Financial Data Science

Confidently analyze, interpret, and act on financial data with this practical introduction to the fundamentals of financial data science.

Master the fundamentals
Step-by-step introductions to core topics will equip you with a solid foundation for applying data science techniques to real-world complex financial problems.

Extract meaningful insights
Learn how to use data to lead informed, data-driven decisions, with over 50 examples and case studies and hands-on MATLAB® and Python code.

Explore the cutting edge
Dive into the latest techniques and tools in machine learning for financial data analysis, including deep learning and natural language processing.

Accessible to readers with or without a specialized background in finance or machine learning, and including coverage of data representation and visualization, data models and estimation, principal component analysis, clustering methods, optimization tools, mean/variance portfolio optimization, and financial networks, this is the ideal introduction for financial services professionals, and graduate students in finance and data science.

Giuseppe C. Calafiore is a Professor of Automatic Control at the Electronics and Telecommunications Department at Politecnico di Torino, Italy, where he coordinates the Control Systems and Data Science group, and a former Professor at VinUniversity, Hanoi, and Visiting Professor at the University of California, Berkeley, where he co-taught graduate courses in financial data science. He is a co-author of *Optimization Models* (2014), and a Fellow of the IEEE.

Laurent El Ghaoui is Vice-Provost of Research and Innovation, and Dean of Engineering and Computer Science, at VinUniversity. He is a former Professor of Electrical Engineering and Computer Science at the University of California, Berkeley, where he taught topics in data science and optimization models within the Haas Business School Master of Financial Engineering programme. He is a co-author of *Optimization Models* (2014).

Giulia Fracastoro is an Assistant Professor at the Electronics and Telecommunications Department at Politecnico di Torino. In 2017, she obtained her Ph.D. in Electronics and Telecommunications Engineering from Politecnico di Torino with a thesis on design and optimization of graph transform for image and video compression. Her main research interests are graph signal processing and neural networks on graph-structured data.

Alicia Y. Tsai is a Research Engineer at Google DeepMind. She obtained her Ph.D. in Computer Sciences from the University of California, Berkeley. Her main research interests are optimization, natural language processing, and machine learning. She is also a founding board member of the Taiwan Data Science Association and the founder of Women in Data Science (WiDS) Taipei.

Combining the theoretical backbone of data science with real-world applications in modern financial markets, this book offers invaluable insights and practical techniques for mastering quantitative finance. A must-read for anyone looking to harness the power of data science in finance.

Kiseop Lee, Purdue University

Covering the essential topics in the fast-growing field of financial data science, this is a valuable resource for professionals, students, and for anyone seeking conceptual depth, technical clarity, and a practical guide to working with real-world financial data.

Paul Mende, MIT Sloan School of Management

A masterful balance of rigorous theoretical techniques and practical financial applications, this is a must-read for anyone seeking both depth and applicability in the evolving field of financial data science.

Dragos Bozdog, Stevens Institute of Technology

An outstanding overview of techniques supporting critical financial engineering tasks, striking a careful balance between theoretical foundations and real-world financial applications. A must-read for anyone looking to understand and contribute to the future of financial data science.

Agostino Capponi, Columbia University

Financial Data Science

Giuseppe Calafiore
Politecnico di Torino, Italy, and VinUniversity, Hanoi

Laurent El Ghaoui
VinUniversity, Hanoi

Giulia Fracastoro
Politecnico di Torino, Italy

Alicia Tsai
University of California, Berkeley

Shaftesbury Road, Cambridge CB2 8EA, United Kingdom

One Liberty Plaza, 20th Floor, New York, NY 10006, USA

477 Williamstown Road, Port Melbourne, VIC 3207, Australia

314–321, 3rd Floor, Plot 3, Splendor Forum, Jasola District Centre,
New Delhi – 110025, India

103 Penang Road, #05–06/07, Visioncrest Commercial, Singapore 238467

Cambridge University Press is part of Cambridge University Press & Assessment,
a department of the University of Cambridge.

We share the University's mission to contribute to society through the pursuit of
education, learning and research at the highest international levels of excellence.

www.cambridge.org
Information on this title: www.cambridge.org/highereducation/isbn/9781009432245

DOI: 10.1017/9781009432283

© Giuseppe C. Calafiore, Laurent El Ghaoui, Giulia Fracastoro, and Alicia Y. Tsai 2026

This publication is in copyright. Subject to statutory exception and to the provisions
of relevant collective licensing agreements, no reproduction of any part may take
place without the written permission of Cambridge University Press & Assessment.

When citing this work, please include a reference to the DOI 10.1017/9781009432283

First published 2026

Cover image: Andriy Onufriyenko/Moment/GettyImages

A catalogue record for this publication is available from the British Library

A Cataloging-in-Publication data record for this book is available from the Library of Congress

ISBN 978-1-009-43224-5 Hardback

Additional resources for this publication at www.cambridge.org/calafiore-fds

Cambridge University Press & Assessment has no responsibility for the persistence or
accuracy of URLs for external or third-party internet websites referred to in this
publication and does not guarantee that any content on such websites is, or will
remain, accurate or appropriate.

For EU product safety concerns, contact us at Calle de José Abascal, 56, 1°,
28003 Madrid, Spain, or email eugpsr@cambridge.org

*To Martina, Francesco, Masha, and Charlotte, in their
inverse order of appearance in my life,
and in loving memory of my father*
G. C.

*To my children Alexandre and Camille, the sweetness
of my life*
L. El G.

To Sofia and Matilde
G. F.

*To my parents and Chien-Wei, who never stopped believing
in me*
A. T.

Contents

1 **Preface** 1
   1.1 Structure of the book 3
   1.2 Scope and target audience 4

2 **Data representation and visualization** 7
   2.1 Vector data representation 8
   2.2 Basic matrix and vector operations 11
   2.3 Data centering and scaling 14
   2.4 Comparing feature vectors 18
   2.5 Data projection and scoring 19
   2.6 Data visualization 24
   2.7 Exercises 27

3 **Data models and estimation** 31
   3.1 A basic estimation model 32
   3.2 Sample covariance and directional variance 38
   3.3 A stochastic model for portfolio analysis and design 40
   3.4 Issues with the sample covariance matrix 45
   3.5 Factor models 48
   3.6 Ledoit and Wolf's shrinkage model 53
   3.7 Sparse covariance and precision matrix estimation 54

3.8 *Confidence ellipsoids and outliers* 57
3.9 *Time-series data models* 61
3.10 *Exercises* 70

Appendix A   73

A.1 *Properties of symmetric matrices* 73
A.2 *Point estimates* 76
A.3 *Identity plus low-rank covariance approximation* 78

4 Principal component analysis and low-rank approximation   81

4.1 *Principal component analysis* 82
4.2 *Low-rank approximations* 87
4.3 *Extensions* 95
4.4 *Exercises* 107

5 Clustering methods   111

5.1 *k-means* 112
5.2 *Spectral clustering* 118
5.3 *Density-based spatial clustering* 119
5.4 *Clusterpath* 120
5.5 *Hierarchical clustering* 121
5.6 *Dimensionality reduction for clustering* 123
5.7 *Exercises* 127

6 Linear regression models   131

6.1 *Linear regression* 131
6.2 *Training regression models* 135
6.3 *Ridge regression, Lasso, Elastic Net, etc.* 137
6.4 *Cross validation and selection of the penalty parameter* 143
6.5 *Exercises* 151

Appendix B   158
   B.1 *Bayesian learning*   158

7 *Linear classifiers*   165
   7.1 *Basics of binary classification*   165
   7.2 *The logistic classifier model*   170
   7.3 *Linear support vector machines*   179
   7.4 *Regularization and robustness*   184
   7.5 *Exercises*   188

8 *Nonlinear classifiers and kernel methods*   193
   8.1 *Motivation*   193
   8.2 *Kernel trick*   196
   8.3 *Nonlinear SVM*   199
   8.4 *Decision trees and random forests*   203
   8.5 *Exercises*   212

9 *Neural networks and deep learning*   215
   9.1 *Neural networks as feature extractors*   216
   9.2 *Feedforward neural networks*   217
   9.3 *Training a neural network*   221
   9.4 *Convolutional neural networks*   228
   9.5 *Recurrent neural networks*   232
   9.6 *Selected modern deep learning approaches*   237
   9.7 *Exercises*   241

10 *Optimization tools*   245
   10.1 *Optimization problems in standard form*   246
   10.2 *Convexity*   248

10.3 Convex problem classes   252

10.4 Robust optimization   260

10.5 Exercises   269

11 Mean/variance portfolio optimization   273

11.1 Prices and returns   274

11.2 Optimal risk/return tradeoffs   277

11.3 Additional criteria and constraints   286

11.4 Sharpe ratio optimization   292

11.5 Alternative measures of risk   295

11.6 Portfolio optimization in practice   301

11.7 Exercises   304

12 Beyond the mean/variance model   311

12.1 Risk contributions and risk budgets   311

12.2 Index tracking   315

12.3 Robust portfolio optimization   318

12.4 Multi-period decision problems   323

12.5 Exercises   333

13 Financial networks   337

13.1 Graph definitions and terminology   338

13.2 Some types of financial networks   342

13.3 Liability networks and default contagion   349

13.4 Centrality measures   353

13.5 Clustering and community detection   363

13.6 Exercises   367

14 Text analytics 371

14.1 Representing words 371
14.2 Distributed word representation 377
14.3 Deep neural networks for text analytics 384
14.4 Evaluating text analytics tasks 388
14.5 Exercises 393

Index 397

1
Preface

FINANCIAL DATA SCIENCE is a rapidly growing field that combines data science, finance, statistics, and machine learning to analyze, interpret and act on financial data. With the advent of big data and the proliferation of data sources, financial data science has become an essential tool for financial institutions, hedge funds, and investors looking to extract insights from financial data for the purpose of making informed decisions. This includes, for instance, making predictions from financial data such as stock prices, trading volumes, and economic indicators, recognizing regular and fraudulent patterns in online transactions, optimizing investment portfolios, understanding the network interconnection structure of financial institutions and its effect on systemic default risk, and so on.

By using data-driven approaches to make decisions, financial institutions can reduce risk, increase efficiency, and improve profitability. Banks can use machine learning to detect fraudulent transactions in real-time, reducing the risk of financial loss. Investment firms can use data science to optimize portfolio management, resulting in higher returns for investors. Overall, data science in finance is transforming the way financial institutions operate, make decisions and improve their profitability by harnessing the power of data. As such, the demand for professionals skilled in financial data science is expected to continue to grow in the coming years.

The field of financial data science has evolved significantly over the years, driven by advancements in technology, the increasing availability of large amounts of financial data, and the growing need for data-driven decision-making in the finance industry. From an historical perspective, some key milestones and influential works have shaped the development of financial data science. Early applications focused on statistical methods and quantitative techniques. Notable contributions include the pioneering work of Harry Markowitz in the 1950s on modern portfolio theory,[1] which laid the foundation for optimizing investment portfolios based on risk and return tradeoffs. This marked the beginning of quantitative finance, where mathematical models and statistical methods were used to analyze financial data and synthesize investment decisions. Then, the advent of computers and the availability of computational power in the 1970s and 1980s revolutionized the field of finance. Researchers started employing computational techniques to analyze financial data and develop pricing models for derivatives and other complex financial instruments. Notable works during this period include the Black–Scholes–Merton model for option pricing[2] and the development of the Monte Carlo simulation method for valuing options.[3] Next, the twenty-first century witnessed an explosion in the volume, variety, and velocity of financial data, leading to the emergence of data science as a powerful tool in finance. With the availability of high-frequency trading data, social media sentiment data, and other alternative data sources, traditional statistical methods proved insufficient to handle the complexity and scale of the data. Financial institutions started adopting data science techniques, such as machine learning, data mining, and natural language processing, to gain insights, make predictions, and automate trading strategies. Notable advancements include the use of machine learning algorithms for credit risk assessment,[4] algorithmic trading strategies based on pattern recognition,[5] and sentiment analysis for predicting market movements.[6] Another significant development was the emergence of open-source platforms and libraries tailored for financial data science, such as Python's pandas, NumPy, and scikit-learn, as well as R's quantmod and caret. These tools have democratized access to advanced data analytics techniques and facilitated the collaboration between researchers and practitioners in the finance industry.

As the field of data science and artificial intelligence is evolving at a fast pace, with new sophisticated tools being proposed continuously, however, it is difficult for authors to explore, organize, and propose to the readers a selection of methodologies and tools that is sufficiently coherent, relevant for practical applications, and meaningful in terms of long-lasting impact on the field. Also, the presentation needs to strike a balance between rigor and simplicity of exposition. The choice made by the authors of the present book is to focus more on well-established models and classical approaches, rather than delve into the complexities of the most recent techniques (say, e.g., attention mechanisms in neural networks and transformers)

[1] Harry M. Markowitz, "Portfolio Selection," *The Journal of Finance*, vol. 7(1), pp. 77–91, 1952.

[2] F. Black and M. Scholes, "The Pricing of Options and Corporate Liabilities," *Journal of Political Economy*, vol. 81(3), pp. 637–654, 1973.

[3] P. P. Boyle, "Options: A Monte Carlo Approach," *Journal of Financial Economics*, vol. 4(3), pp. 323–338, 1977.

[4] E. I. Altman, G. Marco, and F. Varetto, "Corporate Distress Diagnosis: Comparisons Using Linear Discriminant Analysis and Neural Networks (the Italian Experience)," *Journal of Banking & Finance*, vol. 18(3), pp. 505–529, 1994.

[5] F. Lillo and R. N. Mantegna, "Strategy Complexity and Market Efficiency," *The European Physical Journal B*, vol. 15(4), pp. 603–606, 2000.

[6] J. Bollen, H. Mao, and X. Zeng, "Twitter Mood Predicts the Stock Market," *Journal of Computational Science*, vol. 2(1), pp. 1–8, 2011.

which may not even yet be fully understood by the research community. Also, classical financial data (e.g., stock prices) is typically non-stationary, hence only a relatively small and most recent part of the data is usually meaningful for the training of prediction models, and in such situations simple, well-understood models may provide more effective results than over-sophisticated models that need huge amount of training data in order to function reliably.

With this approach in mind, this book aims to provide an introductory guide to some key topics in financial data science, covering a wide range of aspects, including data representation and visualization, data models and estimation, principal component analysis, clustering methods, linear and nonlinear classifiers, neural networks and deep learning, optimization tools, mean/variance portfolio optimization, financial networks, up to an introduction to the rapidly evolving field of text analytics and its applications in finance. By exploring these topics, readers will gain a solid foundation in applying data science techniques to solve complex financial problems and make informed decisions in today's data-driven finance industry.

1.1 Structure of the book

The book is divided into 14 chapters. Chapter 2 introduces the reader to data representation and visualization. This includes topics such as vector data representation, basic matrix and vector operations, data centering and scaling, comparing feature vectors, data projection and scoring, and data visualization. Chapter 3 delves into data models and estimation. This includes a basic estimation model, sample covariance and directional variance, a stochastic model for portfolio analysis and design, issues with the sample covariance matrix, factor models, Ledoit and Wolf's shrinkage model, sparse covariance and precision matrix estimation, confidence ellipsoids and outliers, and time-series data models. Chapter 4 focuses on principal component analysis and low-rank approximation, while Chapter 5 covers clustering methods such as k-means, spectral clustering, density-based spatial clustering, clusterpath, hierarchical clustering, and dimensionality reduction for clustering. Chapter 6 discusses linear regression models, including linear regression, training regression models, Ridge regression, Lasso, Elastic Net, and so on, and cross-validation and selection of the penalty parameter. Chapter 7 covers linear classifiers, such as the basics of binary classification, the logistic classifier model, linear support vector machines, regularization, and robustness. Chapter 8 explores nonlinear classifiers such as decision trees and random forests, nonlinear SVMs, and kernel methods. Chapter 9 focuses on neural networks and deep learning, including neural networks as feature extractors, feedforward neural networks, training a neural network, convolutional neural networks, recurrent neural networks, and selected modern deep learning approaches. Chapter 10 covers optimization tools, including optimization problems in standard form, convexity, convex problem classes, and robust optimization. Chapter 11 focuses on mean/variance portfolio optimization, optimal risk/return tradeoffs, additional criteria and constraints, Sharpe ratio optimization, alternative measures of risk, and portfolio optimization in practice. Chapter 12 goes beyond the mean/variance model and covers risk contributions and risk budgets, index tracking, robust portfolio optimization, and multi-period portfolio allocation decision problems. Chapter 13 introduces financial networks, including graph definitions and terminology, some types of financial networks, liability networks and default contagion, centrality measures, clustering, and community detection. Finally, Chapter 14 explores text analytics, including the basics of text analytics, sentiment analysis, topic modeling, and word embeddings.

Overall, this book provides a rather comprehensive overview of financial data science, covering a wide range of topics and techniques. The presentation is rigorous but kept at an accessible level; technical proofs are provided only when necessary for a better comprehension of the topic and are usually confined to the chapters' appendices. Examples in the text typically use real-world data sets, which are available to the user either from the original sources or in the online resources for the book, at www.cambridge.org/calafiore-fds. The codes for the examples are also available in the online resources, so the reader can reproduce the results hands on. End-of-chapter exercises are also provided to help the reader check their comprehension of each chapter's topics.

1.2 Scope and target audience

By assimilating this book's material, readers will be equipped with the knowledge and skills necessary to navigate the intricate intersection of finance and data science, empowering them to extract meaningful insights from financial data and drive innovation in the field.

This book can be used as a textbook for a Master's level course in data science or machine learning with a financial engineering application flavour, or as a self-learning manual for perspective quantitative finance professionals needing to acquire the basic tools of their trade. Whether the reader is a student, a finance professional, data scientist, or simply has an interest in financial data analysis and modeling, this book can accompany and guide them in the journey of learning financial data science, where the past, present, and future converge to shape the way we analyze, model, and interpret financial phenomena.

2
Data representation and visualization

A CENTRAL PRELIMINARY ASPECT in any data science project is the representation and organization of *data*. In financial applications data will typically be available in the form of tables of numerical values containing, for instance, the collection of daily closing prices of different stocks over a given time span, or the fundamental characteristics (e.g., price/earning ratio, market cap, dividends per share, etc.) of listed companies in a given quarter. Other times data may possess a non-numeric nature: for example, textual data (e.g., from tweets, or news headlines) is abundant and rich in information content, and even image data (e.g., images of car occupancy in parking lots) may sometimes be related to financial information. In the case of non-numeric data, however, the analyst usually applies ad-hoc elaborations on the data so to transform it into a numerical representation. In this chapter we discuss about the standard numerical representation of data in the form of a collection of vectors of variables, or *features*, which form a *data frame*. We shall then illustrate standard preliminary data manipulation techniques, such as centering, normalization, and dropping or imputation of missing values. We next discuss about projecting high-dimensional data along a given direction (line) or onto a given plane for the purpose of visualizing slices of the data via human-understandable one-dimensional or two-dimensional plots. Finally, we examine some standard graphical exploratory analysis tools, such as line charts, histograms, and scatter plots.

2.1 Vector data representation

We introduce data representation with a practical example: a bank wants to quickly profile customers that apply for a home loan, based on customer details provided while filling an online application form. These details are Gender, Marital Status, Education, Number of Dependents, Income, Loan Amount, Credit History, and others. The target to be predicted is whether the loan will be approved or not.[1] The dataset is available in the form of a comma separated (csv) file, in which each row represents a customer (or, in general, a *sample*, a *specimen*, or *observation*), and columns represent characteristics (synonyms are *features* or *variables*) relative to each customer. There are $m = 614$ rows (customers) in the data set. Table 2.1 illustrates the $n = 11$ features that are considered in this data set, for each customer. The last column in the data set represents the actual loan status of each customer, which is a Boolean variable describing whether the loan was eventually approved or not. This type of measured outcome is also called a *label*, *output*, or *target* value corresponding to the sample, and data sets that are provided with such information are called *labeled*.

[1] The loan data set is taken from Kaggle and it is available in this book's online resources in the file loan_data_set.csv.

Feature	Description
Loan_ID	Unique loan ID
Gender	Male/female
Married	Applicant married (Y/N)
Dependents	Number of dependents
Education	Graduate/undergraduate
Self_Employed	Self-employed (Y/N)
ApplicantIncome	Applicant income
CoapplicantIncome	Coapplicant income
LoanAmount	Loan amount in thousands
Loan_Amount_Term	Term of loan in months
Credit_History	Credit history ok? (Y/N)
Property_Area	Urban/semi-urban/rural
Loan_Status	(Target) loan approved (Y/N)

Table 2.1 Description of the features and target for the loan data set.

We observe that in the considered data set some of the features are natively numeric (e.g., the applicant's income in euros), while others are Boolean (e.g., marital status), or categorical (the feature describing the type of area on which the property lies). A first step towards a vector representation of the data is then to assign integer values to the categorical variables, so to reduce all features to numerical format. For instance, we may assign Gender to 1 if female and 0 if male, Education to 1 if graduate and 0 if undergraduate, Self_Employed to 1 if true or 0 otherwise, Credit_History to 1 if good or 0 otherwise, Property_Area to 1 if urban, 2 if semi urban, or 3 if rural. Finally, the target Loan_Status is set to 1 if the loan is approved or 0 otherwise. Proceeding in this way, the kth customer can be represented by a vector $x^{(k)}$ containing $n = 11$ entries corresponding to the relevant features of (Gender, Married, Dependents, Education, Self_Employed, ApplicantIncome, CoapplicantIncome, LoanAmount, Loan_Amount_Term, Credit_History, Property_Area). To the kth customer we also associate the value of the corresponding target variable y_k, which is 1 if the loan is approved or 0 otherwise.[2]

A collection of features x of dimension n will typically be represented by a column of values, that is, by a *vector*

$$x = \begin{bmatrix} x_1 \\ x_2 \\ \vdots \\ x_n \end{bmatrix},$$

[2] The Python code in file `loan_data_set_numeric.py` maps the categorical features into numeric features and saves the result in the numeric data file `loan_data_set_num.csv`.

and denoted by $x \in \mathbb{R}^n$ to specify that its entries are real numbers and that the dimension is n. The transposition operator $\top$ is used to reshape a column vector into a row vector, or vice-versa:

$$x^\top = \begin{bmatrix} x_1 & x_2 & \cdots & x_n \end{bmatrix}, \quad x^{\top\top} = x.$$

Occasionally, it might be convenient to introduce a vector without specifying if it is in row or column shape, in which case we use the notation $x = (x_1, x_2, \ldots, x_n)$.

Now, for the loan data set, we have $m = 614$ feature vectors $x^{(k)} \in \mathbb{R}^n$, with $k = 1, \ldots, m$, each of which corresponds to a customer. By juxtaposing these vectors we obtain a *data matrix* X of n rows and m columns, where each column represents a customer:

$$X = \begin{bmatrix} x^{(1)} & x^{(2)} & \cdots & x^{(m)} \end{bmatrix} \in \mathbb{R}^{n,m}.$$

The element X_{ij} in position row i and column j of X contains the value of the ith feature for the jth customer. The transposition operator acts on a matrix by exchanging the row and column indices of its elements, that is, $X_{ij}^\top = X_{ji}$.

Notice that data files typically contain samples (customers, in this case) in rows and features in columns, while the data matrix X, according to our convention, contains samples in columns and features in rows. Hence, the data matrix is essentially the transpose of the data frame contained in the csv file. Although this is simply a convention, we will stick to this notation and consider data matrices X

in which each column represents the vector of features of a sample. To the data matrix, in the present example, is associated also a *response vector* $y \in \mathbb{R}^m$, which contains the observed output of each customer. The output response vector is used in the training phase of supervised learning algorithms. We next give a few examples of vector data representation.

Example 2.1 (*Financial price and return data*) A typical example of vector representation of data is given by historical sequences of asset prices or returns. If Δ denotes a time duration, for example, one day, one month, or one minute, we let $p(i,t)$ represent the market price of asset i at time $t\Delta$, where t is an integer time index. Collecting the prices of n assets at times $t = 0, \ldots, m$, we define a price data matrix $P = [p^{(0)} \cdots p^{(m)}] \in \mathbb{R}^{n,m+1}$, where $p^{(t)} \doteq [p(1,t) \cdots p(n,t)]^\top$, $t = 0, \ldots, m$, is a column vector containing the prices of all assets at time t.

Also, from asset prices we can compute *simple returns* as

$$r(i,t) \doteq \frac{p(i,t) - p(i,t-1)}{p(i,t-1)}, \quad i = 1, \ldots, n; \, t = 1, \ldots, m,$$

and defining the return vectors $r^{(t)} = [r(1,t) \cdots r(n,t)]^\top$, $t = 1, \ldots, m$, we obtain a return data matrix $R = [r^{(1)} \cdots r^{(m)}] \in \mathbb{R}^{n,m}$ in which each column represents a time point and each rows corresponds to an asset, so that the entry R_{it} in position row i and column t represents the return of asset i at time t. In an analogous way one can define the *log-returns*

$$\rho(i,t) \doteq \log \frac{p(i,t)}{p(i,t-1)} = \log(1 + r(i,t)), \quad i = 1, \ldots, n; \, t = 1, \ldots, m,$$

and hence the log-return vectors $\rho^{(t)}$, $t = 1, \ldots, m$, and the log-return data matrix. Observe that when $|r(i,t)|$ are very small, as usually happens with high-frequency data at minute, hourly, or even daily rate, then $\log(1 + r(i,t)) \simeq r(i,t)$ and hence simple and log-returns are numerically close.

Example 2.2 (*Image data*) Another example of vectorized data representation is given by images. A gray-scale image is a $n_x \times n_y$ array of pixel values, for example, from 0 to 255 where 0 is for black, 255 is for white and intermediate values are gray levels, see an example in Figure 2.1.

The rectangular array representation of an image can be converted to a representation with a vector of size $n = n_x n_y$ obtained by stacking all columns of the image array into a single column. Color (RGB) images can be treated similarly: each color channel is represented by a $n_x \times n_y$ array, one for Red, one for Green, and one for Blue, and the overall information is stacked into a vector of size $n = n_x \times n_y \times 3$. If we have a collection of m images, each coded into an n-dimensional vector as just described, we may construct a data matrix $X \in \mathbb{R}^{n,m}$ in which the columns (samples) represent the images and the rows (features) refer to pixels.

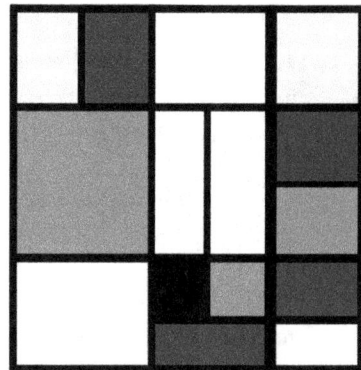

Figure 2.1 An 800×800 grayscale image represented as a matrix of integers, see MATLAB® file mondrian.mat in the online resources.

Example 2.3 (*Textual data*) Yet another example of data vectorization comes from the so-called *bag-of-words* representation of textual information. Given a collection of m documents (say, news headlines):

- We store all the n unique words appearing in the collection in a list called "dictionary." This operation may require word segmentation for certain languages (e.g., Chinese) and stemming.
- For each $j = 1, \ldots, n$, the jth word is represented as a "one-hot" (basis) vector, that is, a vector of dimension n composed of all zeros, except for a one in position j.
- Each document $i = 1, \ldots, m$, is then represented as a vector $x^{(i)}$ obtained as the sum of all the vectors corresponding to the unique words present in the document.

This leads to an *occurence* (presence/absence) representation, see an example in Figure 2.2. The collection of all document vectors $x^{(i)}$, $i = 1, \ldots, m$, can then be represented by means of the term-document matrix $X = [x^{(1)} \cdots x^{(m)}]$, in which columns represent documents and rows refer to dictionary words (features). There are many alternatives to the bag-of-words representation. For instance, one may use vectors containing the frequencies of each word in each document, instead of the simple binary occurrence flags, or more sophisticated approaches such as the term frequency-inverse document frequency (Tf-Idf) encoding, see Chapter 14 for further details about representation of text.

	softbank	billion $B	trade trading traders	gains	stock	option	bet	boom	sit	profit	buying
SoftBank's $4 Billion Trading Gains on U.S. Stock-Option Bet	1	1	1	1	1	1	1				
Options Traders Whipped Up Stock Boom With SoftBank Buying	1		1		1	1		1			1
SoftBank sits on $4B profit on options trade	1	1	1			1				1	

Figure 2.2 Bag-of-words representation of text.

2.2 Basic matrix and vector operations

We assume that the reader is familiar with basic matrix operations and linear algebra. However, we next briefly recall some useful facts. The (standard) *inner product* between two vectors $x \in \mathbb{R}^n$, $y \in \mathbb{R}^n$, also known as the scalar product, is defined as

$$x^\top y \doteq \sum_{i=1}^n x_i y_i.$$

The "size" of a vector can be measured in many ways, the most common being:

- the ℓ_2 ("Euclidean") norm $\|x\|_2 \doteq \sqrt{x^\top x}$, which corresponds to the ordinary notion of length or distance from standard geometry;

- the ℓ_1 ("Manhattan") norm $\|x\|_1 \doteq |x_1| + \cdots + |x_n|$, which corresponds to a notion of distance for an agent moving across an orthogonal grid;

- the ℓ_∞ ("peak") norm $\|x\|_\infty \doteq \max_{1 \leq i \leq n} |x_i|$, which corresponds to the maximum amplitude of a vector's elements.

Corresponding to each of the above distance notions we may define the sets of points whose distance from the origin is no larger than a given radius $\varrho \geq 0$ (norm balls), that is, the sets $\{x: \|x\|_p \leq \varrho\}$, for $p = 1, 2, \infty$, see examples in Figure 2.3.

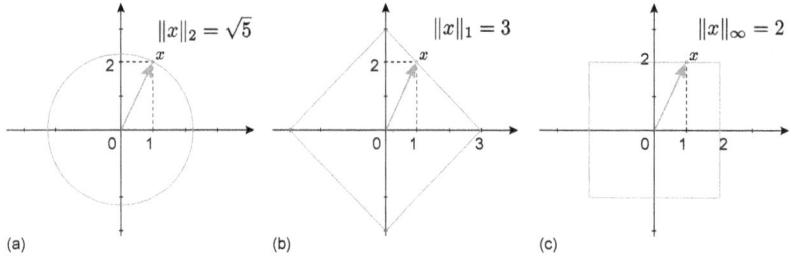

Figure 2.3 The sets $\{x \in \mathbb{R}^2 : \|x\|_2 \leq \sqrt{5}\}$ (a), $\{x \in \mathbb{R}^2 : \|x\|_1 \leq 3\}$ (b), and $\{x \in \mathbb{R}^2 : \|x\|_\infty \leq 2\}$ (c).

The product of a matrix $A \in \mathbb{R}^{n,m}$ and a vector $x \in \mathbb{R}^m$ is denoted by Ax, which is a vector in $\mathbb{R}^n$ that can be represented in two equivalent ways in terms of either the columns or the rows of A. If A is expressed as the juxtaposition of its columns, that is, $A = [c^{(1)} \cdots c^{(m)}]$, then

$$Ax = \sum_{j=1}^{m} c^{(j)} x_j,$$

which indicates that Ax is a vector obtained by taking a linear combination of the columns of A, using the entries in x as coefficients. If instead A is expressed as a stack of its rows, that is,

$$A = \begin{bmatrix} r^{(1)\top} \\ \vdots \\ r^{(n)\top} \end{bmatrix},$$

then

$$Ax = \begin{bmatrix} r^{(1)\top} x \\ \vdots \\ r^{(n)\top} x \end{bmatrix},$$

which indicates that each entry of vector Ax is given by the inner product of the corresponding row of A and x.

The product of a matrix $A \in \mathbb{R}^{n,m}$ and a matrix $B \in \mathbb{R}^{m,p}$ is denoted by AB, which is a matrix in $\mathbb{R}^{n,p}$ that can be defined such that if B is expressed by means of its columns $B = [b^{(1)} \cdots b^{(p)}]$ then

$$AB = A[b^{(1)} \cdots b^{(p)}] = [Ab^{(1)} \cdots Ab^{(p)}],$$

and if A is expressed by its rows as in (2.2) then

$$AB = \begin{bmatrix} r^{(1)\top} \\ \vdots \\ r^{(n)\top} \end{bmatrix} B = \begin{bmatrix} r^{(1)\top} B \\ \vdots \\ r^{(n)\top} B \end{bmatrix}.$$

Similar to vectors, there are size measures also for matrices. The most commonly used are:

- the Frobenius norm $\|X\|_F \doteq \sqrt{\operatorname{Tr}(XX^\top)}$, which corresponds to the ℓ_2 norm of the vectorization of matrix X. Here $\operatorname{Tr}(\cdot)$ denotes the *trace* function, which returns the sum of the diagonal entries of its argument;

- the ℓ_1 and ℓ_∞ matrix norms, which are defined respectively as the ℓ_1 and the ℓ_∞ norm of the vectorization of of matrix X, that is, $\|X\|_1 \doteq \sum_{i,j} |X_{ij}|$ and $\|X\|_\infty \doteq \max_{i,j} |X_{ij}|$;

- operator norms, such as the spectral norm $\|X\|_\sigma \doteq \max\{\|Xw\|_2 : \|w\|_2 = 1\}$.

A particularly important special structure for matrices is the *dyadic* structure. A matrix $X \in \mathbb{R}^{n,m}$ is said to be a *dyad* if it can be factored as

$$X = uv^\top, \quad \text{where } u \in \mathbb{R}^n, v \in \mathbb{R}^m.$$

The element of X in position (i,j) is thus $X_{ij} = u_i v_j$. We also refer to dyads as *outer products*, or *rank-one* matrices. If a data matrix $X = [x^{(1)} \cdots x^{(m)}] \in \mathbb{R}^{n,m}$ is a dyad, that is, if $X = uv^\top$, then every column $x^{(j)}$ is a scalar multiple of the same vector u, since $x^{(j)} = v_j u$, $j = 1,\ldots,m$. This means that all columns (samples) are parallel to each other. The same actually holds for the rows of X, since each row $r^{(i)\top}$ of X is a scalar multiple of the same vector $v^\top$, that is, $r^{(i)\top} = u_i v^\top$, $i = 1,\ldots,n$.

Example 2.4 (*Dyadic approximation of price data matrix*) Let a data matrix $X = [x^{(1)} \cdots x^{(m)}] \in \mathbb{R}^{n,m}$ contain by rows time-series data of the prices of n financial instruments over m periods of time, as discussed in Example 2.1. It is very unlikely that X has rank one. However, there exist numerical techniques that permit to find a rank-one (hence dyadic) approximation to X, see, for example, Section 4.2.3. The interpretation of the dyadic approximation $X \simeq uv^\top$ is that v contains a prototypal time profile for the asset time series, and the coefficients u_i, $i = 1, \ldots, n$, provide the scalings of this time profile, so that the scaled vectors $u_i v$ approximate the corresponding price series contained in the ith row of X.

For purpose of illustration, Figure 2.4(a) shows the daily time series of the scaled US dollar (USD) value of $n = 5$ major cryptocurrencies – Bitcoin (BTC), Ethereum (ETH), XRP, Cardano (ADA), Litecoin (LTC) – over the period from Jan. 2, 2020 to Jan. 30, 2021, for a total of $m = 395$ time points. Figure 2.4(b) shows instead, for comparison, the dyadic approximation of these data.

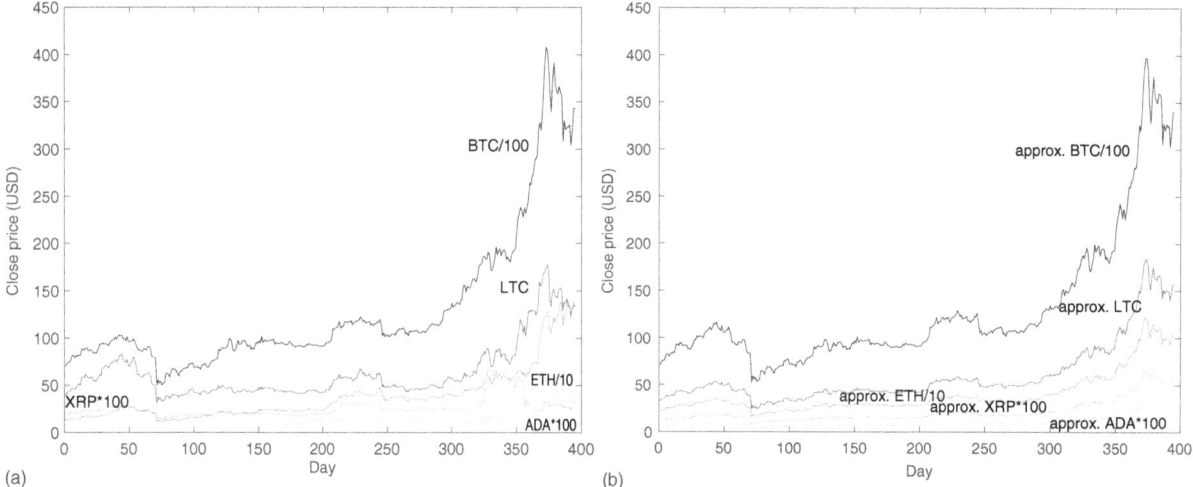

Figure 2.4 (a) Plot of scaled cryptocurrency value in USD from Jan. 2, 2020 to Jan. 30, 2021. (b) Plot of the rank-one approximation of the corresponding data matrix, see the MATLAB® file cryptocurr.m in the online resources.

2.3 Data centering and scaling

Given a data matrix $X = [x^{(1)}\ x^{(2)} \cdots x^{(m)}] \in \mathbb{R}^{n,m}$ we can interpret the data as a cloud of m points in an n-dimensional Euclidean space. A first key parameter a data analyst is interested in is the *location* of the data cloud. A natural way to summarize "where" the data cloud is located is to compute the centroid (barycenter) $\bar{x} \in \mathbb{R}^n$ of the data points, which is simply defined as the average of the data points:

$$\bar{x} \doteq \frac{1}{m} \sum_{k=1}^{m} x^{(k)} = \frac{1}{m} X \mathbf{1}_m,$$

where $\mathbf{1}_m$ denotes a vector of ones of dimension m. Here, the expression $X \mathbf{1}_m$ refers to the multiplication of matrix $X \in \mathbb{R}^{n,m}$ and vector $\mathbf{1}_m \in \mathbb{R}^m$, which results in a column vector of size m in which the ith entry is given by the scalar product of the ith row of X with $\mathbf{1}_m$, as discussed in Section 2.2. Then, we may remove this offset from the data and define the centered data points as

$$\check{x}^{(k)} \doteq x^{(k)} - \bar{x}, \quad k = 1, \ldots, m,$$

and the centered data matrix as

$$\check{X} \doteq [\check{x}^{(1)} \; \check{x}^{(2)} \cdots \check{x}^{(m)}] = X - \bar{x}\mathbf{1}_m^\top.$$

We may now interpret $\check{X}$ as a cloud of m points in n-dimensional space whose barycenter is zero. Further, different features may have different units and scale, hence sometimes it may be useful to rescale each feature x_i by dividing it by a suitable scale parameter l_i, which may be selected as the range of variation of the $\check{x}_i$ variable across samples (rescaling), or as its sample standard deviation (standardization). Defining a diagonal scaling matrix $L \doteq \mathrm{diag}(l_1, \ldots, l_n)$, rescaled feature vectors are obtained by multiplying each centered feature vector $\check{x}^{(k)}$ on the left by L^{-1}. Further, certain learning algorithms require the feature vectors to be also normalized in size, which is obtained by dividing each centered and rescaled feature vector by its norm (typically, the Euclidean or the ℓ_1 norm). This type of normalization is obtained by multiplying each column $\check{x}^{(k)}$ of the centered data matrix by a suitable scalar r_k^{-1}. Defining a diagonal scaling matrix $R \doteq \mathrm{diag}(r_1, \ldots, r_m)$, the centered, rescaled, and normalized feature vectors are thus obtained as

$$\tilde{x}^{(k)} \doteq L^{-1}(x^{(k)} - \bar{x})r_k^{-1}, \quad k = 1, \ldots, m,$$

and the centered, rescaled and normalized data matrix is given by

$$\tilde{X} \doteq [\tilde{x}^{(1)} \; \tilde{x}^{(2)} \cdots \tilde{x}^{(m)}] = L^{-1}(X - \bar{x}\mathbf{1}_m^\top)R^{-1}. \tag{2.1}$$

We remark that centering the data according to the centroid is not the only possible choice. Sometimes, the features are required to be centered and rescaled so to fall in a standard range, such as $[-1, 1]$ or $[0, 1]$. For the case of the $[0, 1]$ range, for instance, the centering of the data shall be made with respect to the vector $\bar{x} = \min_{k=1,\ldots,m} x^{(k)}$, where the minimum is computed entry-wise, and the scalings are given by $l_i = \max_{k=1,\ldots,m} x_i^{(k)} - \min_{k=1,\ldots,m} x_i^{(k)}$, $i = 1, \ldots, n$.

Remark 2.1 (*Centering, scaling, and look-ahead bias*) Care should be exerted when centering and/or scaling data to be used to train and test predictive models. As is discussed extensively in later parts of this book (see, e.g., Section 6.4), the available data is often partitioned into two chunks, one to be used for training the predictive model and one to be used for testing the quality of the model on new, unseen data. The key point here is that, for fair unbiased testing, in the model training phase we should use the training data only, and not see and exploit any information contained in the test data set, for otherwise this will introduce a "look-ahead" bias in the trained model. Thus, if the center and range of the data are computed using the *entire* data set and then used to center and normalize the training data set, we are potentially introducing look-ahead bias since the centered and scaled training data will contain information relative to the test data set too. Data centering and scaling is thus deprecated in all situations in which the described type of look-ahead bias can affect the model training. In some cases, centering and scaling can be achieved intrinsically by the model via the addition of offset terms and weighted penalties, see for instance Remark 6.1.

2.3.1 Dealing with missing data

Oftentimes the available data set is incomplete, in the sense that some of the entries of the data frame are missing or spurious. Spurious data refers to variable values which are clearly wrong (e.g., out of the scale of the same variable for other samples, or of a different data type) and should then be removed from the data set. Missing values are "holes" in the data matrix, corresponding to unavailable variable values. Missing values arise from a variety of reasons, the most common of which is that some variable value corresponding to some sample was not recorded. In the loan data example, for instance, there are several missing values which correspond to situations in which some customer did not fill the required field in the questionnaire. Missing values are typically recorded as NaN (not-a-number) in the data matrix.

There are two main avenues for dealing with missing data: imputation or the removal of samples/variables. Imputation methods work by developing suitable guesses for missing data, and they are most useful when the percentage of missing data is low. Removal methods work instead either sample-wise or feature-wise: if the feature vector of a sample contains many missing values, we may just discard that sample from the data set (sample-wise removal). Similarly, if one specific feature appears to be frequently empty across samples, we may remove that feature from the variables, and thus reduce the variable dimension (feature-wise removal). Clearly, discarding samples or reducing the number of features will, in general, affect negatively the quality of the inference model that we intend to build. Sample or feature removal should therefore employed with care, possibly on data sets in which the removed samples or features are a small percentage of the corresponding total. It should also be remarked that some learning algorithms have internal ways for handling missing data based on the training loss reduction, or may just work by ignoring missing data: in such cases the best option is to do noting on the missing data, and let the training algorithm handle the problem. We next briefly discuss the most common data imputation methods.

Mean, median, or mode imputation In this method, any missing values in a given row (feature) of the data matrix X are replaced with the mean (or median, or mode) of the non-missing values of that row. This is a very simple and computationally efficient method which, in the case of mode (i.e., most frequent value), can also work with categorical features. On the other hand, this method works on each row independently of the others, and hence does not consider the correlations between features.

Regression imputation Mean, median, or mode imputation only look at the distribution of the values of the variable with missing entries. If we know there is a correlation between the missing value and other variables, we can often get better guesses by regressing the missing variable on other variables. What this means in practice is that we postulate a (typically linear) regression model of the form

$$x_i = \sum_{j \neq i} \beta_j x_j + \epsilon_i$$

for the ith feature which is missing in some sample k, where β_j are the regression coefficients and ϵ_i accounts for regression errors. Then, we estimate the regression coefficients β_j using the data from other samples in which the ith feature is not missing, and finally we impute the value of x_i in sample k by using the above regression equation. The imputation can be deterministic (i.e., $x_i = \sum_{j \neq i} \beta_j x_j$), or we may aim to preserve the variability of data by adding random noise to the prediction, that is, $x_i = \sum_{j \neq i} \beta_j x_j + \epsilon_i$, where the variance of the noise ϵ_i is estimated in the regression.

Imputation via k-NN The k-nearest neighbors (k-NN) is a simple and well-known algorithm for classification that uses "feature similarity" for predicting the target output of a new data point. In k-NN a new sample is assigned an output value based on how closely it resembles the points in the training set. This approach can be useful in making predictions about the missing values in a sample by finding the k-closest neighbors to the sample with missing data, and then imputing them based on the non-missing values in the neighbors.

Other imputation methods Many other data imputation methods exist. For example, multivariate imputation by chained equation (MICE) is a technique that works by filling the missing data multiple times; it is implemented in an R package as well as in the `fancyimpute` package in Python. Data imputation can actually be seen in the general context of the so-called *matrix completion* problems, for which there exist a vast literature and sophisticated numerical methods, such as the nuclear norm minimization method, the spectral regularization algorithm, and the low-rank singular value decomposition (SVD).[3] All in all, however, there is no perfect or universal method for data imputation. Each strategy can be more suitable for a certain purpose or data set. For instance, simple imputation methods are often the only choice for very large-scale datasets, while low-rank matrix completion methods based on convex optimization may perform better for small-sized datasets. The most suitable approach for dealing with missing data should be tailored to the problem under study, possibly via trial-and-error iterations.

[3] For further details we direct the reader to M. Laurent, "Matrix Completion Problems," in C. Floudas and P. Pardalos (eds.) *Encyclopedia of Optimization*, New York: Springer, 2008.

2.4 Comparing feature vectors

We next discuss standard techniques for comparing two feature vectors, for the purpose of assessing their similarity. The most natural notion for measuring the similarity (or, actually, the dissimilarity) between two vectors $u, v \in \mathbb{R}^n$ is to compute their Euclidean distance

$$d(u,v) = \|u - v\|_2,$$

see Figure 2.5. From elementary geometry (law of cosines), we have that

$$d^2(u,v) = \|u - v\|_2^2 = \|u\|_2^2 + \|v\|_2^2 - 2\|u\|_2\|v\|_2 \cos \theta. \quad (2.2)$$

Further, by expressing $\|u - v\|_2^2 = (u - v)^\top (u - v) = \|u\|_2^2 + \|v\|_2^2 - 2u^\top v$, we have that

$$d^2(u,v) = \|u\|_2^2 + \|v\|_2^2 - 2u^\top v = \|u\|_2^2 + \|v\|_2^2 - 2\|u\|_2\|v\|_2 \cos \theta,$$

from which we obtain the key identity[4]

$$u^\top v = \|u\|_2\|v\|_2 \cos \theta. \quad (2.3)$$

The inner product $u^\top v$ is thus closely related to the angle θ between vectors u and v. When $|\cos \theta| = 1$ we say that vectors u, v are *aligned*; more specifically, they are said to be *parallel* if $\cos \theta = 1$, and *antiparallel* if $\cos \theta = -1$. Vectors u, v are said to be *orthogonal* if $\cos \theta = 0$, that is, if $u^\top v = 0$.

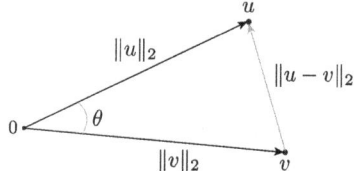

Figure 2.5 The law of cosines for general triangles relates the length of one side of the triangle with the cosine of the opposite angle.

[4] The so-called Cauchy–Schwartz inequality $u^\top v \leq \|u\|_2\|v\|_2$ simply follows from (2.3), by observing that $|\cos \theta| \leq 1$.

By looking at the previous expression for $d(u,v)$ we observe that the distance between u and v can be attributed to two causes: difference in size, and difference in direction, or angle. Indeed, it holds that
$$\|u - v\|_2^2 \geq (\|u\|_2 - \|v\|_2)^2,$$
which tells us that if the sizes of u and v are very different, then the distance $d(u,v)$ will be large, no matter what the angle between u and v is. On the other hand, if u,v have the same size $q \doteq \|u\|_2 = \|v\|_2$, as happens for instance when samples in a data set are normalized in size, then
$$\|u - v\|_2^2 = 2q^2(1 - \cos\theta),$$
and so in this case the distance between u and v is proportional to $(1 - \cos\theta)$.

In certain applications, such as in text analytics, the feature vectors are nonnegative, in which case it holds that $u^\top v \geq 0$, hence from (2.3) it holds that $\cos\theta \geq 0$, and we see from (2.2) that, for fixed size of u, v, the dissimilarity $d(u,v)$ is minimal if $\cos\theta = 1$ (i.e., when u, v are aligned) and it is maximal if $\cos\theta = 0$ (i.e., when u, v are orthogonal). For this reason, in applications where $u, v \geq 0$, feature similarity can be measured simply in terms of the $\cos\theta$, also called *cosine similarity*:
$$\cos\theta = \frac{u^\top v}{\|u\|_2\|v\|_2}.$$

We remark that all the above reasoning holds when the Euclidean norm is taken as a metric for measuring the distance between vectors. This type of choice is very common but it is not mandatory, and there exist specific situations in which other metrics (like, e.g., the ℓ_1 norm) may be more suitable for measuring the distance between vectors.

2.5 Data projection and scoring

Data in dimension $n > 3$ is impossible to plot and visualize directly. For this and other reasons we may be interested in "slicing" the data so to create lower-dimensional views of the data set. In the following we shall assume without loss of generality that our data points $x^{(k)} \in \mathbb{R}^n$, $k = 1, \ldots, m$, have already been centered around their centroid, so that the data set is centered in zero.

2.5.1 Projection along a line

The first problem we consider is computing the projections of the data points along a given line. A line $\mathcal{L}_u$ passing through the origin is described by

$$\mathcal{L}_u = \{x \in \mathbb{R}^n : x = \alpha u, \alpha \in \mathbb{R}\},$$

where $u \in \mathbb{R}^n$, $\|u\|_2 = 1$, is a given unit-norm vector which defines the direction of the line in space. Line $\mathcal{L}_u$ is indeed the locus of all points that are scalar multiples of the u vector, see Figure 2.6.

The projection $x_{\mathcal{L}_u}$ of a point $x \in \mathbb{R}^n$ onto the line $\mathcal{L}_u$ is defined as the point in $\mathcal{L}_u$ at minimum Euclidean distance from x. In formulae:

$$x_{\mathcal{L}_u} \doteq \arg\min_{\xi \in \mathcal{L}_u} \|x - \xi\|_2.$$

The Projection theorem[5] guarantees that such projection exists and it is unique. Moreover, it can be univocally determined by enforcing the necessary and sufficient conditions

$$x_{\mathcal{L}_u} \in \mathcal{L}_u, \text{ and } (x - x_{\mathcal{L}_u}) \perp u,$$

where the last condition means that the residual vector $x - x_{\mathcal{L}_u}$ must be orthogonal to u, see Figure 2.7.

The first condition $x_{\mathcal{L}_u} \in \mathcal{L}_u$ is enforced by imposing that $x_{\mathcal{L}_u} = \alpha u$ for some scalar α that we will determine. The second condition $(x - x_{\mathcal{L}_u}) \perp u$ is enforced by imposing that $(x - x_{\mathcal{L}_u})^\top u = 0$ since, as discussed in Section 2.4, two vectors are orthogonal if and only if their inner product is zero, which, by substituting $x_{\mathcal{L}_u} = \alpha u$ and recalling that $u^\top u = \|u\|_2^2 = 1$, results in the condition

$$(x - \alpha u)^\top u = 0 \quad \Leftrightarrow \quad \alpha = x^\top u.$$

The projection $x_{\mathcal{L}_u}$ is therefore given by

$$x_{\mathcal{L}_u} = (x^\top u) u,$$

where the proportionality coefficient $\alpha = x^\top u$ is called the *score* of x along direction u.

We can now apply the above idea, and compute the projections of the data points $x^{(k)} \in \mathbb{R}^n$, $k = 1, \ldots, m$, along a given direction u. The resulting projections will have the form $(x^{(k)\top} u) u$, $k = 1, \ldots, m$, with scores

$$\alpha_k \doteq x^{(k)\top} u, \quad k = 1, \ldots, m, \tag{2.4}$$

see Figure 2.8 for a two-dimensional illustration.

The directional variance of data Given (centered) data points $x^{(k)} \in \mathbb{R}^n$, $k = 1, \ldots, m$, and a direction $u \in \mathbb{R}^n$, $\|u\|_2 = 1$, we have seen in the previous section how to compute the projections and scores of

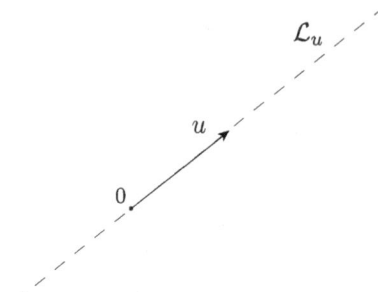

Figure 2.6 Line $\mathcal{L}_u$ is the locus of scalar multiples of the given direction described by the unit-norm vector u.

[5] See, for example, Section 2.3 of G. Calafiore and L. El Ghaoui, *Optimization Models*, Cambridge: Cambridge University Press, 2014.

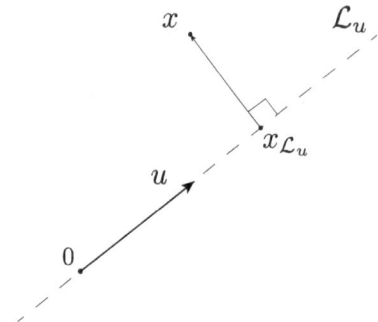

Figure 2.7 Projection of x onto the line $\mathcal{L}_u$.

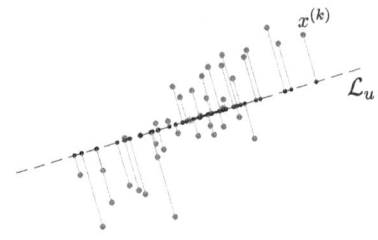

Figure 2.8 Data points and their projections onto a line.

the data along the direction u. The scores α_k in (2.4) form a set of real numbers which represent the distance from zero to the projection of point $x^{(k)}$ onto the line $\mathcal{L}_u$. Depending on the data set and on the direction u we choose, the scores may be more or less concentrated or dispersed around zero. A classical measure of dispersion is the sample variance, that is, the average squared distance of the projected points from zero. The *sample variance* of the projections of the data points along direction u is therefore defined as

$$\sigma_u^2 \doteq \frac{1}{m}\sum_{k=1}^m \alpha_k^2 = \frac{1}{m}\sum_{k=1}^m (x^{(k)\top}u)^2 = \frac{1}{m}\sum_{k=1}^m u^\top x^{(k)} x^{(k)\top} u$$

$$= u^\top \left(\frac{1}{m}\sum_{k=1}^m x^{(k)} x^{(k)\top}\right) u$$

$$= u^\top S u, \qquad (2.5)$$

where

$$S \doteq \frac{1}{m}\sum_{k=1}^m x^{(k)} x^{(k)\top}$$

is the sample covariance matrix of the data. Given the covariance matrix S of the (centered) data set, and given a probing direction u, we can then readily understand how much "variability" is present in the data along the given direction, by evaluating the quadratic form $\sigma_u^2 = u^\top S u$. In Chapter 4 we shall push this discussion further, and study how to actually *find* interesting directions in data space, such as the direction u along which the data variability is maximal, or the direction along which it is minimal.

Deflation and successive projections Suppose we project data points $x^{(k)} \in \mathbb{R}^n$, $k = 1, \ldots, m$, along a given direction u, obtaining the projected points $x_u^{(k)} \doteq (x^{(k)\top}u)u$, $k = 1, \ldots, m$. The projected points $x_u^{(k)}$ capture the data variability along direction u. We may then remove from the original data their components along u, thus *deflating* the data from their variability along this direction. The deflated data set is composed of vectors

$$x_{\neg u}^{(k)} \doteq x^{(k)} - x_u^{(k)} = x^{(k)} - (x^{(k)\top}u)u, \quad k = 1, \ldots, m.$$

Suppose now that we want to probe the variability of the deflated data set along another direction v. Projecting the u-deflated points $x_{\neg u}^{(k)}$ along v we obtain the projections

$$x_{\neg u;v}^{(k)} = \left(x^{(k)\top}v\right)v - \left(x^{(k)\top}u\right)(u^\top v)v, \quad k = 1, \ldots, m.$$

If we instead project the original data points directly along v, we obtain
$$x_v^{(k)} \doteq \left(x^{(k)\top} v\right) v, \quad k = 1, \ldots, m.$$

We observe that $x_v^{(k)} - x_{\neg u;v}^{(k)} = \left(x^{(k)\top} u\right)(u^\top v)v \neq 0$ in general. That is, if we first deflate the data from the u component and then project along v we obtain a different result with respect to projecting the data directly along v. This is due to the fact that the new probing direction v has in general a nonzero score $u^\top v$ along u. When probing data along successive directions it is therefore a good idea to choose the successive directions so to have zero score with respect to the preceding ones. In our case, if we choose v such that $u^\top v = 0$ (i.e., v is orthogonal to u), then we obtain that $x_v^{(k)} = x_{\neg u;v}^{(k)}$. This choice also simplifies the deflation process, since now the data deflated from the components along u and v are simply given by

$$x_{\neg uv}^{(k)} \doteq x^{(k)} - x_u^{(k)} - x_v^{(k)} = x^{(k)} - \left(x^{(k)\top} u\right) u - \left(x^{(k)\top} v\right) v.$$

The process can be continued by considering a third direction w, chosen so to be orthogonal to both u and v, projecting onto w, deflating the data and so on.

2.5.2 Projection onto a subspace

The line $\mathcal{L}_u$ passing through the origin considered in Section 2.5.1 is a one-dimensional subspace of $\mathbb{R}^n$. More precisely, it is the subspace generated as the *span* of a single vector $u \in \mathbb{R}^n$. It seems then natural to generalize the concept to projections onto subspaces generated as the span of a collection of $d \leq n$ independent vectors. Specifically, we let $u^{(j)} \in \mathbb{R}^n$, $j = 1, \ldots, d$, be a given collection of orthonormal vectors, that is, such that for $i, j = 1, \ldots, d$, it holds that

$$u^{(j)\top} u^{(i)} = \begin{cases} 1 & \text{if } i = j, \\ 0 & \text{otherwise,} \end{cases}$$

and we define

$$\mathcal{U} \doteq \{x \in \mathbb{R}^n : x = \alpha_1 u^{(1)} + \cdots + \alpha_d u^{(d)}; \alpha_i \in \mathbb{R}, i = 1, \ldots, d\}.$$

Letting $U \doteq [u^{(1)} \cdots u^{(d)}] \in \mathbb{R}^{n,d}$ and $\alpha^\top = [\alpha_1 \cdots \alpha_d]$, we write in more compact matrix notation

$$\mathcal{U} = \{x \in \mathbb{R}^n : x = U\alpha; \alpha \in \mathbb{R}^d\}.$$

Given a vector $x \in \mathbb{R}^n$, its projection onto $\mathcal{U}$ is defined as

$$x_\mathcal{U} = \arg\min_{\zeta \in \mathcal{U}} \|\zeta - x\|_2.$$

Again, the Projection theorem guarantees that the projection exists and it is unique, and it is univocally determined by enforcing the following necessary and sufficient conditions:

$$x_\mathcal{U} \in \mathcal{U}, \quad \text{and} \quad (x - x_\mathcal{U}) \perp \mathcal{U}.$$

The first condition requires that $x_\mathcal{U} = U\alpha$ for some $\alpha \in \mathbb{R}^d$ to be determined. The second condition requires that the inner product between $x - x_\mathcal{U}$ and each column of U is zero, that is, considering that $U^\top U = I_d$,

$$(x - U\alpha)^\top U = 0 \quad \Leftrightarrow \quad \alpha = U^\top x.$$

From this we conclude that the projection $x_\mathcal{U}$ is given by

$$x_\mathcal{U} = UU^\top x,$$

where the scores of x along the directions $u^{(1)}, \ldots, u^{(d)}$ are given by

$$\alpha_1 = {u^{(1)}}^\top x, \ \alpha_2 = {u^{(2)}}^\top x, \ldots, \alpha_d = {u^{(d)}}^\top x.$$

A data set $X \in \mathbb{R}^{n,m}$ is thus readily projected onto the subspace span by the columns of U by the formula

$$X_\mathcal{U} = (UU^\top)X.$$

Projection on a plane A plane passing through zero is a two-dimensional subspace that can be described by means of the span of two orthogonal directions $u^{(1)}, u^{(2)} \in \mathbb{R}^n$. The projection of a (centered) data set $X \in \mathbb{R}^{n,m}$ onto this plane can be represented graphically in terms of the scores (α_1, α_2) of the data points along the two directions $u^{(1)}, u^{(2)} \in \mathbb{R}^n$, where the scores are computed as $\alpha_1(k) = {u^{(1)}}^\top x^{(k)}$, $\alpha_2(k) = {u^{(2)}}^\top x^{(k)}$, for $k = 1, \ldots, m$. The scores plot is a two-dimensional plot which is easy to visualize graphically.

Recalling our loan data example, for instance, we may consider as direction $u^{(1)}$ the sum of the applicant income (variable n. 6) and the co-applicant income (variable n. 7), and as direction $u^{(2)}$ the requested loan amount (variable n. 8), that is,

$$u^{(1)} = \frac{1}{\sqrt{2}} \begin{bmatrix} 0 \\ 0 \\ 0 \\ 0 \\ 0 \\ 1 \\ 1 \\ 0 \\ 0 \\ 0 \\ 0 \end{bmatrix}, \quad u^{(2)} = \begin{bmatrix} 0 \\ 0 \\ 0 \\ 0 \\ 0 \\ 0 \\ 0 \\ 1 \\ 0 \\ 0 \\ 0 \end{bmatrix}.$$

Projecting the centered and standardized data set onto the plane generated by $u^{(1)}, u^{(2)}$, we obtain the score plot shown in Figure 2.9.

2.6 Data visualization

Data visualization constitutes a fundamental step towards the understanding of a data set, especially in the early phase of the so-called exploratory data analysis, in which the data scientist tries to understand the peculiar characteristics of the data set under study, in order to decide which more advanced tools to use next. We next illustrate some of the standard data visualization techniques that are commonly used in a financial data science context.

Line graphs A line graph uses points connected by line segments from left to right to show changes in value. The horizontal axis depicts a continuous progression, often that of time, while the vertical axis reports values for a quantity of interest across that progression. A typical example of line graph is the graph of the evolution of a stock price as a function of time, see an example in Figure 2.10.

Candlestick charts Candlestick charts are frequently used for displaying price data evolution for stocks. Since price data in the stock market is available in an asynchronous way (i.e., a price is formed at any instant, whenever a bid offer and ask are met), this data is made synchronous and regular in time by collecting the price event data over given fixed periods of time, for example, one minute or one hour, and then plotting some statistics of the collected data. A typical candle plot displays a vertical box or bar with the price at the beginning of the candle period (the so-called open price) and the price at the end of the candle period (the so-called close price), and superim-

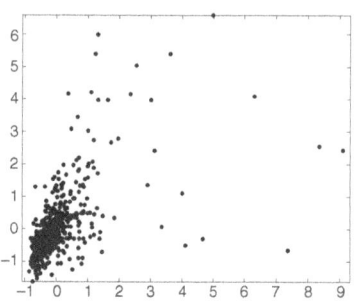

Figure 2.9 Two-dimensional score plot for the loan data set, see file `ex_scores_loanexample.m` in the online resources.

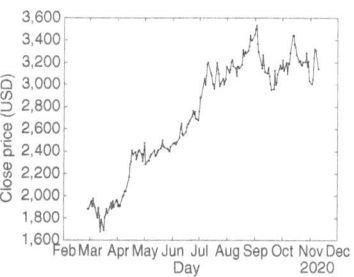

Figure 2.10 Line graph showing the daily close price of the Amazon stock (AMZN) in the period from Feb. 27 to Nov. 9, 2020 (179 price data points), see file `plot_stock.m` in the online resources.

poses to the box a vertical line extending from the minimum (low) to the maximum (high) of the prices in the candle period. Also, the convention is that if the closing price is greater than the opening price, the body (the region between the open and close price) of the box is unfilled, otherwise the body is filled. Figure 2.11 gives an example of a candlestick chart.

Figure 2.11 A 5 minute candlestick chart of the Apple (AAPL) stock price on Nov. 25, 2024 (Yahoo finance data and graph).

Box plots A box plot is a standardized way of displaying the distribution of data based on a summary of the empirical data statistics. Typically, a box plot shows the median value of the samples, the first quartile Q_1 and the third quartile Q_3, the minimum range defined as $Q_1 - 1.5(Q_3 - Q_1)$, the maximum range defined as $Q_3 + 1.5(Q_3 - Q_1)$, as well as extreme values of the sample (i.e., those above the maximum range and below the minimum range), considered as outliers. Other choices are possible, such as displaying the mean of the data, the plus/minus one standard deviation range, and the maximum and minimum of the data. Box plots give a readily understandable summary about how tightly the data is grouped, about how the data is symmetrical or skewed, about outliers, and so on. Figure 2.12 gives an example of box plot, based on the loan data described in Section 2.1; this box plot summarizes the distribution of the applicant's income versus gender.

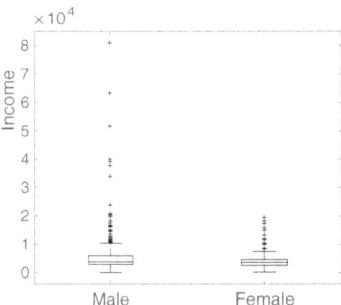

Figure 2.12 Box plot of applicant's income vs. applicant's gender (loan data set), see file ex_boxplot.m in the online resources.

Histograms Histograms are a standard tool for displaying the frequency distribution in a variable. In a histogram, a number of bins of given center and width are created, and then the scalar values of the variable of interest are assigned to the bins and counted. The histogram plot then shows the absolute or relative counts of the bins by means of a bar chart. Figure 2.13 gives an example of histogram of the distribution of the requested loan amounts, based on the loan data described in Section 2.1.

Scatterplots Scatterplots display the relation between pairs of variables, by plotting on a plane one variable versus the other. They are often used to highlight correlation between variables. In the case of n variables $x_1, \ldots, x_n$, a scatter plot matrix is often used, which displays an $n \times n$ array of two-dimensional plots, where the plot in position (i, j) in the plot array displays x_j as a function of x_i.

As an example, we consider a data set containing fundamental financial information about most of the companies listed in the Standard & Poor 500 index. The complete data set is available in the online resources in file fundamentals_SP500.csv, which contains data for $m = 453$ companies, each described by a vector of features that

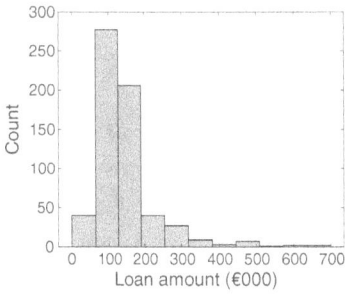

Figure 2.13 Histogram of the requested loan amounts (loan data set), see file ex_boxplot.m in the online resources.

include information such as the industrial sector code, the number of employees (NEMPLY), the number of shares (NSHARES), the market capitalization (MKTCAP), revenue (TTMREV), earnings before interest, taxes, depreciation and amortization (TTMEBITD), dividends per share (TTMDIVSHR), price/earnings ratio (PEEXCLXOR), and many others. Figure 2.14 shows the scatterplot array for the fundamental S&P 500 data, restricted for simplicity to the four features (NEMPLY, MKTCAP, TTMDIVSHR, PEEXCLXOR), centered and standardized.

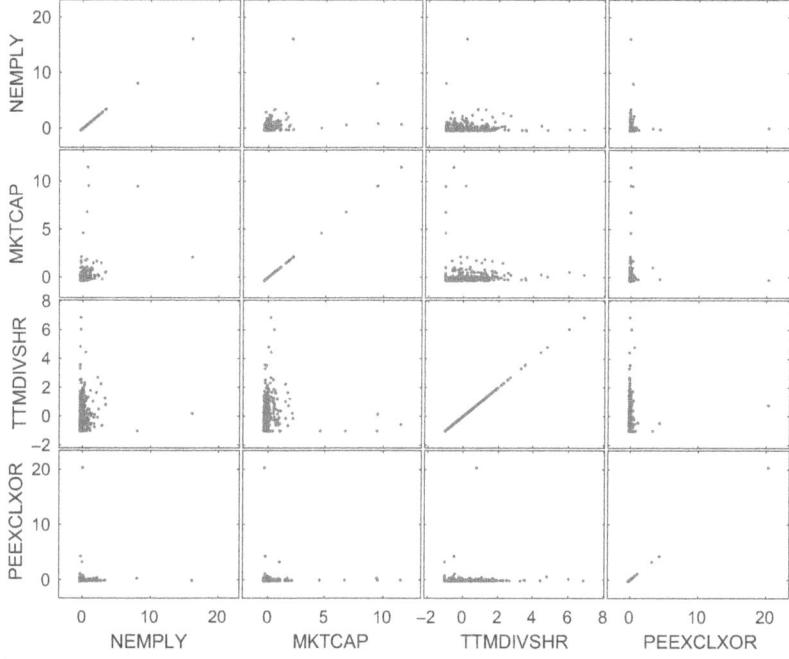

Figure 2.14 Scatterplot of the features (NEMPLY, MKTCAP, TTMDIVSHR, PEEXCLXOR) from the standardized S&P 500 fundamentals data set; see MATLAB® file ex_scatterplot.m in the online resources.

Heatmaps The idea behind heatmap plots is to replace numbers with colors. Heatmaps are widely used with geospatial data, and are commonly applied for describing density or intensity of variables, for visualizing patterns, variance, and for spotting anomalies. Often, heatmaps are used to visually display the correlation pattern among variables. For example, the heatmap of the Pearson correlation matrix for the loan data is shown in Figure 2.15.

Other visualizations What we illustrated in the previous paragraphs is only a summary of some of the most frequently used data visualization approaches. This list is by no means complete, so the reader is invited to explore other methods that can be useful for their project of interest, such as, for instance, pie charts, network graphs, three-dimensional scatter plots, radar charts, and so on.

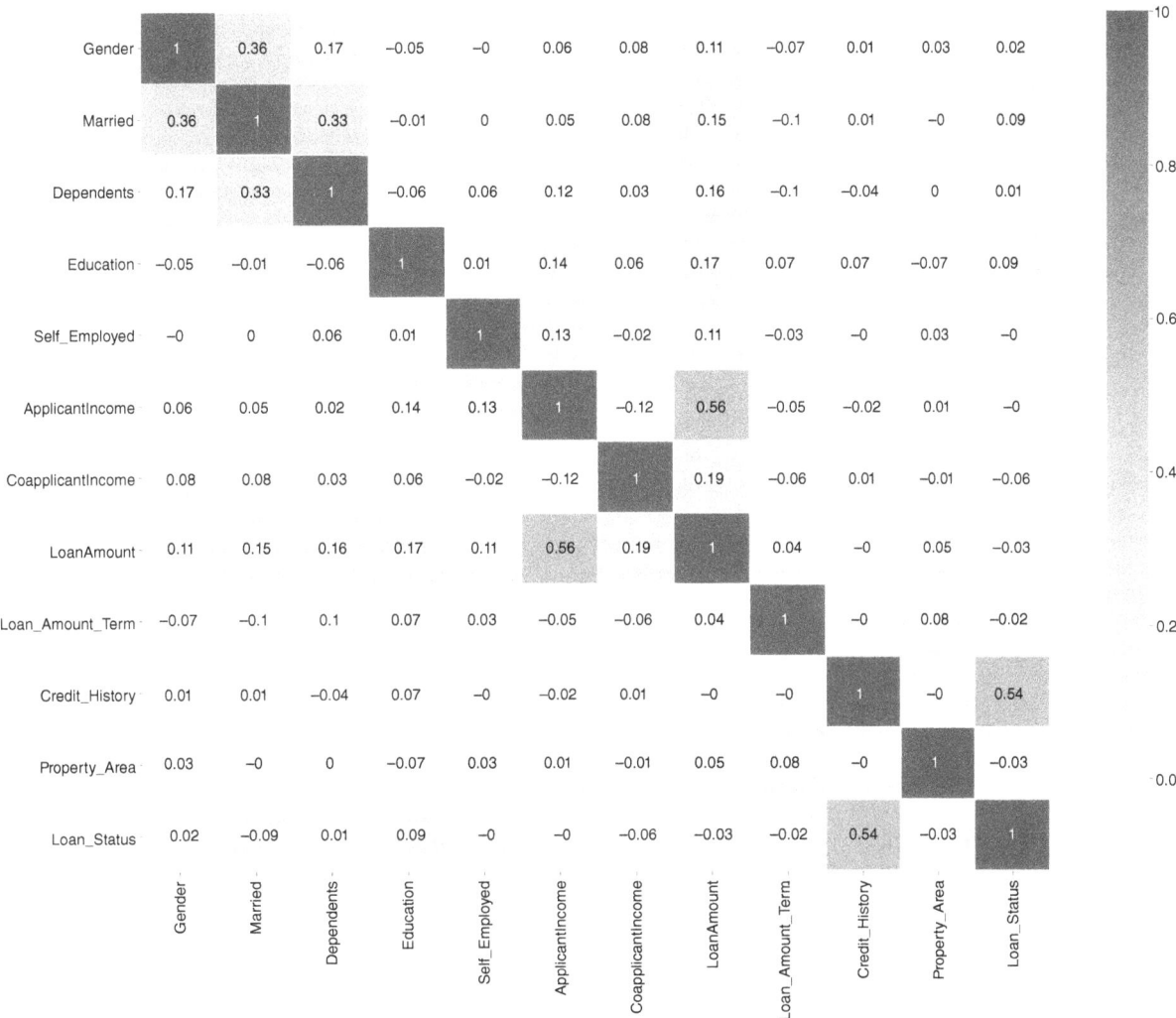

Figure 2.15 Heatmap of the Pearson correlation matrix for the loan data; see Python file `loan_data_set_numeric.py` in the online resources.

2.7 Exercises

Exercise 2.1 *Scatterplots and box plots*

In Section 2.6 we introduced scatterplots and box plots. In this exercise we will use them to analyze a data set containing information about the spending habits of customers of a wholesale distributor. The data set contains information about 440 customers and each of them is described using eight features: the geographical region, the retail channel, and the annual spending amounts of various product categories (namely: fresh products, milk, grocery products, frozen products, detergents and paper products, and delicatessen).

28   2 DATA REPRESENTATION AND VISUALIZATION

1. Download the file customers.csv in the online resources.[6] Discard the information regarding the geographical region and the retail channel, since they are uninformative for our purpose.

   [6] This data set is also available at https://archive.ics.uci.edu/dataset/292/wholesale+customers.

2. Draw a box plot of the features representing the annual spending amounts of the various product categories, in order to display the data distribution of the various features.

3. Draw a 6 × 6 scatter plot matrix of the various features. Are there any pairs of features which exhibit some degree of correlation?

Exercise 2.2 *Data projection*

1. Download from the book's online resources the file SP500_2015.csv, which contains the daily price data of the companies included in the Standard & Poor 500 index for the year 2015.

2. Apple (AAPL), Microsoft (MSFT), Amazon (AMZN), Johnson & Johnson (JNJ), JP Morgan Chase (JPM), Facebook (FB), Exxon Mobil (XOM), Alphabet C (GOOG), and Alphabet A (GOOGL) are some of the most important companies included in this index. Build a data matrix by considering only the data of these companies (sorted in alphabetic order of their tickers).

3. Draw a line graph of the daily prices of these stocks.

4. Let us suppose that we have a portfolio with the following asset allocation (negative for shorting):

   $$v = [1\ 0\ -1\ 0\ \tfrac{1}{2}\ 0\ \tfrac{3}{4}\ 0\ \tfrac{1}{4}]^\top.$$

   Compute the projection of the data matrix along the line $\mathcal{L}_v = \{x \in \mathbb{R}^n : x = \alpha v,\ \alpha \in \mathbb{R}\}$. This projection represent a synthetic asset created by mixing together the original stocks. Draw a line graph of the daily prices of this synthetic asset.

Exercise 2.3 *Directional variance*

We consider the same data set introduced in the previous exercise.

1. As in the previous exercise, we restrict our study to a limited number of companies. Build a data matrix by considering only the data of the following stocks (sorted in alphabetic order of their tickers): American Water (AWK), Devon Energy (DVN), Edison International (EIX), Marathon Oil (MRO), Sysco (SYY), Welltower (WELL), Williams Companies (WMB), and Wynn Resorts (WYNN).

2. Using the data of these stocks, compute their average returns and variance.

3. Compute the directional variances of the daily returns along the directions given by the following vectors:

$$v_1 = [0\ 1\ 0\ 1\ 0\ 0\ 1\ 1]^\top,$$

$$v_2 = [1\ 0\ 1\ 0\ 1\ 1\ 0\ 0]^\top.$$

These two directions define two different subsets of the stocks selected in point 1. Which direction provides the highest variance? Which provides the highest average return?

Exercise 2.4 *Bar graph*
A simple but effective way of visualizing data that vary over time is to draw a bar graph. A bar graph is a graph that presents data with rectangular bars with heights or lengths proportional to the values that they represent.

1. Download from the online resources the file norway_new_car_sales_by_make.csv from the online resources, which contains the monthly car sales in Norway for the period 2007–2017 by make.

2. Draw a bar graph representing the number of cars sold by Volkswagen, Toyota, Peugeot, Tesla, and Fiat in Feb. 2014, Feb. 2015, and Feb 2016.

3. Can you extract any information about car sales in Norway for the years 2017–2018 from these data?

Exercise 2.5 *Website traffic*
The file website_data.csv in the online resources contains data about the number of visits to a small website.[7] The website is mainly visited by university students.

[7] The dataset can also be downloaded from http://openmv.net/info/website-traffic.

1. Download the data set and draw a line graph of the daily website traffic. Can you detect any significant trend from this visualization?

2. Draw a box plot that shows the variability in website traffic for each day of the week. Can you detect any significant trend from this visualization?

3
Data models and estimation

PRACTICAL QUANTITATIVE tools for analysis, design, or prediction are habitually based on working assumptions and simple statistical models. In a financial context, for instance, classical models for asset returns assume that returns follow some stochastic process and, under further hypotheses such as stationarity, parameters of the model such as the expectation and covariance can be estimated by using statistical estimation techniques, on the basis of available *samples* or observations of realizations.

In this chapter, we introduce a basic statistical models for static and dynamic data generation, and discuss classical techniques based on the Bayesian approach for the estimation of the parameters of the model. We discuss in particular the estimation of covariance and correlation matrices, which are key objects used in computational finance for exploratory data analysis, risk analysis, hedging, portfolio optimization, and outlier detection. Standard estimation techniques, such as those based on the maximum likelihood principle, however, hinge upon asymptotic efficiency theories, which means in practice that they work well when the number m of samples to be used for the estimation is significantly larger than the dimension n of the sample vectors. Unfortunately, this situation is seldom verified in a financial context, where one needs for instance to estimate covariance or correlations among a large number n of assets (say, in the range of hundreds, or even thousands) based on a limited number m of time samples, and where hence the situation $m < n$ is not uncommon. We shall discuss various covariance estimation methods, starting from the naive sample estimate and then moving to factor models, shrinkage estimators, and sparse covariance estimation. We also discuss related problems such as the computation of confidence regions and outlier detection.

3.1 A basic estimation model

In order to pose the parameter estimation problem on a sound statistical basis, we need to assume that the data we observe are generated by some underlying stochastic model, the so-called *data generation mechanism* (DGM). This mechanism is assumed to exist but it is otherwise unknown; we can see and collect the data generated by this mechanism, but cannot access the mechanism itself. Sometimes we may assume that the DGM has some postulated structure, while the *parameters* entering this model structure remain unknown. The parameters of the DGM may be, for instance, the mean vector and covariance matrix of a probability distribution function assumed in the DGM, and our purpose is to estimate these unknown parameters by exploiting (a) observations, that is, available samples (realizations) generated by the DGM, and possibly (b) prior knowledge we may have on the value of these parameters.

We shall henceforth consider an independent identically distributed (i.i.d.) DGM of the form

$$x^{(i)} \sim f_\theta(x), \quad \text{for } i = 1, \ldots, m, \tag{3.1}$$

which means that the m samples of the feature vector $x \in \mathbb{R}^n$ are generated independently according to the same probability distribution f_θ, which depends on an unknown parameter θ (this parameter may actually denote a collection of parameters, such as the mean and the covariance matrix of the distribution), that needs be estimated. We next introduce the essential aspects of a general methodology for estimation known as *Bayesian inference*.

3.1.1 Bayesian estimation

Let $\theta \in \mathbb{R}^d$ be a vector of parameters we are interested in. For instance, θ may contain the elements of the mean vector of a distribution, and/or the elements of its covariance matrix. In a Bayesian setting, our assumptions and a-priori knowledge about θ, before observing any data, are expressed in the form of a *prior probability distribution* $p(\theta)$ on θ, also called *prior belief*, or simply *prior*. This prior summarizes all we know about the unknown parameter prior to seeing the data, for instance from previous experiments or expert advice. If we do know nothing, we may select a so-called *flat prior*, that is, a uniform probability distribution $p(\theta)$ that does not favor any specific value.

We then let $\mathcal{D} = \{x^{(i)}\}_{i=1,\ldots,m}$ denote a set of observed data generated according to (3.1). The statistical model relating $\mathcal{D}$ to θ is expressed by the conditional distribution $p(\mathcal{D}|\theta)$, which is called the *likelihood*. Knowing the likelihood means knowing the structure of the DGM, that is, the class of probability distribution functions that is generating the data, albeit we do not know the specific parameters θ of such distribution. Bayes' rule for for probability distributions then states that

$$p(\theta|\mathcal{D}) = \frac{p(\mathcal{D}|\theta)p(\theta)}{p(\mathcal{D})}, \qquad (3.2)$$

where $p(\theta|\mathcal{D})$ is the *posterior distribution*, representing the updated state of knowledge about θ, after we see the data in $\mathcal{D}$, and $p(\mathcal{D})$ is the marginal distribution of the data, which can be expressed in terms of the prior and the likelihood via the continuous version of the total probability rule

$$p(\mathcal{D}) = \int p(\mathcal{D}|\theta)p(\theta)\mathrm{d}\theta.$$

Since $p(\mathcal{D})$ only acts as a normalization constant in (3.2), we can state Bayes' rule as

$$\text{Posterior} \propto \text{Likelihood} \times \text{Prior},$$

where $\propto$ means "proportional to." The posterior distribution of θ can be used to infer information about θ, that is, to find a suitable point estimate $\hat{\theta}$ of θ, as discussed in Section 3.1.2.

The Bayesian approach is very suitable for on-line, recursive inference, since it can be applied recursively: as new data is gathered the current posterior becomes the new prior, and a new posterior is computed based on the new data, and so on recursively, as illustrated in Figure 3.1.

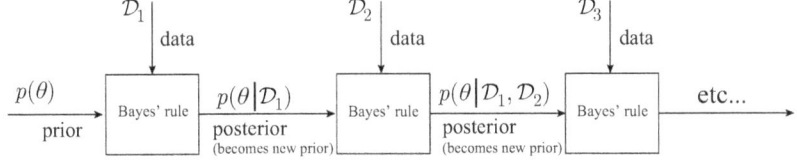

Figure 3.1 Recursive application of Bayes' rule.

3.1.2 Loss functions and point estimates

A point estimate $\hat{\theta}$ of θ is a vector that minimizes the expected value of a given *loss function* under the posterior density of θ. A loss function $\mathcal{L}$ measures the average cost in predicting θ via $\hat{\theta}$. Typical loss functions are the following ones (assuming for simplicity θ to be scalar):

- quadratic:
$$\mathcal{L}(\theta - \hat{\theta}) = (\theta - \hat{\theta})^2;$$

- absolute value:
$$\mathcal{L}(\theta - \hat{\theta}) = |\theta - \hat{\theta}|;$$

- hit-or-miss: for given $\delta > 0$,
$$\mathcal{L}(\theta - \hat{\theta}) = \begin{cases} 0 & \text{if } |\theta - \hat{\theta}| \leq \delta, \\ 1 & \text{if } |\theta - \hat{\theta}| > \delta; \end{cases}$$

- Huber loss: for given $\delta > 0$,
$$\mathcal{L}(\theta - \hat{\theta}) = \begin{cases} \frac{1}{2}(\theta - \hat{\theta})^2 & \text{if } |\theta - \hat{\theta}| \leq \delta, \\ \delta(|\theta - \hat{\theta}| - \delta/2) & \text{if } |\theta - \hat{\theta}| > \delta. \end{cases}$$

A Bayesian point estimator is defined as the point that minimizes the expected loss, under the posterior conditional density:

$$\begin{aligned}\hat{\theta} &= \arg\min_{\hat{\theta}} \int \mathcal{L}(\theta - \hat{\theta}) p(\theta|\mathcal{D}) \mathrm{d}\theta \\ &= \arg\min_{\hat{\theta}} \mathrm{E}\{\mathcal{L}(\theta - \hat{\theta})|\mathcal{D}\}.\end{aligned} \quad (3.3)$$

Given a prior $p(\theta)$ and a likelihood function $p(\mathcal{D}|\theta)$, we compute the posterior $p(\theta|\mathcal{D}) = \text{const.} \times p(\mathcal{D}|\theta)p(\theta)$ and, based on this posterior and of the loss function of choice, we can obtain various point estimators, among which (see Appendix A.2 for a derivation of the following facts):

- $\hat{\theta}_{\text{MMSE}}$ is the *mean* of the posterior $p(\theta|\mathcal{D})$. It is the value that minimizes the expected quadratic loss.

- $\hat{\theta}_{\text{MMAE}}$ is the *median* of the posterior $p(\theta|\mathcal{D})$. It is the value that minimizes the expected absolute loss.

- $\hat{\theta}_{\text{MAP}}$ is the *maximum* (i.e., peak, or mode) of the posterior $p(\theta|\mathcal{D})$. It is the value that minimizes the expected hit-or-miss loss.

- $\hat{\theta}_{\text{ML}}$ is the *maximum* of the *likelihood* function $p(\mathcal{D}|\theta)$.

Example 3.1 (*Estimation of the success probability*) Suppose we are interested in a repetitive phenomenon whose outcome can be either "success," which we tag with the number 1, or "failure," which we tag with the number 0. Such a phenomenon can be, for instance, the price of an asset going up (1) or down (0) one day with respect to the preceding day, the outcome of a bet on a possibly unfair dice toss (say, success if the result is above 4, or failure otherwise), the result of a biased coin toss, and so on. The point here is that we do not know precisely the probability $\theta \in [0,1]$ of success, or the probability $1-\theta$ of failure, so we want to estimate such probability from available data. For instance, we suspect that a coin is biased, hence we believe that $\theta \neq 0.5$, and we want to estimate the actual value of θ by tossing the coin many times and by using the resulting data for estimating θ.

If we make the hypothesis that the outcomes of the phenomenon are independent (i.e., success/failure in one trial does not depend on successes/failures in preceding trials) then we are in the framework of (3.1), where $x^{(i)}$, $i = 1, \ldots, m$, are the binary (0/1) outcomes in m realizations of the phenomenon under consideration, and f_θ is a Bernoulli probability distribution, that is,

$$f_\theta(x) = \begin{cases} \theta & \text{if } x = 1, \\ 1-\theta & \text{if } x = 0, \end{cases}$$

which can be compactly expressed as $f_\theta(x) = \theta^x (1-\theta)^{1-x}$, for $x \in \{0,1\}$. Since the observations $x^{(i)}$, $i = 1, \ldots, m$, are independent, the likelihood function $p(\mathcal{D}|\theta)$ is

$$p(\mathcal{D}|\theta) = \prod_{i=1}^{m} f_\theta(x^{(i)}) = \theta^S (1-\theta)^{m-S},$$

where $S \doteq \sum_{i=1}^{m} x^{(i)}$ is the number of observed successes in the m trials. Note that the likelihood is proportional to a Binomial distribution. We next need to introduce a distribution quantifying the prior knowledge on θ. In the context of Bayesian learning with Binomial likelihood it is customary to use a Beta distribution for quantifying the prior:

$$p(\theta) = \text{Beta}_{\alpha,\beta}(\theta) \doteq \frac{1}{B(\alpha,\beta)} \theta^{\alpha-1}(1-\theta)^{\beta-1}, \quad \theta \in [0,1],$$

where $\alpha > 0$, $\beta > 0$ are parameters of the distribution and $B(\alpha, \beta)$ is a normalization constant. The distribution $\text{Beta}_{\alpha,\beta}(\theta)$ has mean value $\alpha/(\alpha+\beta)$ and has a peak in $(\alpha-1)/(\alpha+\beta-2)$ when $\alpha, \beta > 1$, see Figure 3.2.

For $\alpha = \beta = 1$ we simply obtain the uniform distribution on $[0,1]$, which can be used as a flat prior in case we have no prior idea about the location of θ. For $\alpha = \beta = 2$ we obtain instead a Beta distribution with mean and peak in $\theta = 0.5$, which can be used, for instance, if we have some prior hints that the success probability should be around 0.5, as in the case of a coin toss.

Given the prior and likelihood, the posterior is computed according to Bayes' rule as

$$p(\theta|\mathcal{D}) \propto p(\mathcal{D}|\theta)p(\theta) = \theta^S(1-\theta)^{m-S}\theta^{\alpha-1}(1-\theta)^{\beta-1}$$
$$= \theta^{(S+\alpha)-1}(1-\theta)^{(m-S+\beta)-1}.$$

The posterior density is thus still of Beta type. It is then immediate to obtain the following point estimates of θ:

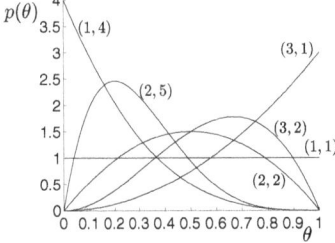

Figure 3.2 Examples of shapes of the $\text{Beta}_{\alpha,\beta}(\theta)$ pdf for various (α, β) pairs.

- the minimum mean square error (MMSE) estimate (mean of the posterior) is

$$\hat{\theta}_{\text{MMSE}} = \frac{S+\alpha}{\alpha+\beta+m};$$

- the maximum a-posteriori (MAP) estimate (peak of the posterior) is

$$\hat{\theta}_{\text{MAP}} = \frac{S+\alpha-1}{\alpha+\beta+m-2};$$

- the maximum likelihood estimate, which coincides with the MAP estimate for flat prior (i.e., $\alpha = \beta = 1$), is

$$\hat{\theta}_{\text{ML}} = \frac{S}{m}.$$

Observe that the ML estimate corresponds to the natural sample average estimate (number of observed successes over total number of trials). The ML estimate neglects the prior and only uses data evidence to estimate the parameter. While the ML approach is optimal for large sample size (i.e., for $m \to \infty$), it may lead to misleading estimates for small sample size, due to unlikely but possible events of "bad" samples. For instance, if we are estimating the probability of heads of a biased coin and observe a sequence of $m = 10$ tails values (an unlikely but possible event), then the ML estimate would be $\hat{\theta}_{\text{ML}} = 0$, which would lead to a model by which we will predict always tails in the future. If we considered prior information, instead, for instance by using $\alpha = \beta = 2$ in the prior, then the same observed data would lead us to the more prudent estimate $\hat{\theta}_{\text{ML}} = 1/(m+2)$. In general, the attractive feature of the Bayesian approach is to provide a sound framework for suitably blending the evidence coming from data with useful prior information, when available.

3.1.3 Estimation of mean and covariance in a Gaussian model

We next consider the case when the distribution f_θ in the DGM (3.1) is a multivariate Gaussian distribution, that is,

$$f_\theta(x) = \frac{1}{(2\pi)^{n/2}\sqrt{\det \Sigma}} \exp\left(-\frac{1}{2}(x-\mu)^\top \Sigma^{-1}(x-\mu)\right),$$

where $\mu \in \mathbb{R}^n$ is the mean of the distribution, Σ is the covariance matrix (an $n \times n$ symmetric matrix, see Appendix A.1 for definitions, notation and facts about symmetric matrices), assumed to be invertible, and $\theta = (\mu, \Sigma)$ contains the unknown parameters of the distribution, which are to be estimated. Since the observations $x^{(i)}$, $i = 1, \ldots, m$, are independent, the likelihood function $p(\mathcal{D}|\theta)$ is

$$p(\mathcal{D}|\theta) = \prod_{i=1}^{m} f_\theta(x^{(i)})$$

$$= \left(\frac{1}{(2\pi)^{n/2}\sqrt{\det \Sigma}}\right)^m \exp\left(-\frac{1}{2}\sum_{i=1}^{m}(x^{(i)}-\mu)^\top \Sigma^{-1}(x^{(i)}-\mu)\right).$$

Next, assume we have no prior knowledge on θ, whence the prior belief $p(\theta)$ is taken as constant. Then, the posterior is identical to the likelihood, and since mean, mode, and median coincide for the Gaussian distribution, we have that the MMSE, MMAE, and MAP estimators in this case all coincide with the maximum likelihood estimator, that is, the value of θ that maximizes the likelihood $p(\mathcal{D}|\theta)$. Observe that since the natural logarithm is a strictly monotone increasing function, maximizing the likelihood is equivalent to maximizing the logarithm of the likelihood, or log-likelihood

$$\ell(\theta) \doteq \ln p(\mathcal{D}|\theta)$$
$$= -\frac{m}{2}\ln(2\pi)^n + \frac{m}{2}\ln\det\Sigma^{-1} - \frac{1}{2}\sum_{i=1}^{m}(x^{(i)}-\mu)^\top \Sigma^{-1}(x^{(i)}-\mu),$$

which is in turn equivalent to minimizing the function

$$q(\mu, P) \doteq -\ln\det P + \frac{1}{m}\sum_{i=1}^{m}(x^{(i)}-\mu)^\top P(x^{(i)}-\mu)$$

with respect to (μ, P), where $P = \Sigma^{-1}$ is the so-called *precision matrix*. The minimum of q with respect to μ is obtained by equating to zero the gradient

$$\nabla_\mu q(\mu, \Sigma) = P\sum_{i=1}^{m}\left(\mu - x^{(i)}\right),$$

which results in an estimator $\hat{\mu}$ for μ,

$$\hat{\mu} = \frac{1}{m}\sum_{i=1}^{m} x^{(i)} \doteq \bar{x},$$

that coincides with the centroid, or sample mean, $\bar{x}$ of the data points. Substituting back into q, we obtain the negative log-likelihood as a function of P only:

$$\begin{aligned} q(P) &= -\ln\det P + \frac{1}{m}\sum_{i=1}^{m}(x^{(i)}-\bar{x})^\top P(x^{(i)}-\bar{x}) \\ &= -\ln\det P + \mathrm{Tr}(SP), \end{aligned} \qquad (3.4)$$

where we defined the *sample covariance* matrix

$$S \doteq \frac{1}{m}\sum_{i=1}^{m}(x^{(i)}-\bar{x})(x^{(i)}-\bar{x})^\top. \qquad (3.5)$$

It can be proved that $q(P)$ is a convex function of P over the domain where P is positive definite. Further, matrix differential calculus shows that $\nabla_P(-\ln\det P) = -P^{-1}$ and $\nabla_P(\mathrm{Tr}(SP)) = S$, therefore by setting the gradient of $q(P)$ to zero we obtain the condition

$S - P^{-1} = 0$, from which it follows that, if S is positive definite,[1] then the minimum of $q(P)$ is achieved at the optimal solution $P = S^{-1}$. Since the precision matrix P is the inverse of the covariance Σ, the maximum likelihood estimate of Σ is

$$\hat{\Sigma} = S,$$

that is, the maximum likelihood estimate of the covariance matrix is given by the sample covariance matrix S. It is worth to observe that equation (3.5), which is commonly used for computing the data covariance, is a rigorously valid estimate of the actual covariance of the DGM only under several assumptions that we made so far, namely: that the DGM is i.i.d., that the underlying probability distribution is normal (Gaussian), that we use a Bayesian approach with a flat prior, that we use MAP approach for point estimation, and that S itself is positive definite. Changing one or more of these assumptions will likely change the answer about what is the corresponding point estimate for the covariance matrix. This should help us understand that (3.5) is far from being *the* undisputed way to estimate a covariance matrix. In Sections 3.5–3.7 we shall discuss alternative covariance estimation methods that address some of the shortcomings of the plain sample covariance estimate. Most of these developments make use of basic properties of symmetric matrices, which are recalled in Appendix A.1.

[1] See Appendix A.1.3 for definitions, properties, and notation related to symmetric positive definite and semidefinite matrices.

3.2 Sample covariance and directional variance

Assume that data samples $x^{(1)}, \ldots, x^{(m)}$ from the DGM (3.1) are i.i.d. realizations of a random variable x having probability distribution f_θ with (unknown) expected value $\mu \in \mathbb{R}^n$ and covariance matrix $\Sigma \in \mathbb{S}_+^n$. Collecting these samples in the data matrix

$$X = [x^{(1)} \cdots x^{(m)}], \qquad (3.6)$$

we can proceed as in Section 2.3 and express the sample mean of the data as

$$\bar{x} \doteq \frac{1}{m} \sum_{i=1}^{m} x^{(i)} = \frac{1}{m} X \mathbf{1}_m, \qquad (3.7)$$

and the centered data matrix as

$$\check{X} \doteq [\check{x}^{(1)} \; \check{x}^{(2)} \cdots \check{x}^{(m)}] = X - \bar{x}\mathbf{1}_m^\top, \qquad (3.8)$$

where

$$\check{x}^{(i)} \doteq x^{(i)} - \bar{x}, \quad i = 1, \ldots, m,$$

are the centered data samples. Then, the sample covariance matrix (3.5) can be written compactly as

$$S = \frac{1}{m} \sum_{i=1}^{m} (x^{(i)} - \bar{x})(x^{(i)} - \bar{x})^\top = \frac{1}{m}\check{X}\check{X}^\top. \qquad (3.9)$$

As we have seen in Section 3.1.3, under suitable assumptions, $\bar{x}$ is the MAP estimate of the true unknown expectation $\mu = \mathrm{E}\{x\}$, and S is the MAP estimate of the true unknown covariance matrix $\Sigma = \mathrm{E}\{(x-\mu)(x-\mu)^\top\}$.

Given a direction represented by a vector $u \in \mathbb{R}^n$ such that $\|u\|_2 = 1$, the *directional variance* of the random variable x along direction u is defined as the variance of the score $\alpha_u = u^\top x$ of the projection of x along u, see Section 2.5. The score α_u is a scalar random variable with

$$\mu_u \doteq \mathrm{E}\{\alpha_u\} = \mathrm{E}\{u^\top x\} = u^\top \mathrm{E}\{x\} = u^\top \mu,$$
$$\sigma_u^2 \doteq \mathrm{var}\{\alpha_u\} = \mathrm{E}\{(u^\top x - u^\top \mu)^2\} = \mathrm{E}\{u^\top(x-\mu)(x-\mu)^\top u\}$$
$$= u^\top \Sigma u.$$

Since μ and Σ are typically unknown, so are also the mean μ_u and variance σ_u^2 of the directional score of x along u. However, we can construct point estimates of these quantities by substituting μ and Σ with their point estimates, given respectively, for instance, by the sample mean $\bar{x}$ and the sample covariance matrix S. We have in particular for the sample directional variance, using (3.9), that

$$\hat{\sigma}_u^2 \doteq u^\top S u = \frac{1}{m} u^\top \check{X}\check{X}^\top u = \frac{1}{m}\|\check{X}^\top u\|_2^2.$$

By taking $u = e^{(i)}$, where $e^{(i)}$ is vector of all zeros except for a one in position i, we obtain the directional variance along the ith coordinate axis, which corresponds to

$$\sigma_i^2 = e^{(i)^\top} \Sigma e^{(i)} = \Sigma_{ii}, \quad i = 1, \ldots, n,$$

that is, the ith diagonal element of the covariance matrix. We define the *total variance* of x as the sum of the directional variances along all coordinate axes:

$$\mathrm{tvar}(x) = \sum_{i=1}^{n} e^{(i)^\top} \Sigma e^{(i)} = \sum_{i=1}^{n} \Sigma_{ii} = \mathrm{Tr}(\Sigma).$$

Again, point estimates of the directional variance along coordinate axes and of the total variance are obtained by substituting Σ with its point estimate in the above formulae.

While the diagonal elements of Σ represent the directional variance along the coordinate axes, the off-diagonal elements of Σ_{ij}, $i \neq j$, are related to the *correlation coefficients* between the i,j components of the random vector x. In particular, the *Pearson correlation matrix* Q of x is defined such that

$$Q_{ij} \doteq \frac{\Sigma_{ij}}{\sigma_i \sigma_j}, \quad i,j = 1, \ldots, n,$$

or, in matrix notation,

$$Q = \mathrm{diag}\left(\sigma_1^{-1}, \ldots, \sigma_n^{-1}\right) \Sigma \, \mathrm{diag}\left(\sigma_1^{-1}, \ldots, \sigma_n^{-1}\right).$$

It descends from the definition that Q is PSD (positive semidefinite) and such that $Q_{ii} = 1$, for $i = 1, \ldots, n$, and $|Q_{ij}| \leq 1$ for all $i \neq j$.

3.3 A stochastic model for portfolio analysis and design

We next illustrate the ideas of covariance and directional variance with an applicative example dealing with the analysis of a financial portfolio. Consider n assets and a fixed time period Δ, say one day, or one month. Let r_i, $i = 1, \ldots, n$, denote the *return* of an investment in asset i over the given period Δ. This means that if we invest $w_i \geq 0$ dollars in asset i now, then after a time duration Δ we receive back an amount $(1 + r_i)w_i$. A *portfolio* $w = (w_1, \ldots, w_n)$ is a vector containing the dollar amounts invested at the beginning of the period among the n available assets. If we denote by W the total amount invested in the different assets, $W_{\mathrm{ini}} = \sum_{i=1}^n w_i$, then the amount we may receive back at the end of the period is

$$W_{\mathrm{end}} = \sum_{i=1}^n (1 + r_i) w_i = W_{\mathrm{ini}} + w^\top r,$$

where $r = (r_1, \ldots, r_n)$ is the vector of asset returns. The overall return of our portfolio is therefore

$$\rho(w) = \frac{W_{\mathrm{end}} - W_{\mathrm{ini}}}{W_{\mathrm{ini}}} = \frac{1}{W_{\mathrm{ini}}} w^\top r.$$

Clearly, in reality, the value of r is not known exactly. An usual working hypothesis made by practitioners is to assume that r is a random n-dimensional vector with some mean μ and covariance matrix Σ:

$$\mu \doteq \mathrm{E}\{r\}, \quad \Sigma \doteq \mathrm{var}\{r\} = \mathrm{E}\{(r - \mu)(r - \mu)^\top\}.$$

These parameters are themselves unknown in practice, but at least one may estimate them from historical data by using, for instance, some of the techniques discussed in this chapter. So, if $\mathrm{E}\{r\} = \mu$ and $\mathrm{var}\{r\} = \Sigma$, then we have for the (random) portfolio return $\rho(w)$ that

$$\bar\rho(w) \doteq \mathrm{E}\{\rho(w)\} = \frac{1}{W_{\mathrm{ini}}} \mathrm{E}\{w^\top r\} = \frac{1}{W_{\mathrm{ini}}} w^\top \mu,$$

$$\sigma_w^2 \doteq \mathrm{var}\{\rho(w)\} = \frac{1}{W_{\mathrm{ini}}^2} \mathrm{var}\{w^\top r\} = \frac{1}{W_{\mathrm{ini}}^2} w^\top \Sigma w.$$

Assuming for simplicity and without loss of generality that the initial investment is one unit, that is, $W_{\mathrm{ini}} = 1$, we have that the expected return of our portfolio is $\bar\rho(w) \doteq w^\top \mu$, and its variance is $\sigma_w^2 \doteq w^\top \Sigma w$. We observe that the portfolio variance is proportional to the directional variance of r, along the portfolio direction $w/\|w\|_2$. The portfolio variance σ_w^2 is often taken as a measure of the *riskiness* of the portfolio, while $\bar\rho(w)$ expresses the expected profitability of our investment. A portfolio allocation w is thus typically evaluated along these two axes: the expected profitability (expected return), and its associated risk (variance of the portfolio return). Clearly, an investor ideally has preference towards receiving a large return with a small associated risk. Such an ideal situation is hardly realizable in practice, since higher expected returns are typically associated with higher risk, and lower risk investments typically yield lower returns. A "good" portfolio is thus one that realizes a reasonable *tradeoff* between expected return and risk.

We exemplify the above approach with a simple example. Consider a universe of three stocks, say American Airlines (AAL), Best Buy (BBY), and Microsoft (MSFT). As of Feb. 3, 2021, a rough estimate for the daily expected returns and covariance for these stocks is

$$\hat\mu = \begin{bmatrix} 0.2574 & 0.1605 & 0.0996 \end{bmatrix}^\top / 100,$$

$$\hat\Sigma = \begin{bmatrix} 1.5149 & -0.0722 & 0.0046 \\ -0.0722 & 0.4257 & 0.1109 \\ 0.0046 & 0.1109 & 0.3699 \end{bmatrix} / 1000.$$

Let's believe, for the purpose of this example, that this estimated expected value and covariance matrix are the true ones. If an investor a seeks only to maximize their expected return, then the best choice would be to invest all of the budget $W_{\mathrm{ini}} = 1$ on the first asset, AAL, setting $w^{(a)} = (1, 0, 0)$, since this asset has the highest expected return $\bar\rho(w^{(a)}) = \hat\mu_1 = 0.2574\%$. However, the risk associated with this investment is high, being

$$\hat\sigma_{w^{(a)}}^2 \doteq w^{(a)\top} \hat\Sigma w^{(a)} = \hat\Sigma_{11} = 1.5149 \times 10^{-3}.$$

An investor b who is relatively more averse to risk may choose to invest all of their budget on the third asset, MSFT, setting $w^{(b)} = (0, 0, 1)$, since this asset has the lowest risk $\hat{\sigma}^2_{w^{(b)}} = \hat{\Sigma}_{33} = 0.3699 \times 10^{-3}$. However, the expected return associated with this investment is low, being $\bar{\rho}(w^{(b)}) = \hat{\mu}_3 = 0.0996\%$. A third investor c may instead decide to allocate half of their budget on AAL and half on MSFT, setting $w^{(c)} = (0.5, 0, 0.5)$. Investor c then has

$$\bar{\rho}(w^{(c)}) = w^{(c)\top} \hat{\mu} = 0.1785\%,$$
$$\hat{\sigma}^2_{w^{(c)}} = w^{(c)\top} \hat{\Sigma} w^{(c)} = 0.4735 \times 10^{-3}.$$

So, choice c is better than a in terms of risk, but worse in terms of expected return, while choice c is better than b in terms of expected return, but worse in terms of risk. Finally, consider an investor d who would allocate their budget 40% on AAL, 40% on BBY, and 20% on MSFT, which corresponds to the portfolio vector $w^{(c)} = (0.4, 0.4, 0.2)$. In such case, we have

$$\bar{\rho}(w^{(d)}) = w^{(d)\top} \hat{\mu} = 0.1871\%,$$
$$\hat{\sigma}^2_{w^{(d)}} = w^{(d)\top} \hat{\Sigma} w^{(d)} = 0.3207 \times 10^{-3}.$$

We may then conclude that portfolio d is better than portfolio c in terms of *both* expected return and risk. We thus say that portfolio c is *dominated* by d or, equivalently, that d dominates c. Portfolio d also dominates portfolio b, since again it has both higher expected return and lower risk. It is interesting to observe that portfolio d actually yields a risk level which is lower than the lowest level achievable by any single asset alone (portfolio b). This fact shows that portfolio *diversification*, under the current working hypotheses, may effectively reduce the risk of the investment. Figure 3.3 shows the three individual assets (AAL, BBY, MSFT) plotted as circles on a risk/return plane, together with the a, b, c, d portfolios. In the figure, risk is expressed in terms of the standard deviation of the portfolio return, that is, the square root of $\hat{\sigma}^2_w$.

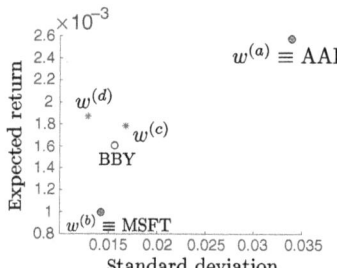

Figure 3.3 Assets and portfolios on a risk/return plane.

This example should also suggest that, if an actual investment decision is to be made based on a mean/variance model, the investor needs to pay great attention on how the mean and covariance parameters are estimated since, obviously, wrong or unreliable estimates may lead to wrong portfolio allocations. Just to highlight the problem, think of plain sample estimates of the mean and covariance of the returns, which are in practice computed by averaging over m return samples collected over a look-back window of data recorded in the past: changing the width m of the look-back window changes the estimates, and unfortunately there is no universal rule for establishing which the "right" width is.

In the next example, we explore in further depth the mechanism by which diversification decreases risk, and discuss the role of correlation.

Example 3.2 (*A two-asset portfolio*) We consider a problem of portfolio allocation with two assets. The random return vector is $r = (r_1, r_2)$, with mean $\mu = (\mu_1, \mu_2)$ and covariance matrix

$$\Sigma = \begin{bmatrix} \sigma_1^2 & \sigma_{12} \\ \sigma_{12} & \sigma_2^2 \end{bmatrix} \succeq 0.$$

We assume that at least one of the two assets has nonzero variance, say $\sigma_1 > 0$. Then, from the PSD condition on Σ it follows that $|\sigma_{12}| \leq \sigma_1 \sigma_2$. Both μ and Σ are assumed to be known, although in reality all we'll have, at best, is a good estimate of them. We further assume that $\mu_1 \geq 0$, $\mu_2 \geq 0$, and we assume without loss of generality that the expected returns are ordered so that $\mu_1 \geq \mu_2$. We also assume that $\sigma_1^2 \geq \sigma_2^2$, since the physiological situation is that the higher is the expected return the higher is the risk.

We let our portfolio allocation be given by vector $w = (\lambda, 1 - \lambda)$, with $\lambda \in [0,1]$. The expected portfolio return is thus a weighted average of μ_1, μ_2:

$$\bar{\rho}(\lambda) = w^\top \mu = \lambda \mu_1 + (1-\lambda)\mu_2, \quad \lambda \in [0,1],$$

where $\bar{\rho}(\lambda)$ is affine and monotonic non-increasing with $\lambda \in [0,1]$. The variance of the portfolio return is instead given by

$$\sigma_w^2(\lambda) = w^\top \Sigma w = (\sigma_1^2 + \sigma_2^2 - 2\sigma_{12})\lambda^2 - 2(\sigma_2^2 - \sigma_{12})\lambda + \sigma_2^2, \quad \lambda \in [0,1].$$

Notice that the fact that Σ is PSD implies that $[1\ -1]\Sigma[1\ -1]^\top = \sigma_1^2 + \sigma_2^2 - 2\sigma_{12} \geq 0$, whence $\sigma_w^2(\lambda)$ is a convex parabola. The vertex (minimum) of this parabola is at

$$\lambda_{\text{vert}} = \frac{\sigma_2^2 - \sigma_{12}}{(\sigma_2^2 - \sigma_{12}) + (\sigma_1^2 - \sigma_{12})}.$$

Notice that $\Sigma \succeq 0$ and the position that $\sigma_2^2 \leq \sigma_1^2$, imply that $|\sigma_{12}| \leq \sigma_1 \sigma_2 \leq \sigma_1^2$. If $-\sigma_1^2 \leq \sigma_{12} < \sigma_2^2$ we have that $\lambda_{\text{vert}} \in [0,1]$ and hence the optimal $\lambda \in [0,1]$ is given by λ_{vert}. If else $\sigma_2^2 \leq \sigma_{12} \leq \sigma_1^2$ then $\lambda_{\text{vert}} \leq 0$ and hence $\sigma_w^2(\lambda)$ is increasing in $[0,1]$, thus achieving the constrained minimum at $\lambda = 0$, see the example in Figure 3.4.

The minimum risk level achievable is therefore

$$\sigma_{\min}^2 = \min_{\lambda \in [0,1]} \sigma_w^2(\lambda) = \begin{cases} \dfrac{\sigma_1^2 \sigma_2^2 - \sigma_{12}^2}{\sigma_1^2 + \sigma_2^2 - 2\sigma_{12}} & \text{if } \sigma_{12} < \sigma_2^2, \\ \sigma_2^2 & \text{if } \sigma_{12} \geq \sigma_2^2. \end{cases}$$

We may observe that in the "high positive correlation" situation in which $\sigma_{12} \geq \sigma_2^2$, the optimal (minimum risk) portfolio is all concentrated on the second asset (since $\lambda = 0$) and the corresponding risk cannot be reduced below the σ_2^2 level. On the other hand, the situation in which $\sigma_{12} < \sigma_2^2$ allows to achieve risk levels that are below the σ_2^2 level, while also increasing the expected return of the investment. Further, the rate

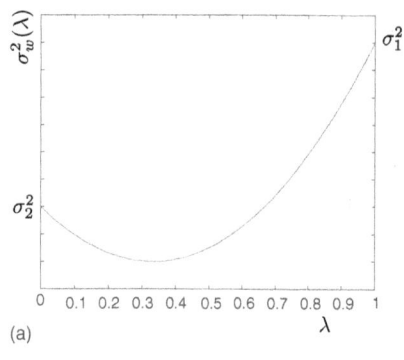
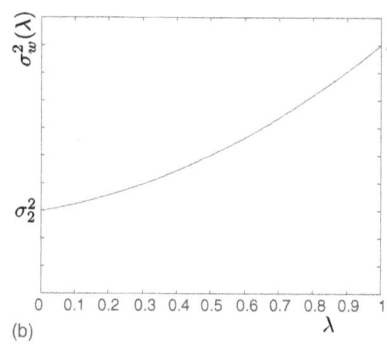

Figure 3.4 Plots of $\sigma_w^2(\lambda)$: (a) the case $\sigma_{12} < \sigma_2^2$; (b) the case $\sigma_{12} \geq \sigma_2^2$.

of decrease of the risk at $\lambda = 0$ is proportional to $-\sigma_2^2 + \sigma_{12}$, hence the smaller is σ_{12} the steeper is the decrease of the risk curve near $\lambda = 0$. Also, we have that the risk reduction

$$\delta(\sigma_{12}) \doteq \sigma_2^2 - \sigma_{\min}^2 = \frac{(\sigma_2^2 - \sigma_{12})^2}{(\sigma_2^2 - \sigma_{12}) + (\sigma_1^2 - \sigma_{12})}$$

is monotone decreasing in σ_{12}, for $\sigma_{12} < \sigma_2^2$, which implies that the deepest reduction in risk is achieved for negative σ_{12}. We may conclude from this simple example that negative correlation is desirable for the purpose of obtaining effective risk minimization from portfolio diversification.

We further observe that the objective in portfolio selection may be in general different than simply minimizing the risk. In fact, an investor also typically has the the goal of achieving a desired expected return, while controlling risk. Such mixed goals can be treated mathematically by minimizing over $\lambda \in [0, 1]$ an objective of the form

$$c(\lambda) = \sigma_w^2(\lambda) - 2\gamma \bar{\rho}(\lambda),$$

where $\gamma \geq 0$ is a parameter that weights the relative importance of the risk term and of the return term. Letting $\delta \doteq \mu_1 - \mu_2 \geq 0$, and under the condition

$$\sigma_{12} \leq \min(\sigma_1^2 - \gamma\delta, \sigma_2^2 + \gamma\delta),$$

we have that the constrained minimizer of $c(\lambda)$ is

$$\lambda^*(\gamma) = \frac{\sigma_2^2 - \sigma_{12} + \gamma\delta}{\sigma_1^2 + \sigma_2^2 - 2\sigma_{12}},$$

which corresponds to an optimal portfolio having risk and expected return, respectively:

$$\sigma^2(\gamma) = \frac{\sigma_1^2 \sigma_2^2 - \sigma_{12}^2 + \gamma^2 \delta^2}{\sigma_1^2 + \sigma_2^2 - 2\sigma_{12}},$$

$$\bar{\rho}(\gamma) = \frac{\sigma_2^2 - \sigma_{12} + \gamma\delta}{\sigma_1^2 + \sigma_2^2 - 2\sigma_{12}} \mu_1 + \frac{\sigma_1^2 - \sigma_{12} - \gamma\delta}{\sigma_1^2 + \sigma_2^2 - 2\sigma_{12}} \mu_2.$$

We can now plot $\bar{\rho}(\gamma)$ versus $\sigma^2(\gamma)$ for a suitable range of the tradeoff parameter $\gamma \geq 0$ to trace the optimal risk/return tradeoff frontier, see Figure 3.5 for a numerical example. Figure 3.6 shows a collection of optimal frontiers, each of which corresponds to a different level of cross-covariance σ_{12}; it is apparent in this plot that smaller and negative values of the correlation lead to improved frontiers that *dominate* the frontiers for higher correlations.

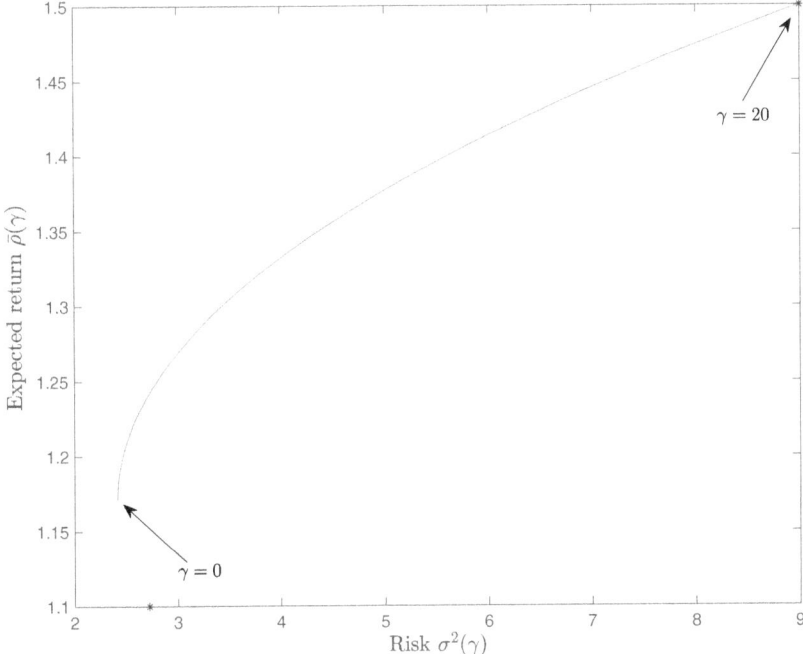

Figure 3.5 Optimal frontier plot of $\bar{\rho}(\gamma)$ vs. $\sigma^2(\gamma)$, for $\mu_1 = 1.5$, $\mu_2 = 1.1$, $\sigma_1^2 = 9$, $\sigma_2^2 = 2.7225$, $\sigma_{12} = 1$.

Further information about portfolio construction and analysis is contained in Chapter 11, which is devoted to an in-depth discussion of techniques for designing portfolio allocations to obtain desirable tradeoffs between expected return and risk.

3.4 Issues with the sample covariance matrix

The sample mean $\bar{x}$ in (3.7) and sample covariance matrix S in (3.9) are a staple of classical statistical inference. The main reason for this is that, due to the law of large numbers, these maximum likelihood estimates converge (in some suitable probabilistic sense) to the true and unknown quantities as the number m of samples tends to infinity, that is,

$$\bar{x} \stackrel{m \to \infty}{\longrightarrow} \mu, \quad S \stackrel{m \to \infty}{\longrightarrow} \Sigma.$$

It has to be understood, however, that this fact is true only if the data dimension n stays fixed, and the number of samples m goes to

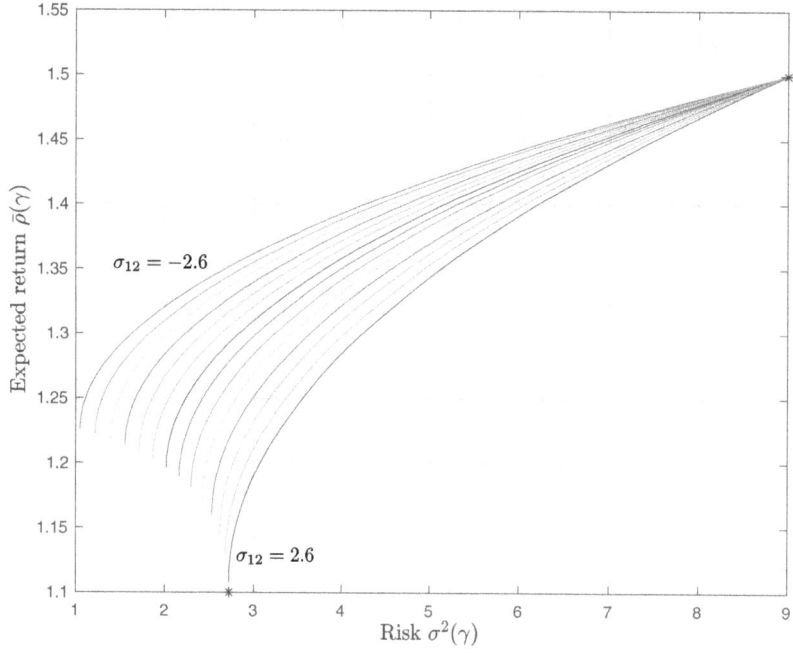

Figure 3.6 Optimal frontiers of $\bar{\rho}(\gamma)$ vs. $\sigma^2(\gamma)$, for $\mu_1 = 1.5$, $\mu_2 = 1.1$, $\sigma_1^2 = 9$, $\sigma_2^2 = 2.7225$, and σ_{12} ranging from -2.6 to 2.6 in 14 equi-spaced values.

infinity. Or, in practice, sample estimates work well when $m \gg n$. Eigenvalues of the covariance matrix are objects of particular importance in data science, for example because they are related to the *explained variance* in principal component analysis (PCA), where one searches for a good low-dimensional approximation to the data by projecting the data onto a suitable k-dimensional subspace, see, for example, Chapter 4. If one compares the eigenvalues $\lambda_i(\Sigma)$, $i = 1, \ldots, n$ of the true covariance matrix with the eigenvalues $\ell_i(S)$, $i = 1, \ldots, n$ of the sample covariance matrix computed on the basis of m samples, then when n is fixed and $m \to \infty$ a classical result[2] guarantees that

$$(\ell_i - \lambda_i) \xrightarrow{m \to \infty} \mathcal{N}\left(0, 2\lambda_i^2/m\right),$$

which means that the difference between eigenvalues $\ell_i - \lambda_i$ tends in distribution to a normal random variable with zero mean and variance $2\lambda_i^2/m$. Hence indeed, for fixed n and growing m, the eigenvalues of the sample covariance matrix tend to the eigenvalues of the actual covariance matrix. However, the situation is quite different when m grows, but the ratio n/m tends to a fixed limit $\gamma \in (0, \infty)$, that is, in the "large m, large n" case.[3] To see this fact, let us consider the case of a random vector $x \in \mathbb{R}^n$ with zero mean $\mu = 0$ and covariance matrix $\Sigma = I_n$. Since the covariance is the identity, all its eigenvalues are equal to one, that is, $\lambda_i(\Sigma) = 1$, $i = 1, \ldots, n$. We

[2] See T. W. Anderson, "Asymptotic Theory for Principal Component Analysis," *The Annals of Mathematical Statistics*, vol. 34, 1963.

[3] For an in-depth discussion of this case, see N. El Karoui, "Spectrum Estimation for Large Dimensional Covariance Matrices Using Random Matrix Theory," *The Annals of Statistics*, vol. 36, pp. 2757–2790, 2008.

now experiment by drawing m random samples $x^{(j)}$, $j = 1,\ldots,m$, according to $\mathcal{N}(0, I_n)$, constructing the sample covariance S based on these samples, and then looking at the sorted eigenvalues of S, in comparison with the eigenvalues of Σ, which are all one. This type of comparison is shown in Figure 3.7, for different n/m ratios. It is clear from these plots that for small n/m the eigenvalues of the sample covariance are close to the ideal value (i.e., all ones), while as n/m gets larger the sample covariance eigenvalues are no longer reliable estimates of the eigenvalues of the true covariance matrix.

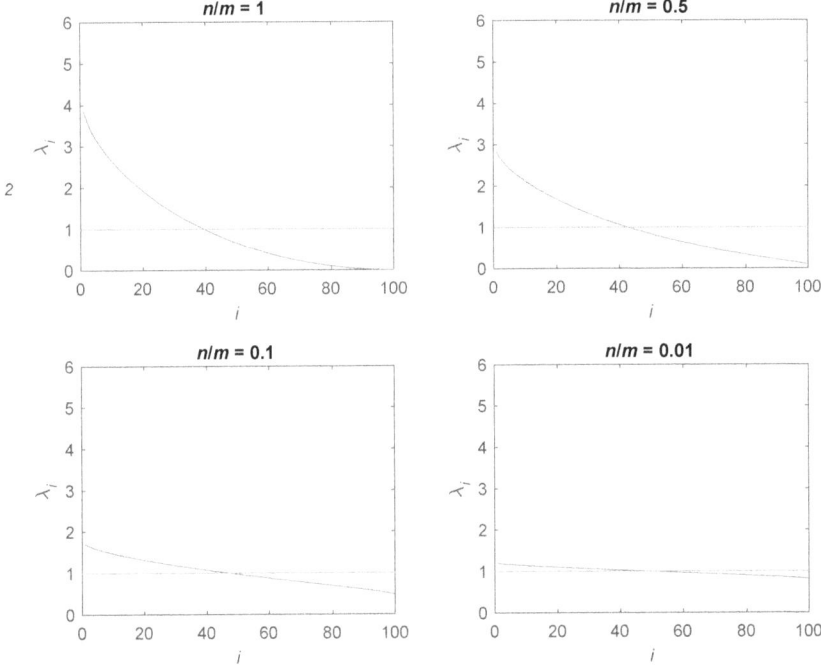

Figure 3.7 Sorted eigenvalues of an $n \times n$, $n = 100$, sample covariance matrix, estimated from m samples $\mathcal{N}(0, I_n)$, at various levels of the n/m ratio; see file mar_pastur.m in the online resources.

Another problem that arises from the sample covariance matrix is that it may not reflect the *structure* present in the true unknown covariance matrix. For instance, the true Σ may be low rank, but due to noise in the data the estimate S is not. Or, the inverse covariance $P = \Sigma^{-1}$ may be sparse,[4] but again due to noise in the data this property may not be preserved in the sample covariance. Both the recovery of low-rank structure in the covariance matrix and of sparsity in the precision matrix are discussed in later sections.

We conclude this section with a note on how to assess the quality of an estimation method, using a cross-validation stability principle. Since in practice we do not have infinitely many samples available, we must make the best possible use of the m samples we have. A general procedure is therefore the following one. Given a sample es-

[4] The inverse of the covariance matrix is called the *precision matrix*. Sparsity in the precision matrix is a characteristic related to conditional independence among features, see Section 3.7.

timate $\hat{Z}$ of a random quantity Z, computed using m samples, let $k = 1$ and fix an integer $N \geq k$:

1. Randomly remove $\gamma\%$ of the data points (say, $\gamma = 10$).

2. Compute a new estimate $\hat{Z}^{(k)}$ based on the reduced sample set.

3. If $k = N$ finish, otherwise set $k = k+1$ and repeat from 1.

At the end of the above loop we have N new estimates $\hat{Z}^{(k)}$, $k = 1, \ldots, N$, that we can use to compute errors with respect to the initial estimate, that is,

$$\delta_k \doteq \|\hat{Z}^{(k)} - \hat{Z}\|, \quad k = 1, \ldots, N.$$

In the case when $\hat{Z}$ is a vector, we can use one of the standard vector norms for computing the errors, and similarly we can use matrix norms such as the Frobenius norm for computing the errors in the matrix case, for example, when $\hat{Z}$ is a sample covariance matrix. We may then compute statistics of the errors such as the average error, or the errors histogram, in order to evaluate the estimation quality.

3.5 Factor models

3.5.1 "Identity plus low-rank" model

We next consider first the case in which the samples are generated by a low-rank covariance model, plus $\mathcal{N}(0, I_n)$ noise, that is, we assume a DGM of the form

$$x = Lz + \sigma e + \mu, \tag{3.10}$$

where $\mu \in \mathbb{R}^n$, $L \in \mathbb{R}^{n,k}$ is the so-called *factor loading* matrix, with $k \leq n$, $z \sim \mathcal{N}(0, I_k)$ contains the random *factors*, $e \sim \mathcal{N}(0, I_n)$ is a random noise vector, and $\sigma \geq 0$ is the noise level. The rationale behind this model is that, ideally, without noise (i.e., for $\sigma = 0$), x would have a covariance matrix $LL^\top \succeq 0$ of rank $\leq k$. However, the presence of noise in the samples "hides" this structure and makes the covariance matrix of x be full rank for $\sigma > 0$, that is,

$$\Sigma \doteq \mathrm{var}\{x\} = LL^\top + \sigma^2 I_n. \tag{3.11}$$

Now, if we collect samples $x^{(1)}, \ldots, x^{(m)}$ of x, and construct the sample covariance matrix S from these samples, we are not guaranteed that S has the structure in (3.11), nor do we have actual estimates for L and σ. The approach we follow is therefore to find numerically some estimates $\hat{L} \in \mathbb{R}^{n,k}$ and $\hat{\sigma} \in \mathbb{R}$ such that the matrix

$$\hat{\Sigma} \doteq \hat{L}\hat{L}^\top + \hat{\sigma}^2 I_n$$

is a good approximation of the sample covariance matrix S. Measuring the approximation error by means of the Frobenius norm of the residual, we have to solve the following optimization problem:

$$(\hat{L}, \hat{\sigma}) = \arg \min_{L \in \mathbb{R}^{n,k}, \sigma \in \mathbb{R}} \|S - (LL^\top + \sigma^2 I_n)\|_F. \quad (3.12)$$

Remarkably, this problem has a "closed-form" solution, which can be derived as described in Appendix A.3. Precisely, letting $S = U\Lambda U^\top$ be the eigenvalue decomposition of S, with $\Lambda = \text{diag}(\lambda_1, \ldots, \lambda_n)$, $\lambda_1 \geq \cdots \geq \lambda_n \geq 0$, and $U = [u_1, \ldots, u_n]$ orthogonal, we have that the optimal $\hat{L}$ is given by

$$\hat{L} = \left[\sqrt{\lambda_1 - \hat{\sigma}^2} u_1 \cdots \sqrt{\lambda_k - \hat{\sigma}^2} u_k \right],$$

where

$$\hat{\sigma}^2 = \frac{\sum_{i=k+1}^n \lambda_i}{n-k}.$$

The overall estimated covariance is therefore

$$\hat{\Sigma} = \hat{L}\hat{L}^\top + \hat{\sigma}^2 I_n = \sum_{i=1}^k (\lambda_i - \hat{\sigma}^2) u_i u_i^\top + \hat{\sigma}^2 I_n$$

$$= \sum_{i=1}^k \lambda_i u_i u_i^\top - \hat{\sigma}^2 \sum_{i=1}^k u_i u_i^\top + \hat{\sigma}^2 I_n$$

$$= \sum_{i=1}^k \lambda_i u_i u_i^\top + \hat{\sigma}^2 \sum_{i=k+1}^n u_i u_i^\top,$$

where the last passage follows from the fact that U is orthogonal, whence $\sum_{i=1}^n u_i u_i^\top = I_n$. It is interesting to observe that $\hat{\Sigma}$ is obtained from S by replacing its $n-k$ smallest eigenvalues with their average $\hat{\sigma}^2$.

3.5.2 "Diagonal plus low-rank" model

We now generalize slightly the previous model by allowing different levels of noise on the components of x. The DGM takes the form

$$x = Lz + \text{diag}(\sigma_1, \ldots, \sigma_n) e + \mu, \quad (3.13)$$

where all terms are defined as in Section 3.5.1, and $\sigma_1, \ldots, \sigma_n \geq 0$ are the noise levels on the components of x. The covariance matrix of x takes the "diagonal plus low-rank" form

$$\Sigma \doteq \text{var}\{x\} = LL^\top + D, \quad D \doteq \text{diag}\left(\sigma_1^2, \ldots, \sigma_n^2\right). \quad (3.14)$$

The estimation problem now amounts to extracting an estimate with the structure

$$\hat{\Sigma} \doteq \hat{L}\hat{L}^\top + \hat{D}, \quad \hat{D} = \mathrm{diag}\left(\hat{\sigma}_1^2, \ldots, \hat{\sigma}_n^2\right)$$

from the available sample covariance S. This problem is posed as

$$(\hat{L}, \hat{D}) = \arg\min_{L \in \mathbb{R}^{n,k}, D \succeq 0 \text{ diagonal}} \|S - (LL^\top + D)\|_F. \quad (3.15)$$

This time, the problem has no closed-form or easily computable globally optimal solution. However, we can apply an heuristic alternate minimization algorithm to obtain a locally optimal solution. The idea is to alternate the minimization of the objective $\|S - (LL^\top + D)\|_F$ over diagonal D and over $Q = LL^\top$. We now detail the two steps involved in this approach.

Q-step: given $D = \mathrm{diag}(d_1, \ldots, d_n) \succeq 0$, we consider the problem

$$L(D) = \arg\min_{L \in \mathbb{R}^{n,k}} \|S - (LL^\top + D)\|_F. \quad (3.16)$$

Since $LL^\top$ is PSD and has rank $\leq k$, we rewrite this problem as a rank-constrained one:

$$Q(D) = \arg\min_{Q \in \mathbb{S}_+^n} \|S(D) - Q\|_F, \text{ s.t.: } \mathrm{rank}(Q) \leq k,$$

where $S(D) \doteq S - D$. Applying a positive-semidefinite version of the Eckart–Young–Mirsky theorem (see Appendix A.3) we obtain that

$$Q(D) = \sum_{i=1}^{k} \lambda_i^+(S(D)) u_i u_i^\top, \quad (3.17)$$

where $\lambda_1(S(D)) \geq \cdots \geq \lambda_n(S(D))$ are the eigenvalues of $S(D)$, $u_1, \ldots, u_n$ are the corresponding eigenvectors, and

$$\lambda_i^+(S(D)) \doteq \max(\lambda_i(S(D)), 0), \quad i = 1, \ldots, n.$$

D-step: letting $Q \doteq Q(D)$, with $Q(D)$ computed in (3.17), we consider the problem

$$D(Q) = \arg\min_{D = \mathrm{diag}(d_1, \ldots, d_n) \succeq 0} \|S(Q) - D\|_F$$

$$= \arg\min_{D = \mathrm{diag}(d_1, \ldots, d_n) \succeq 0} \|S(Q) - D\|_F^2, \quad (3.18)$$

where $S(Q) \doteq S - Q$. The objective function in the above problem can be rewritten as

$$\|S(Q) - D\|_F^2 = \sum_{i=1}^{n}(S_{ii}(Q) - d_i)^2 + \sum_{i \neq j} S_{ij}(Q)^2,$$

where $S_{ij}(Q)$ denotes the element in position row i and column j of matrix $S(Q)$. The minimum is then attained for $d_i = S_{ii}(Q)$, if $S_{ii}(Q) \geq 0$, or $d_i = 0$ otherwise, that is,

$$D(Q) = \text{diag}\left(S_{11}^+(Q), \ldots, S_{nn}^+(Q)\right), \qquad (3.19)$$

where $S_{ii}^+(Q) = \max(S_{ii}(Q), 0)$, $i = 1, \ldots, n$. We can summarize the overall approach in the following meta-algorithm.

Algorithm 3.1 (Diagonal plus low-rank model estimation) Given a sample covariance matrix $S \in \mathbb{S}_+^n$ and a desired rank upper bound $k \leq n$, the algorithm returns a matrix $\hat{L} \in \mathbb{R}^{n,k}$ and a matrix $\hat{D} = \text{diag}(d_1, \ldots, d_n) \succeq 0$ such that $\hat{\Sigma} = \hat{L}\hat{L}^\top + \hat{D}$ is a good approximation of S.

1. (Initialization) Solve problem (3.12), and let $D \leftarrow \hat{\sigma} I_n$.
2. (Q-step) Compute Q according to (3.17).
3. (D-step) Compute D according to (3.19).
4. (Exit condition) Evaluate objective $\|S - (Q + D)\|_F$: if relative improvement with respect to the previous iteration is below a given threshold then proceed to step 5, otherwise repeat from step 2.
5. Return $\hat{D} \leftarrow D$ and the $\hat{L}$ factor of $Q = \hat{L}\hat{L}^\top$:

$$\hat{L} = \left[\sqrt{\lambda_1^+(S(D))}u_1 \cdots \sqrt{\lambda_k^+(S(D))}u_k\right],$$

where $\lambda_i^+(S(D))$, $i = 1, \ldots, k$, are as in (3.17).

3.5.3 Benefits of factor models

Factor models bring primarily an advantage of interpretability of the data model under study. Suppose for instance that $x \in \mathbb{R}^n$ is a random vector representing the returns of a collection of n assets, where n is relatively large, say in the order of hundreds. If we can fit to x a factor model of low rank $k \ll n$ and small error $\|D\|_F$, then this means that each entry x_i, $i = 1, \ldots, n$, can be "explained" approximately as a linear combination of the same few k factors $z_1, \ldots, z_k$. The entry L_{ij} of the factor loading matrix represents the sensitivity of the ith asset return to the movements of the jth factor.

Besides interpretability, factor models may also bring computational benefits. Consider for instance a basic portfolio risk/return tradeoff problem, in which one needs to compute the portfolio that minimizes a combination of portfolio variance and expected return,[5] that is, using the notation set in Section 3.3,

$$\min_{w \in \mathbb{R}^n} f(w) \doteq w^\top \Sigma w - 2\lambda \mu^\top w, \qquad (3.20)$$

[5] This problem is treated extensively in Chapter 11.

where $w \in \mathbb{R}^n$ is the portfolio composition vector, here assumed without shorting restrictions (i.e., the portfolio positions w_i can be either *long* $w_i > 0$, *short* $w_i < 0$, or neutral $w_i = 0$), $\lambda \geq 0$ is a tradeoff parameter, Σ is the covariance matrix of the random return vector $r \in \mathbb{R}^n$, and μ is its expected value.

Since $\Sigma \succeq 0$, the problem (3.20) is a convex quadratic minimization problem, for which a solution can be obtained in closed form by computing the gradient of $f(x)$ and equating it to zero:

$$\nabla f(w) = \Sigma w - \lambda \mu = 0. \quad (3.21)$$

If we assume that $\Sigma \succ 0$, then (3.21) has the unique solution

$$w^* = \lambda \Sigma^{-1} \mu,$$

which represents the optimal portfolio allocation. From a computational point of view, solving the system of linear equations in (3.21) with generic coefficient matrix $\Sigma \in \mathbb{S}_+^n$ requires $O(n^3)$ operations. Let us now see what happens if we exploit a diagonal plus low-rank structure in Σ. If $\Sigma = LL^\top + D$, with D diagonal and $L \in \mathbb{R}^{n,k}$, $k \ll n$, then equation (3.21) reads

$$(LL^\top + D)w = \lambda \mu.$$

Defining $y \doteq L^\top w$ we rewrite this equation as

$$\begin{bmatrix} D & L \\ L^\top & -I_k \end{bmatrix} \begin{bmatrix} w \\ y \end{bmatrix} = \begin{bmatrix} \lambda \mu \\ 0 \end{bmatrix}.$$

From the first row of the above equations we obtain $w = D^{-1}(\lambda \mu - Ly)$ which, substituted in the second row, yields an equation in $y \in \mathbb{R}^k$

$$(L^\top D^{-1} L + I_k)y = \lambda L^\top D^{-1} \mu. \quad (3.22)$$

We can therefore proceed as follows:

- invert diagonal D, which takes n operations;

- form matrix $(L^\top D^{-1} L + I_k)$ and solve (3.22) for y, which requires $O(nk^2 + k^3)$ operations;

- retrieve w as $w = D^{-1}(\lambda \mu - Ly)$, which requires $O(nk)$ operations.

Overall, with the above method we obtain the optimal solution in $O(nk^2)$ operations, which is much lower than $O(n^3)$, when $k \ll n$.

3.6 Ledoit and Wolf's shrinkage model

We discussed in Section 3.4 that in situations where the dimension n is comparable with the number of samples m the plain sample covariance matrix can be a poor estimator of the true covariance. When the dimension n is larger than the number m of observations available, the sample covariance matrix is indeed not even invertible. When the ratio n/m is less than one but not negligible, the sample covariance matrix is invertible but numerically ill-conditioned, which means that inverting it amplifies estimation errors dramatically. For large n it is difficult to find enough observations to make n/m negligible, and therefore it is important to develop a well-conditioned estimator for large-dimensional covariance matrices. One such estimator, known as *shrinkage estimator*, has been proposed by Ledoit and Wolf.[6] The proposed estimator $\hat{\Sigma}$ is optimal in the sense that it minimizes the quadratic loss $\mathrm{E}\{\|\hat{\Sigma} - \Sigma\|_F^2\}$ in a so-called *general asymptotic* regime, that is, when m goes to infinity and n/m remains bounded.

[6] See O. Ledoit and M. Wolf, "A Well-Conditioned Estimator for Large-Dimensional Covariance Matrices," *Journal of Multivariate Analysis*, vol. 88, pp. 365–411, 2004.

The structure of Ledoit and Wolf's shrinkage estimator is very simple, since the estimate is formed by taking a linear combination of the sample covariance S and the identity matrix, namely

$$\hat{\Sigma} = \rho_1 I_n + \rho_2 S. \tag{3.23}$$

It can be proved that $\mathrm{E}\{\|\hat{\Sigma} - \Sigma\|_F^2\}$ is minimized with the choice of parameters

$$\rho_1 = \frac{\beta^2}{\delta^2}\bar{\sigma}^2, \quad \rho_2 = \frac{\alpha^2}{\delta^2},$$

where

$$\bar{\sigma}^2 \doteq \frac{1}{n}\mathrm{Tr}\,\Sigma, \quad \alpha^2 \doteq \|\Sigma - \bar{\sigma}^2 I\|_F^2,$$
$$\beta^2 \doteq \mathrm{E}\{\|S - \Sigma\|_F^2\}, \quad \delta^2 \doteq \mathrm{E}\{\|S - \bar{\sigma}^2 I\|_F^2\},$$

and that the optimal residual is $\mathrm{E}\{\|\hat{\Sigma} - \Sigma\|_F^2\} = \frac{\alpha^2 \beta^2}{\delta^2}$. Notice that $\bar{\sigma}^2$ is the average variance of the n components of the random vector, and that (3.23) can be rewritten as

$$\hat{\Sigma} = \rho(\bar{\sigma}^2 I_n) + (1-\rho)S, \tag{3.24}$$

where

$$\rho \doteq \frac{\beta^2}{\delta^2}, \quad 1 - \rho = \frac{\alpha^2}{\delta^2}.$$

This formulation highlights the fact that the shrinkage estimator forms a convex combination of S and $\bar{\sigma}^2 I_n$, and the amount of shrinkage is controlled by the parameter $\rho \in [0,1]$, which represents a normalized measure of the error of the sample covariance matrix S. Intuitively, if S is relatively accurate, then we apply a little shrinking (small ρ), while if S is relatively inaccurate, then we apply a larger shrinking (larger ρ).

While all the above is appealing, the careful reader should have noticed that the estimator in (3.23) or in (3.24) cannot be used in practice, since it requires hindsight knowledge of the parameters $\bar{\sigma}^2$, α^2, β^2, and δ^2, which depend on the unknown Σ, and hence are themselves unknown. Ledoit and Wolf, however, prove that under a general asymptotic regime the shrinkage intensity ρ tends to a limiting constant that can be estimated consistently, and the same can be done for the other quantities of interest. Therefore, in practice, we use the following sample estimates of the parameters of interest:

$$\hat{\sigma}^2 \doteq \frac{1}{n} \operatorname{Tr} S, \quad \hat{\delta}^2 \doteq \|S - \hat{\sigma}^2 I\|_F^2,$$

$$\hat{b}^2 \doteq \frac{1}{m^2} \sum_{k=1}^{m} \|(x^{(k)} - \bar{x})(x^{(k)} - \bar{x})^\top - S\|_F^2,$$

$$\hat{\beta}^2 \doteq \min(\hat{b}^2, \hat{\delta}^2), \quad \hat{\alpha}^2 \doteq \hat{\delta}^2 - \hat{\beta}^2,$$

and then obtain the shrinkage intensity as $\hat{\rho} \doteq \hat{\beta}^2/\hat{\delta}^2$ and build the shrinkage estimator as

$$\hat{\Sigma} = \hat{\rho}(\hat{\sigma}^2 I_n) + (1 - \hat{\rho})S. \tag{3.25}$$

Besides (3.25), other approaches also exist for the computation of the shrinkage estimator. For instance, a so-called *Empirical Bayesian* approach would prescribe computing the parameters ρ_1, ρ_2 of the shrinkage estimator (3.23) simply as

$$\rho_1 = \frac{mn - 2m - 2}{nm^2} \det(S)^{1/n}, \quad \rho_2 = \frac{m}{m+1}.$$

Also, in model (3.25) one could select the $\hat{\rho}$ parameter empirically, on the basis of cross-validation experiments. Further, the identity matrix in (3.25) can be replaced with another given positive-definite matrix, and this allows mixing, for instance, of heterogeneous views on markets, such as price-based and based on news or expert advice.

3.7 Sparse covariance and precision matrix estimation

Any symmetric $n \times n$ matrix A has a natural interpretation in terms of an underlying undirected graph in which there are n vertices and the value of $A_{ij} = A_{ji}$ refers to the level of mutual interaction between vertex i and vertex j, see Section 13.2.1 for a more in-depth discussion of correlation networks. In the case of a covariance matrix Σ, the off-diagonal elements Σ_{ij} represent the covariance of variables x_i and x_j, that is,

3.7 SPARSE COVARIANCE AND PRECISION MATRIX ESTIMATION

$$\Sigma_{ij} = E\{(x_i - \mu_i)(x_j - \mu_j)\} = E\{x_i x_j\} - \mu_i \mu_j,$$

and the variables x_i, x_j are said to be *uncorrelated* if $\Sigma_{ij} = 0$. A sparse covariance matrix Σ has many zero entries, which therefore correspond to many pairs of variables that are mutually uncorrelated. When estimating Σ from data, however, the presence of noise may lead to a sample covariance estimate S which is *not* sparse and hence to a loss of structural information about the underlying correlation graph. Sparse covariance estimation techniques aim at finding covariance estimates that are sparse, and thus at recovering the structure of the correlation graph.

Interestingly, also sparsity in the inverse of the covariance matrix (the so-called precision matrix, $P = \Sigma^{-1}$) has a statistical interpretation in terms of *conditional independence* between variables. Given random variables w, y, z, we say that w and y are conditionally independent given z if the conditional density $g(w, y|z)$ of (w, y) given z can be factored as

$$g(w, y|z) = g_1(w|z) g_2(y|z),$$

or, equivalently, the joint density $f(w, y, z)$ of (w, y, z) is written as

$$f(w, y, z) = g_1(w|z) g_2(y|z) h(z),$$

where $h(z)$ is the marginal density of z. The interpretation is that variables w, y are independent, once the value of the third variable z is given. More generally, if $x = (x_1, \ldots, x_n)$ is a random vector with joint pdf $p(x)$, then two variables x_i, x_j are conditionally independent given the other variables x_k, $k \neq i, j$, if the density can be factored as

$$p(x) = p_i(x_i) p_j(x_j),$$

where p_i, p_j depend parametrically also on the other variables x_k, $k \neq i, j$. The above condition states equivalently that the log-density can be expressed as the sum

$$\log p(x) = \log p_i(x_i) + \log p_j(x_j).$$

Interestingly, if $p(x)$ is Gaussian (say, for notational simplicity, with zero mean),

$$p(x) = \frac{1}{(2\pi)^{n/2} \det \Sigma^{1/2}} \exp\left(-\frac{1}{2} x^\top \Sigma^{-1} x\right),$$

and we let $P = \Sigma^{-1}$ denote the precision matrix, then we have that

$$\log p(x) = \text{const.} - \frac{1}{2} x^\top P x$$
$$= \text{const.} - \frac{1}{2} \left(x_i^2 P_{ii} + x_j^2 P_{jj} + 2 x_i x_j P_{ij} \right.$$
$$\left. + \text{ linear terms in } (x_i, x_j) \times \text{ terms containing } (x_k)_{k \neq i,j} \right),$$

and we see that the coefficient of $x_i x_j$ in $\log p(x)$ is P_{ij}. Therefore, $\log p(x)$ can be expressed as the sum $\log p_i(x_i) + \log p_j(x_j)$ if and only if $P_{ij} = 0$. We conclude that, for a Gaussian random vector $x \in \mathbb{R}^n$, the entries x_i, x_j are conditionally independent if and only if the (i,j)th element of the precision matrix is zero. An underlying conditional independence structure among the variables thus entails a sparse precision matrix, but such structure may be obscured by noise in the plain sample estimate, hence special estimation methods should be employed in order to recover such structure. Sparse precision matrix estimation and sparse covariance matrix estimation methods are discussed in the next sections.

3.7.1 Sparse precision matrix estimation

One way in which we can obtain a sparse estimate of the precision matrix is via a suitable regularization of the maximum likelihood criterion in (3.4). Indeed, under the Gaussian assumption, the plain maximum likelihood estimate was obtained by minimizing

$$q(P) = -\ln \det P + \text{Tr}(SP),$$

and the minimizer was proved to be $P^* = S^{-1}$, that is, the inverse of the sample covariance matrix. If we want sparsity, we may include a term in $q(P)$ which penalizes nonzero elements, that is,

$$\tilde{q}_\lambda(P) \doteq -\ln \det P + \text{Tr}(SP) + \lambda \|P\|_0,$$

where $\lambda \geq 0$ is a penalty parameter, and $\|P\|_0$ represents the *cardinality* of its argument, that is, the number of nonzero entries. Minimization of $\tilde{q}_\lambda(P)$, however, is numerically hard, since this function is nonconvex, due to the $\|P\|_0$ term. A standard and effective approach is then to relax the problem by substituting the $\|P\|_0$ term with a convex proxy $\|P\|_1 = \sum_{ij} |P_{ij}|$, thus obtaining the ℓ_1-regularized objective[7]

$$q_\lambda(P) \doteq -\ln \det P + \text{Tr}(SP) + \lambda \|P\|_1. \tag{3.26}$$

Minimizing $q_\lambda(P)$ under the constraint $P \succeq 0$ is a convex optimization problem. For small-/medium-sized problem instances, it can be

[7] The use of the ℓ_1 norm as a replacement of the cardinality is a standard "trick" in machine learning. It is used, for instance, in the Lasso regression as well as in many other contexts as a sparsity-inducing term, see for instance the monograph by T. Hastie, R. Tibshirani, and M. Wainwright, *Statistical Learning with Sparsity: The Lasso and Generalizations*, Boca Raton, FL: CRC Press, 2015.

solved by means of general-purpose parser-solvers, such as CVX, a MATLAB® software for disciplined convex programming, see https://cvxr.com/cvx/. For realistically sized problems, ad-hoc algorithms have been developed that exploit the specific structure of the problem, such as a block-wise coordinate descent algorithm,[8] or the *Graphical Lasso* algorithm.[9]

[8] See O. Banerjee, L. El Ghaoui, and A. D'Aspremont, "Model Selection Through Sparse Maximum Likelihood Estimation for Multivariate Gaussian or Binary Data," *Journal of Machine Learning Research*, vol. 9, pp. 485–516, 2008.

[9] See J. Friedman, T. Hastie, and R. Tibshirani, "Sparse Inverse Covariance Estimation with the Graphical Lasso," *Biostatistics*, vol. 9(3), pp. 432–441, 2007.

3.7.2 *Sparse covariance matrix estimation*

While the approach in the previous section was aimed at identifying a sparse *inverse* covariance matrix, thus revealing the structure of the underlying conditional independence graph among the variables (the so-called Markov network), other approaches consider sparsity directly in the covariance matrix, which is related to the structure of the covariance graph, that is, the graphical model for marginal independencies among variables. Assuming a Gaussian joint pdf for the variables, the negative log-likelihood (3.4) to be minimized is written in terms of $\Sigma = P^{-1}$ as

$$g(\Sigma) = \ln \det \Sigma + \mathrm{Tr}(S\Sigma^{-1}).$$

Similar to the previous case, we can add a sparsity-inducing term to the above cost, obtaining[10]

$$g_\lambda(\Sigma) \doteq \ln \det \Sigma + \mathrm{Tr}(S\Sigma^{-1}) + \lambda \|W \odot \Sigma\|_1, \quad (3.27)$$

where $\lambda \geq 0$ is a penalty parameter, W is a matrix of weights,[11] and $\odot$ denotes element-wise multiplication. The problem of minimizing $g_\lambda(\Sigma)$ under the constraint $\Sigma \succeq 0$, however, is a hard nonconvex one. Nevertheless, there is a "difference of convex" structure in the objective, since $\mathrm{Tr}(S\Sigma^{-1}) + \lambda \|W \odot \Sigma\|_1$ is convex in $\Sigma \succeq 0$, and $\ln \det \Sigma$ is concave, hence $-\ln \det \Sigma$ is convex. This type of structure lends itself to a (possibly suboptimal) solution approach based on an iterative majorize–minimize algorithm, a version of which is detailed for the problem at hand in the cited paper by Bien and Tibshirani.

[10] This approach was proposed and developed in J. Bien and R. Tibshirani, "Sparse Estimation of a Covariance Matrix," *Biometrika*, vol. 98(4), pp. 807–820, 2011.

[11] Common choices for W are a matrix of all ones $W = \mathbf{1}\mathbf{1}^\top$, or a version with zeros on the diagonal to avoid shrinking diagonal elements of Σ, that is, $W = \mathbf{1}\mathbf{1}^\top - I$.

3.8 *Confidence ellipsoids and outliers*

We next discuss how we can exploit, in a Gaussian setting, the covariance estimate in order to determine ellipsoidal regions of confidence for a random vector. Then, we discuss about methods for detecting *outliers*, that is, data points that are "dissimilar" from the rest of a given data set.

3.8.1 Confidence ellipsoids

Preliminarily, we state some facts about the representation of n-dimensional ellipsoids and how they are related to PSD matrices and quadratic forms. Given a PSD matrix $Q \in \mathbb{S}^n$ and a vector $c \in \mathbb{R}^n$, the set

$$\mathcal{E} = \{x \in \mathbb{R}^n : x = c + Qz, \|z\|_2 \leq 1\} \qquad (3.28)$$

represents a bounded *ellipsoid* with center c and orientation and shape described by the Q matrix. In particular, the directions of the semi-axes of the ellipsoid are given by the eigenvectors of Q, which hence define the orientation of $\mathcal{E}$ in space, and the shape of the ellipsoid is defined by the lengths of the semi-axes, which are given by the eigenvalues of Q, $\lambda_1(Q) \geq \cdots \geq \lambda_n(Q) \geq 0$. Letting $P \doteq QQ^\top$, we have that $\lambda_i(P) = \lambda_i(Q)^2$, and the function $\operatorname{Tr} P = \sum_{i=1}^n \lambda_i(P)$ is a useful measure of the "size" of the ellipsoid, which corresponds to the sum of the squares of the semi-axes lengths. Another frequently used measure of size is the *volume* of $\mathcal{E}$, which is given by

$$\operatorname{vol}(\mathcal{E}) = \beta_n \sqrt{\det P} = \beta_n \det Q,$$

where β_n is the volume of the unit hypersphere in $\mathbb{R}^n$. If Q is PD, hence invertible, an equivalent representation of the ellipsoid (3.28) is

$$\mathcal{E} = \{x \in \mathbb{R}^n : (x - c)^\top P^{-1} (x - c) \leq 1\}, \qquad (3.29)$$

or, equivalently yet,

$$\mathcal{E} = \{x \in \mathbb{R}^n : \|Ax - b\|_2 \leq 1\}, \quad A \doteq Q^{-1}, \ b \doteq Q^{-1}c. \qquad (3.30)$$

The problem we consider is the following one: suppose we know that some random vector $x \in \mathbb{R}^n$ (e.g., the vector of asset returns) has a multivariate normal distribution with known mean μ and covariance matrix $\Sigma \succ 0$. We wish to determine an *ellipsoid of confidence* for x, that is, a set $\mathcal{E}_\gamma$ of the form (3.29) that contains x with a given probability $\gamma \in (0,1)$, that is, such that

$$\operatorname{Prob}\{x \in \mathcal{E}_\gamma\} = \gamma.$$

In practice, we need to determine the appropriate center $c \in \mathbb{R}^n$ and shape matrix $Q \in \mathbb{S}^n$ (or squared shape matrix $P \in \mathbb{S}^n$) of the ellipsoid. This problem is a classical one, and the solution is known to be given by[12]

$$c = \mu,$$
$$Q = \sqrt{\chi_n^2(\gamma)} \Sigma^{1/2},$$

[12] See for instance V. Chew, "Confidence, Prediction, and Tolerance Regions for the Multivariate Normal Distribution," *Journal of the American Statistical Association*, vol. 61(315), pp. 605–617, 1966.

where $\chi_n^2(\gamma)$ is the quantile function for probability $\gamma \in (0,1)$ of the Chi-squared distribution with n degrees of freedom. Equivalently,

$$\mathcal{E}_\gamma = \{x \in \mathbb{R}^n : (x - \mu)^\top \Sigma^{-1}(x - \mu) \leq \chi_n^2(\gamma)\}. \tag{3.31}$$

The situation becomes unfortunately more complicated in the realistic case when μ and Σ are unknown, but can be estimated from samples.[13] A rough-cut approach would be to simply use the sample mean $\bar{x}$ and an estimated covariance matrix $\hat{\Sigma}$ (obtained, for instance, via one of the methods described in the preceeding sections) respectively in place of μ and Σ in the above formulae. This type of approach may work when the number m of samples is way larger than the dimension n.

3.8.2 Detection of outliers

Given a data point $z \in \mathbb{R}^n$, we seek to quantify how much *dissimilar* this point is with respect of a given data set $X = [x^{(1)} \cdots x^{(m)}]$. Many approaches are possible; here, we briefly discuss a probabilistic approach and a geometric one, and discuss their similarities.

In a probabilistic approach, under the assumption that the data are generated by a Gaussian DGM, we may use the concept of ellipsoid of confidence in (3.31) and consider the quantity

$$d^2(z) \doteq (z - \hat{\mu})^\top \hat{\Sigma}^{-1} (z - \hat{\mu}), \tag{3.32}$$

where $\hat{\mu}$ is the sample mean, and $\hat{\Sigma}$ is an estimate of the covariance matrix, such as, say, the sample covariance matrix. The quantity $d(z)$ is known as the *Mahalanobis distance* from z to the sample mean. In fact, $d(z)$ is a weighted Euclidean norm

$$d(z) = \|A(z - \hat{\mu})\|_2,$$

where A is a matrix square-root factor of $\hat{\Sigma}^{-1}$, that is, $A^\top A = \hat{\Sigma}^{-1}$. We saw in (3.31) that, for a given level of probability $\gamma \in (0,1)$, the sublevel set $\mathcal{E}_\gamma = \{z : d^2(z) \leq \chi_n^2(\gamma)\}$ is such that the probability of a normal random vector with mean $\hat{\mu}$ and covariance $\hat{\Sigma}$ being contained in $\mathcal{E}_\gamma$ is γ. Consequently, the probability of the random vector being larger than $\chi_n^2(\gamma)$, in the Mahalanobis distance sense, is $1 - \gamma$. Selecting a threshold level γ, say $\gamma = 0.95$, we may decide that vectors z such that $d(z) > \chi_n^2(\gamma)$ have low a-priori probability (i.e., probability $1 - \gamma = 0.05$), hence we classify them as *outliers*. Also, we can use $d(z^{(j)})$, $j = 1, \ldots, p$, to rank observations $z^{(j)}$: those with low $d(z^{(j)})$ value are closer to the center (in the Mahalanobis distance metric) and belong to regions of higher probability, while those with

[13] For a treatment of confidence regions with unknown mean and covariance in the Gaussian case, see for instance X. Dong and T. Mathew, "Central Tolerance Regions and Reference Regions for Multivariate Normal Populations," *Journal of Multivariate Analysis*, vol. 134, pp. 50–60, 2015. If one releases also the normality assumption, then approaches based on the Chebyshev inequality can be followed, see for instance B. Stellato, B. Van Parys, and P. J. Goulart, "Multivariate Chebyshev Inequality with Estimated Mean and Variance," *The American Statistician*, vol. 71, 2015.

high $d(z^{(j)})$ value are farther from the center and belong to regions of lower probability.

In a geometric approach, instead, we measure the dissimilarity of z from the data set $X = [x^{(1)} \cdots x^{(m)}]$ in terms of how far (in Euclidean norm) $z - \hat{\mu}$ is from the span of $x^{(1)} - \hat{\mu}, \ldots, x^{(m)} - \hat{\mu}$. In formulae, we consider the centered data matrix $\check{X}$, defined as in (3.8), and let

$$D(z) \doteq \min_w \|\check{X}w - (z - \hat{\mu})\|_2^2 + \lambda\|w\|_2^2,$$

where $\lambda \geq 0$ is a parameter that penalizes large coefficients in w. A small value of $D(z)$ indicates that $(z - \hat{\mu})$ can almost be expressed as a linear combination $\check{X}w$ of the centered data point, with small weights w. There is an explicit expression for $D(z)$: by solving the convex quadratic minimization over w, we obtain, after a few manipulations,

$$D(z) = (z - \hat{\mu})^\top \left(I + \lambda^{-1}\check{X}\check{X}^\top\right)^{-1} (z - \hat{\mu}).$$

Recalling from (3.9) that $S = (1/m)\check{X}\check{X}^\top$, we rewrite $D(z)$ as

$$D(z) = (z - \hat{\mu})^\top (I + m/\lambda S)^{-1} (z - \hat{\mu}), \qquad (3.33)$$

where S is the sample covariance matrix. Comparing (3.33) with (3.32), we see that by setting

$$\hat{\Sigma} = \rho_1 I + \rho_2 S, \quad \rho_2 = \rho_1 \frac{m}{\lambda}$$

as a shrinkage estimate of the covariance matrix, it holds that

$$D(z) = \rho_1 (z - \hat{\mu})^\top \hat{\Sigma}^{-1} (z - \hat{\mu}) = \rho_1 d^2(z).$$

This means that, with the specific choice of $\hat{\Sigma}$ above, the geometric measure $D(z)$ and the (squared) Mahalanobis distance $d^2(z)$ are proportional to each other, and hence provide identical rankings when applied to the evaluation of multiple observations $z^{(j)}$, $j = 1, \ldots, p$.

Example 3.3 (*Asset returns and outliers*) We consider $n = 2$ assets, namely Best Buy (BBY) and Microsoft (MSFT). Their daily returns over a period of 140 days ending Feb. 12, 2021, are shown as dots in Figure 3.8.

Taking as covariance estimate the plain sample covariance, that is, $\hat{\Sigma} = S$, and assuming $\gamma = 0.98$, whence $\chi_n^2(\gamma) = 7.824$, we plotted the γ-confidence ellipsoid, assuming a Gaussian model with mean $\hat{\mu}$ and covariance $\hat{\Sigma}$. We may then tag as outliers the points that are at a squared Mahalanobis distance greater than 7.824 from the center $\hat{\mu}$, that is, those that fall outside of the γ-confidence ellipsoid. In this example, six such points are detected and highlighted with circles in Figure 3.8.

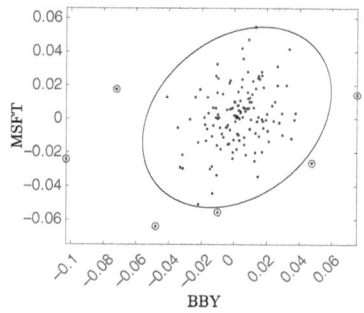

Figure 3.8 One hundred and forty daily returns of BBY and MSFT with their estimated 0.98-confidence ellipsoid and outliers, see ex_outliers.m in the online resources.

3.9 Time-series data models

The basic data model in (3.1) considered so far is a static one, in the sense that the data index i does not carry a dynamic meaning related to time. Indeed, due to the i.i.d. assumption, the samples $x^{(i)}$ can be rearranged in any order, since no causality constraint or time dependence relation holds among them. It is then clear that when dealing with data in which the dynamic aspects are important the DGM (3.1) is inappropriate, and different models that take into explicit account of time causalities and dependencies should be employed instead. A scalar, discrete-time *time series* is a collection of random values $\{y_t\}$, for $t = \ldots, -1, 0, 1, \ldots$, in which the ordering given by the *time* variable t is generally relevant. With some abuse of terminology, we shall refer to as a time series both the stochastic process $\{y_t\}$ and some realization of such process for which we collect m observed data, say $y_1, \ldots, y_m$. The typical problems of interest when dealing with time series are, for instance, making predictions about future values given an observed stream of past values, or estimating some parameters of the underlying distribution of the time series. Both these key tasks would be hopeless if the stochastic law governing the process changed in an unpredictable way with time, since then past data would carry no information about the future behavior of the process. For this reason we shall usually assume that the process under consideration is *weakly stationary* (or covariance stationary), by this meaning that the (unconditional) mean of the process is independent of t and that the *autocovariance* between variable y_t and variable y_{t+k} is only a function of the *lag* k (and not of t), that is, for all t

$$E\{y_t\} = \mu,$$
$$E\{(y_t - \mu)(y_{t+k} - \mu)\} = \gamma_k, \quad \forall k.$$

For brevity, we shall use the term stationary to denote a weakly (covariance) stationary process. Notice that the autocovariance at lag $k = 0$ is simply the unconditional variance of y_t,

$$\gamma_0 = E\{(y_t - \mu)^2\} \doteq \sigma_y^2,$$

which is constant for a stationary time series. The *autocorrelation function* (ACF) of the stationary process $\{y_t\}$ is defined as

$$\rho_k \doteq \frac{\gamma_k}{\gamma_0}, \quad \forall k,$$

and it holds that $\rho_k = \rho_{-k}$, that is, the ACF is symmetric around zero, so it is usually computed only for $k \geq 0$.

Given a batch of observed data $y_1, \ldots, y_m$ from a stationary time series y_t, the mean μ and variance σ_y^2 can be estimated via their sample counterparts

$$\hat{\mu} = \frac{1}{m} \sum_{i=1}^{m} y_i,$$

$$\hat{\sigma}_y^2 = \frac{1}{m} \sum_{i=1}^{m} (y_i - \hat{\mu})^2,$$

and the autocovariance function is estimated by

$$\hat{\gamma}_k \doteq \frac{1}{m} \sum_{i=1}^{m-k} (y_t - \hat{\mu})(y_{t+k} - \hat{\mu}), \quad k = 0, 1, \ldots, K,$$

where $K < m$, and typically $K \simeq m/4$. Correspondingly, the sample autocorrelation function is

$$\hat{\rho}_k \doteq \frac{\hat{\gamma}_k}{\hat{\gamma}_0}, \quad k = 0, 1, \ldots, K.$$

Under appropriate hypotheses, the estimated quantities converge in probability to their exact counterparts as $m \to \infty$.

Example 3.4 (*White noise and random walk*) A zero-mean *white noise* process is a time series $\{\epsilon_t\}$ such that $E\{\epsilon_t\} = 0$ for all t, $E\{\epsilon_t^2\} = \sigma^2$ for all t, and $E\{\epsilon_t \epsilon_\tau\} = 0$ for all $t \neq \tau$. Therefore, white noise is a stationary process uncorrelated in time, that is, such that $\gamma_k = \rho_k = 0$ for all $k > 0$. Sometimes white noise is considered under the stronger condition where $\epsilon_t, \epsilon_\tau$ are independent for all $t \neq \tau$, in which case we talk about *independent white noise*. If further ϵ_t is Gaussian, that is, $\epsilon_t \sim \mathcal{N}(0, \sigma^2)$ for all t, then we have a *Gaussian white noise* process. In general a white noise process may also have a nonzero mean $E\{\epsilon_t\} = \mu$ for all t.

When we need to verify whether an observed sequence is a white noise sequence, we can check the amplitude of the sample autocorrelation coefficients $\hat{\rho}_k$. In particular, under the null hypothesis that the samples are indeed uncorrelated, the standard error of $\hat{\rho}_k$ is proportional to $1/\sqrt{m}$, so that if $|\hat{\rho}_k| \leq \alpha/\sqrt{m}$ (typically with $\alpha = 2$) we may consider it to be approximately zero. Figure 3.9(a) shows an example of $m = 220$ samples of a Gaussian white noise sequence with mean $\mu = 0.01$ and variance $\sigma^2 = 4 \times 10^{-4}$, and Figure 3.9(b) shows the first $K = 20$ values of the sample autocorrelation function, with $2/\sqrt{m}$ confidence bounds.

A *random walk* process is a time series $\{y_t\}$ for which it holds that

$$y_t = y_{t-1} + \mu + \epsilon_t, \tag{3.34}$$

where $\{\epsilon_t\}$ is a white noise process with zero mean and variance σ^2, and μ is the so-called *drift*. Supposing that the value at $t = 0$ is fixed to some value y_0, we obtain via recursion the value at generic $t \geq 1$ as

$$y_t = t\mu + y_0 + \epsilon_t + \epsilon_{t-1} + \cdots + \epsilon_1.$$

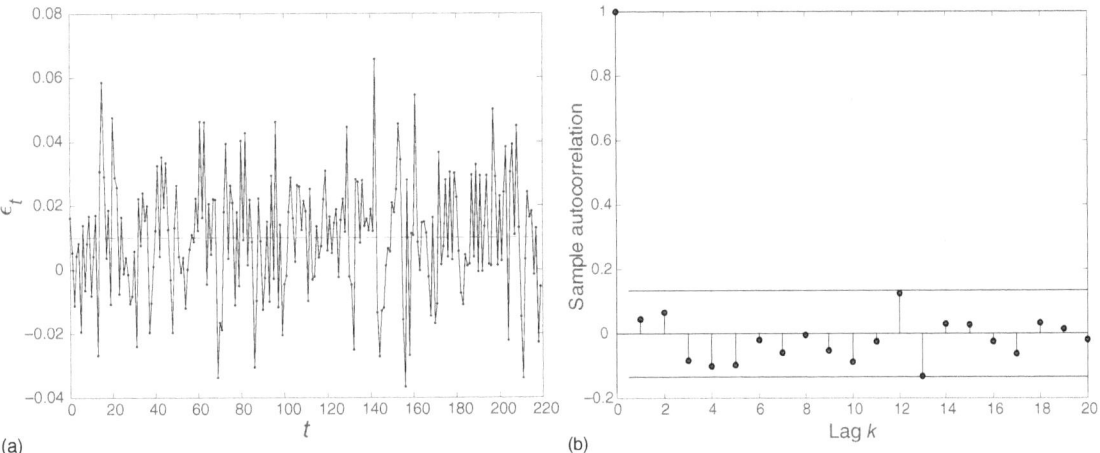

Figure 3.9 A white noise sequence (a) and its sample ACF (b).

By taking the expectation we have that

$$E\{y_t\} = t\mu + y_0,$$

and

$$E\{(y_t - E\{y_t\})^2\} = E\{(\epsilon_t + \epsilon_{t-1} + \cdots + \epsilon_1)^2\} = t\sigma^2,$$

which shows that the random walk is *not* covariance stationary, since both the mean and the variance vary with time. Figure 3.10 shows $m = 220$ samples from a realization of a random walk process with drift $\mu = 0.01$ and variance $\sigma^2 = 0.0034$.

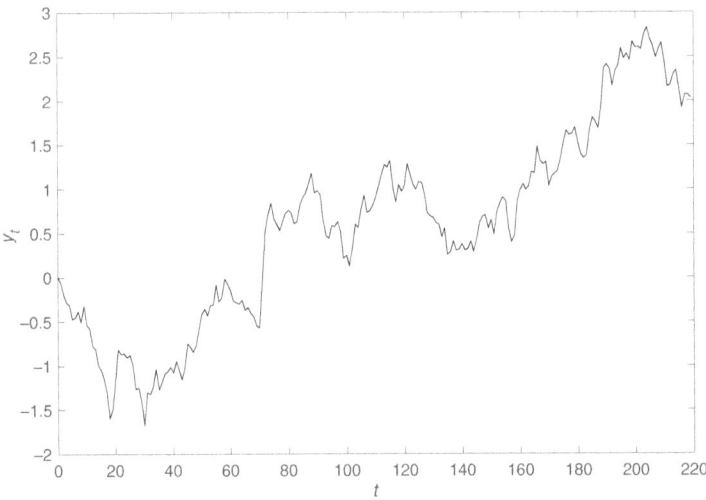

Figure 3.10 A random walk sequence with drift $\mu = 0.01$ and variance $\sigma^2 = 0.0034$.

The random walk model (3.34) is frequently used for modeling the time evolution of the logarithm of a stock price, that is, $y_t = \log p_t$, where p_t is the price of a stock at time t, is typically assumed to follow equation (3.34), where μ represents the time trend of the (log) price. Assuming the initial price $p_0 = 1$, we have that the expected log price trends as $t\mu$ in time, while also its variance grows as $t\sigma^2$. The random walk model (3.34) for stock prices also postulates *unpredictability* of future prices given the past. Indeed, we have that the value of y_t given the whole past path $I_{t-1} = \{y_0, \ldots, y_{t-1}\}$ is the same as the value of y_t given only y_{t-1} (Markov property). Formally this means that $y_t|I_{t-1} = y_t|y_{t-1} = (y_{t-1} + \mu) + \epsilon_t$, which is a random variable with mean $y_{t-1} + \mu$ and variance σ^2. Predicting y_t on the basis of past information is thus like predicting white noise: we can predict the future value based on the conditional expectation

$$E\{y_t|I_{t-1}\} = y_{t-1} + \mu,$$

but the actual random value $y_t|I_{t-1}$ will fluctuate around this prediction with variance

$$\mathrm{var}\{y_t|I_{t-1}\} = \mathrm{var}\{y_t|y_{t-1}\} = \sigma^2,$$

hence conditioning on the past gives no advantage for the purpose of prediction. Further, observe that if y_t denotes a log price, then recalling the definition of log-return, $\varrho_t = \log(p_t/p_{t-1}) = y_t - y_{t-1}$, we have that if y_t is a random walk then ϱ_t obeys the equation

$$\varrho_t = \mu + \epsilon_t, \quad t \geq 0,$$

which means that the log-return is stationary and follows a white noise process with mean μ and variance σ^2. Conversely, assuming a white noise model for the log-return implies that the log price process follows a random walk model.

3.9.1 Moving average processes

A *moving average* (MA) time series $\{y_t\}$ follows a model of the form

$$y_t = \mu + \epsilon_t + \theta_1 \epsilon_{t-1} + \cdots + \theta_q \epsilon_{t-q}, \quad \forall t, \tag{3.35}$$

where $\{\epsilon_t\}$ is a zero-mean white noise process with variance σ^2, the parameters of the model are collected in vector $\theta = (\theta_1, \ldots, \theta_q)$, and q is the order of the model, which is referred to as an MA(q) process. By direct calculation, we obtain that

$$E\{y_t\} = \mu,$$
$$\mathrm{var}\{y_t\} = (1 + \|\theta\|_2^2)\sigma^2 \doteq \gamma_0$$

and it holds for the autocovariance that

$$\gamma_k = \begin{cases} \sigma^2 \sum_{j=0}^{q-k} \theta_{k+j}\theta_j & \text{for } k = 1, \ldots, q, \\ 0 & \text{for } k > q, \end{cases}$$

where we set $\theta_0 \doteq 1$. The MA(q) process is covariance stationary, for any values of the θ parameters. An MA(q) process is easily recognizable from its ACF, since its coefficients γ_k are identically zero for $k > q$. For $q \to \infty$ we have an MA(∞) process of the form

$$y_t = \mu + \sum_{j=0}^{\infty} \psi_j \epsilon_{t-j}, \quad \forall t. \tag{3.36}$$

This type of process is covariance stationary provided that $\sum_{j=0}^{\infty} \psi_j^2 < \infty$, that is, when the sequence of coefficients is square summable.[14] Notice that in MA processes the so-called *innovation* process $\{\epsilon_t\}$ is white noise but need not be an *independent* white noise process, in general. Nonindependent innovations are characteristic of the so-called conditionally heteroskedastic time series that we discuss in Section 3.9.5.

[14] A sufficient condition for square summability is absolute summability, that is, $\sum_{j=0}^{\infty} |\psi_j| < \infty$, a condition that is often assumed in the MA model's context.

3.9.2 Autoregressive processes

An *autoregressive* process of order p, denoted by AR(p), follows a model of the form

$$y_t = \eta + \epsilon_t + \sum_{j=1}^{p} \phi_j y_{t-j}, \quad \forall t, \tag{3.37}$$

where $\{\epsilon_t\}$ is a zero-mean white noise process with variance σ^2, and the parameters of the model are collected in vector $\phi = (\phi_1, \ldots, \phi_p)$. This kind of stochastic model introduces dynamics that may render the system "unstable." Defining the lag operator D such that $D^j[y_t] = y_{t-j}$ we rewrite (3.37) as

$$\Psi(D)[y_t] = \eta + \epsilon_t,$$

where

$$\Psi(D) \doteq 1 - \sum_{j=1}^{p} \phi_j D^j \tag{3.38}$$

is a polynomial of degree p in D. It is a well-known fact in time series theory[15] that the AR(p) process (3.37) is covariance stationary if the roots of $\Psi(D)$ all have modulus larger than unity. In such case, the AR(p) process has a MA(∞) equivalent representation of the form

[15] See for instance Chapter 3 of J. D. Hamilton, *Time Series Analysis*, Princeton, NJ: Princeton University Press, 1994.

$$y_t = \mu + \Psi(D)^{-1}[\epsilon_t],$$

where $\mu = \eta/\Psi(1)$ and $\Psi(D)^{-1} = \psi_0 + \psi_1 D + \psi_2 D^2 + \psi_3 D^3 + \cdots$.

3.9.3 Autoregressive moving average processes

An *autoregressive moving average* process ARMA(p, q) includes an autoregressive component and a moving average component, and follows a model of the form

$$y_t = \eta + \sum_{j=1}^{p} \phi_j y_{t-j} + \sum_{j=0}^{q} \theta_j \epsilon_{t-j}, \quad \forall t, \tag{3.39}$$

where $\{\epsilon_t\}$ is a zero mean white noise process, and we fix $\theta_0 = 1$. Using the delay operator form, we may write

$$\Psi(D)[y_t] = \eta + N(D)[\epsilon_t],$$

where $\Psi(D)$ is given in (3.38), and $N(D) \doteq \sum_{j=0}^{q} \theta_j D^j$. Also in this case, if all the roots of $\Psi(D)$ have modulus larger than unity then the ARMA(p,q) is covariance stationary.

3.9.4 ARIMA(d, p, q) processes

The random walk process $\{y_t\}$ discussed in Example 3.4 is an example of an AR(1) process which is non stationary. However, we have seen that by taking the first-order difference $\{\delta_t^1\}$ of the process, where $\delta_t^1 \doteq y_t - y_{t-1}$, we obtain white noise, which is a stationary process. The interest of ARIMA models, where the "I" stands for "integrated," is to model non-stationary time series having the characteristic of becoming stationary once one takes the dth-order difference

$$\delta_t^d \doteq (1-D)^d[y_t].$$

Specifically, an ARIMA(d, p, q) model on $\{y_t\}$ is an ARMA(p, q) model on the dth-order difference time series $\delta_t^d = (1-D)^d[y_t]$. In most practical applications, first-order differencing ($d = 1$) and occasionally second differencing ($d = 2$) is enough for achieving stationarity.

In practice, the problem of determining the parameters of a time-series model (for instance, an ARIMA(d, p, q) model) from a given number m of observed samples $y_1, \ldots, y_m$ is solved numerically using the maximum likelihood principle via software packages that are available in all of the major computing platforms, such as R, MATLAB®, or Python (e.g., via the statsmodels library).

3.9.5 ARCH models

Autoregressive conditional heteroskedastic (ARCH) models are commonly used in finance for modeling time series that display volatility variation over time and volatility clustering, that is, the observed phenomenon for which large changes in price tend to be followed by large changes and small changes tend to be followed by small changes. Figure 3.11, for example, shows the daily log-returns of the AAPL stock during the five-year period 2017–2022: clusters of high and low volatility periods are clearly visible.

For introducing ARCH models we first consider a standard AR(p) model[16] of the form

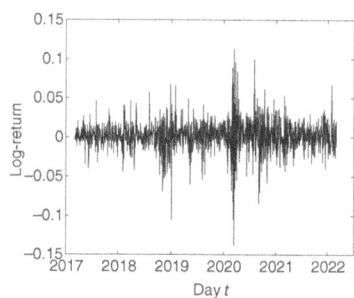

Figure 3.11 AAPL stock daily log-returns: 1,258 sample points from March 6, 2017 to March 2, 2022; daily price data in AAPL_5y.csv in the online resources.

[16] We also allow $p = 0$ in the model, in which case we obtain an AR(0) that is simply a white noise process with mean η.

$$y_t = \eta + u_t + \sum_{j=1}^{p} \phi_j y_{t-j},$$

where $\{u_t\}$ is a zero mean white noise with variance σ^2. We assume that the roots of $\Psi(D)$ in (3.38) are all outside the unit circle, so that the model is covariance stationary. In such case, the (unconditional) mean of y_t is constant:

$$E\{y_t\} = \eta/\Psi(1),$$

while the conditional mean, for given past $I_{t-1} = \{y_0, \ldots, y_{t-1}\}$, is

$$E\{y_t|I_{t-1}\} = \eta + \sum_{j=1}^{p} \phi_j y_{t-j},$$

so that $u_t = y_t - E\{y_t|I_{t-1}\}$ represents the error we make in predicting y_t by means of $E\{y_t|I_{t-1}\}$. While the unconditional variance of this error u_t is constant and equal to σ^2, we can assume that u_t^2 follows itself an AR(b) process, that is,

$$u_t^2 = \zeta + \epsilon_t + \sum_{j=1}^{b} \alpha_j u_{t-j}^2, \quad (3.40)$$

where $\{\epsilon_t\}$ is a zero mean white noise process with variance λ^2. A white noise process $\{u_t\}$ satisfying (3.40) is said to be an ARCH(b) process. Notice that $\{u_t\}$ is white noise (hence it is uncorrelated in time) but it is not independent, since its square $\{u_t^2\}$ follows an AR process, which entails that the correlation of u_t^2 and u_{t+k}^2 is nonnull in general. Typically, since u_t^2 needs be nonnegative at all t, we assume that $\zeta > 0$ and $\epsilon_t \geq -\zeta$, and $\alpha_j \geq 0$ for all j. Under these additional requirements, the condition $\sum_{j=1}^{b} \alpha_j D^j < 1$ guarantees that all the roots of the polynomial $1 - \sum_{j=1}^{b} \alpha_j D^j$ lie outside of the unit circle, hence the squared AR process (3.40) is covariance stationary. Observe that the unconditional variance of u_t is

$$\sigma^2 = E\{u_t^2\} = \zeta / \left(1 - \sum_{j=1}^{b} \alpha_j\right),$$

while the conditional variance, given the past, is

$$E\{u_t^2 | u_0^2, \ldots, u_{t-1}^2\} = \zeta + \sum_{j=1}^{b} \alpha_j u_{t-j}^2.$$

Example 3.5 As already mentioned, a common model for the log-return y_t of a stock is an AR(0) model (white noise) of the form $y_t = \eta + u_t$. Considering for instance the AAPL stock log-returns shown in

Figure (3.11), we see that the sample autocorrelation function of such returns appears to be very small (i.e., practically zero) for all nonzero lags (Figure 3.12(a)) while the sample autocorrelation function of the *squared* returns shows a clear presence of correlation (Figure 3.12(b)), thus indeed suggesting that the time series of log-return can be considered uncorrelated white noise, but not independent white noise, and that an ARCH model might be suitable for modeling this noise.

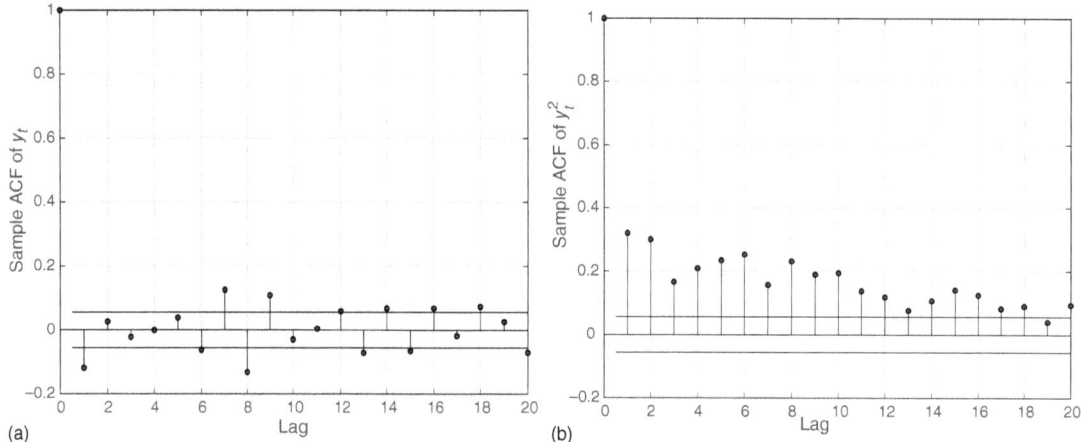

Figure 3.12 Sample ACF of (a) the AAPL log-returns and (b) the squared log-returns; see aapl_ACF.m in the online resources.

An alternative representation of an ARCH(b) process is via the multiplicative expression

$$u_t = \sqrt{h_t} v_t, \qquad (3.41)$$

where $\{v_t\}$ is *independent* white noise with zero mean and unit variance, and h_t evolves according to

$$h_t = \zeta + \sum_{j=1}^{b} \alpha_j u_{t-j}^2. \qquad (3.42)$$

3.9.6 GARCH models

Generalized autoregressive conditional heteroskedastic (GARCH) models[17] are derived as an extension of ARCH models by considering a non-independent white noise process $\{u_t\}$ in the form (3.41) where now h_t evolves according to

$$h_t = \kappa + \sum_{j=1}^{r} \gamma_j h_{t-j}^2 + \sum_{j=1}^{b} \alpha_j u_{t-j}^2. \qquad (3.43)$$

This type of model is referred to as a GARCH(r, b) model; GARCH($0, b$) reduces to ARCH(b), and GARCH($0, 0$) is simply a zero mean

[17] See T. Bollerslev, "Generalized Autoregressive Conditional Heteroskedasticity," *Journal of Econometrics*, vol. 31(3), pp. 307–327, 1986, for the original reference on GARCH models.

independent white noise process. Positivity of h_t is guaranteed by assuming that $\kappa > 0$ and $\gamma_j \geq 0$, $\alpha_j \geq 0$ for all j. Further, if

$$\sum_{j=1}^{p}(\gamma_j + \alpha_j) < 1, \quad p \doteq \max(r, b),$$

then $\{u_t^2\}$ is covariance stationary, and in this case the unconditional mean of u_t^2 is

$$E\{u_t^2\} = \sigma^2 = \kappa / \left(1 - \sum_{j=1}^{p}(\gamma_j + \alpha_j)\right).$$

Further analysis of ARCH/GARCH models goes beyond the scope of this book, so we direct the interested readers to the wide specialistic literature on this topic. Most mainstream computational platforms such as R, Python, or MATLAB®, however, provide packages for estimation of the parameters of ARCH/GARCH models from observations of the time series, and for using the estimated models in forecasting. For instance, continuing Example 3.5, we can fit a GARCH(1,1) models to the observed log-return data, obtaining the model $u_t = \sqrt{h_t} v_t$, with

$$h_t = \kappa + \gamma_1 h_{t-1} + \alpha_1 u_{t-1}^2, \quad (3.44)$$

where the estimated parameters (estimation was performed via the MATLAB® Econometric Toolbox) are

$$\kappa = 1.537 \times 10^{-5}, \quad \gamma_1 = 0.8272, \quad \alpha_1 = 0.13268.$$

Figure 3.13 shows the sample ACF of a realization of (3.44). Comparing it with Figure 3.12(b) we see that this simple GARCH(1,1) essentially captures the correlation structure of the squared returns.

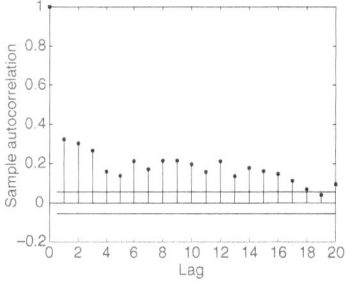

Figure 3.13 sample ACF of a realization of (3.44); see file ex_garch.m in the online resources.

3.10 Exercises

Exercise 3.1 *Covariance matrix of return time series*
Consider a portfolio $w = (w_1, \ldots, w_n)$ and a set of m vectors $r^{(1)}, \ldots, r^{(m)} \in \mathbb{R}^n$ of asset returns.

1. If S is the empirical covariance matrix of the asset returns, what is the financial interpretation of the quantity $w^\top S w$? Justify your answer.

2. Explain why the matrix S is positive semidefinite.

3. Consider a two-asset market. Asset 1 has a variance of 2; asset 2, of 3. Let Σ be the (2×2) covariance matrix of the market. The covariance between the two assets is defined as β, so that

$$\Sigma = \begin{pmatrix} 2 & \beta \\ \beta & 3 \end{pmatrix}.$$

   Find an upper bound on the magnitude $|\beta|$ of the covariance between asset 1 and 2. *Hint:* Express the fact that the portfolio with weights, for example, $w = (-\beta/2, 1)$ must have nonnegative variance.

4. Is the bound you have found previously the best (i.e., the lowest)? Justify your answer.

Exercise 3.2 *Outlier value detection in time series*
In the context of time-series data, "temporal outliers" are values in the time series that do not agree in some sense with the general time evolution of the data.

1. A common approach to outlier detection in time series is to use prediction models. These approaches define a way of building a prediction model, and then compute an "outlier score" for each point in the time series according to its deviation from the predicted value. Suppose we are given a time-series training data set $(y_t)_{1 \leq t \leq T}$, and would like to use for outlier detection an autoregressive (AR) prediction model of the form

$$\hat{y}_{t+1} = \phi_1 y_t + \cdots + \phi_q y_{t-q+1},$$

   where $\phi \in \mathbb{R}^q$ is the parameter vector of the model, to be chosen appropriately by fitting the model to the data. We make the assumption that, over any subsequence, the number of outlier values is small with respect to the length of the subsequence. Discuss a technique to fit this AR model to the time-series data and to use its prediction for detecting outliers.

2. Test the AR method developed in the previous point on synthetic time-series data. Precisely, generate an AR model, add some noise to it, then corrupt it with a few outliers; then check how well the method identifies the outlier values.

Exercise 3.3 *Sparse precision matrix estimation*
In this exercise, the focus is on sparse precision matrix estimation.

1. Consider the currency daily data given in the file INTLFXD_csv.zip in the online resources. Use the "graph LASSO" estimator,[18] which solves the ℓ_1-regularized objective (3.26), to compute a covariance matrix whose inverse is sparse, with Python's default setting for the sparsity parameter. For this part and for the remainder of the problem, remove Malaysia and Venezuela from the data for the purposes of numerical stability of the graphical lasso estimator and subsequent analysis. Scale the data to have zero mean and unit variance.

    [18] At https://people.ece.ubc.ca/xiaohuic/code/glasso/glasso.htm for MATLAB®; for Python, use https://scikit-learn.org/stable/auto_examples/covariance/plot_sparse_cov.html.

2. What is the interpretation of a zero element in the estimated matrix? Conversely, how can we interpret the links connecting a given node (a currency) to its neighbors? Can we get an idea of the correlation structure itself, just by looking at the graph?

3. We now analyze the stability of the edges detected by the estimation method. Plot the number of edges as function of the sparsity parameter. Discuss the results.

4. Discuss how the inverse covariance matrix change over time, when the computation is based on yearly time windows.

5. Try to visualize the graphs you obtained corresponding to the sparse inverse covariance's nonzero elements, using plotly[19] or with the corresponding MATLAB® routines.[20]

    [19] At https://plot.ly/python/network-graphs/.
    [20] At www.mathworks.com/help/matlab/graph-and-network-algorithms.html.

Exercise 3.4 *Mean square error estimate*
Suppose that $x^{(1)}, \ldots, x^{(n)}$ are n i.i.d. realizations of a random variable x having probability distribution

$$p(x|\theta) = \begin{cases} \theta x^{\theta-1} & \text{if } 0 < x < 1, \\ 0 & \text{otherwise.} \end{cases}$$

Suppose also that the value of parameter θ is unknown ($\theta > 0$) and that the prior distribution is a gamma distribution with parameters α and β ($\alpha > 0$ and $\beta > 0$):

$$p(\theta) = \frac{\theta^{\alpha-1} e^{-\beta\theta} \beta^{\alpha}}{\Gamma(\alpha)},$$

where $\Gamma(\alpha)$ is the gamma function: $\Gamma(\alpha) = \int_0^\infty t^{\alpha-1} e^{-t} dt$, for all α where the real part is strictly positive, and $\Gamma(\alpha) = (\alpha - 1)!$ for all positive integers. Determine the posterior distribution of θ and hence obtain the minimum mean square error estimate of θ.

Exercise 3.5 *ATM arrival times*

Let $y^{(1)}, y^{(2)}, \ldots, y^{(n)}$ be realizations of a sequence of i.i.d. random variables representing inter-arrival times of customer to an ATM machine. The random variables have the following probability distribution:

$$p(y|\lambda) = \begin{cases} \lambda e^{-\lambda y} & \text{if } y > 0, \\ 0 & \text{otherwise,} \end{cases}$$

where λ is an unknown, positive parameter with a gamma prior distribution with parameters m and β (see Exercise 3.4 for a definition of the gamma distribution).

1. Show that the posterior distribution of λ given $y^{(1)}, y^{(2)}, \ldots, y^{(n)}$ is a gamma distribution with parameters $n + m$ and $\beta + t$ where $t = \sum_{i=1}^{n} y^{(i)}$.

2. Show that the probability distribution of the $(n+1)$th inter-arrival time y given the first n inter-arrival times $y^{(1)}, y^{(2)}, \ldots, y^{(n)}$ is

$$p(y|y^{(1)}, y^{(2)}, \ldots, y^{(n)}) = \frac{(n+m)(\beta+t)^{n+m}}{(y+\beta+t)^{n+m+1}}.$$

3. Let us consider the following observations for the first seven inter-arrival times (expressed in minutes):

$$\mathcal{D} = \{y^{(1)}, y^{(2)}, \ldots, y^{(n)}\} = \{3, 14, 3, 27, 10, 25, 19\}.$$

Compute the maximum likelihood estimate of λ. Then, suppose that the prior distribution of λ is a gamma distribution with parameters $m = 6$ and $\beta = 1$, compute the minimum mean square error estimate of λ.

Appendix A

A.1 Properties of symmetric matrices

A.1.1 Symmetric, diagonal and orthonormal matrices

A matrix $A \in \mathbb{R}^{n,n}$ is *symmetric* if $A = A^\top$, that is, $A_{ij} = A_{ji}$ for all $i,j = 1,\ldots,n$. We denote the space of symmetric $n \times n$ matrices by $\mathbb{S}^n$. Symmetric matrices are used, for instance, for describing variance-covariance information, as seen in this chapter, for defining quadratic functions of the type $\sum_{i,j} A_{ij} x_i x_j = x^\top A x$ for $x \in \mathbb{R}^n$, or in the representation of undirected graphs, see, for example, Figure A.1.

A *diagonal* matrix is a special type of symmetric matrix for which all elements outside the diagonal are zero. A diagonal matrix D is written as

$$D = \begin{bmatrix} d_1 & & \\ & \ddots & \\ & & d_n \end{bmatrix},$$

which we also write as $D = \mathrm{diag}(d)$, where $d \in \mathbb{R}^n$ is the vector of diagonal elements. When a vector $x \in \mathbb{R}^n$ is multiplied on the left by a diagonal matrix D the result is a vector with scaled entries:

$$Dx = \begin{bmatrix} d_1 x_1 \\ \vdots \\ d_n x_n \end{bmatrix}.$$

When a diagonal matrix D of dimension $n \times n$ multiplies on the left a matrix $A \in \mathbb{R}^{n,m}$ the effect is to multiply the rows of A by the elements on the diagonal of D, that is,

$$DA = \mathrm{diag}(d) \begin{bmatrix} r^{(1)\top} \\ \vdots \\ r^{(n)\top} \end{bmatrix} = \begin{bmatrix} d_1 r^{(1)\top} \\ \vdots \\ d_n r^{(n)\top} \end{bmatrix}.$$

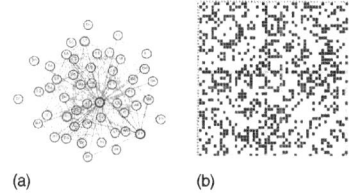

Figure A.1 An undirected graph (a) and its symmetric matrix representation (b).

Similarly, when D multiplies on the right a matrix $B \in \mathbb{R}^{m,n}$ the effect is to multiply the columns of B by the elements on the diagonal of D, that is,

$$BD = \begin{bmatrix} b^{(1)} & \cdots & b^{(n)} \end{bmatrix} \text{diag}(d) = \begin{bmatrix} d_1 b^{(1)} & \cdots & d_n b^{(n)} \end{bmatrix}.$$

An important special diagonal matrix is the *identity* matrix $I = \text{diag}(1,\ldots,1)$, where the size of the matrix is typically inferred from context, or specified with the notation I_n to denote an identity matrix of size $n \times n$.

An $n \times n$ matrix $U = [u^{(1)}, \ldots, u^{(n)}]$ is said to be *orthogonal* (or unitary) if for all $i, j = 1, \ldots, n$, it holds that $u^{(i)^\top} u^{(j)} = \delta_{ij}$, where $\delta_{ij} = 1$ for $i = j$ and zero otherwise. More compactly, we write

$$U^\top U = U U^\top = I,$$

where I is the $n \times n$ identity matrix. Orthonormal matrices leave norms of vectors and angles between vectors invariant: if $x, y \in \mathbb{R}^n$, then

$$(Ux)^\top (Uy) = x^\top U^\top U y = x^\top y, \quad \text{and } \|Ux\|_2 = \|x\|_2.$$

A.1.2 *Eigenvalue decomposition for symmetric matrices*

We are now in position to state a key result that provides a general decomposition for symmetric matrices.

Theorem A.1 (Spectral theorem) *Any symmetric matrix $A \in \mathbb{S}^n$ can be decomposed as*

$$A = U \Lambda U^\top = \sum_{i=1}^n \lambda_i u^{(i)} u^{(i)^\top},$$

where $\Lambda = \text{diag}(\lambda_1, \ldots, \lambda_n)$, being $\lambda_1 \geq \cdots \geq \lambda_n$ the eigenvalues of A (which are all real numbers), and $U = [u^{(1)} \cdots u^{(n)}]$ is an orthogonal matrix, being $u^{(1)}, \ldots, u^{(n)}$ the eigenvectors of A.

The extreme (i.e., maximum and minimum) eigenvalues of A have an interesting interpretation in terms of the maximum and minimum value attained by the quadratic form $x^\top A x$ when x ranges over the unit Euclidean ball, as stated by the following corollary.

Corollary A.1 (Variational characterization of extreme eigenvalues) *Let $\lambda_1 \geq \cdots \geq \lambda_n$ be the ordered eigenvalues of $A \in \mathbb{S}^n$, and let $u^{(1)}, \ldots, u^{(n)}$ be the corresponding set of eigenvectors. Then,*

$$\lambda_1 = \lambda_{\max}(A) \doteq \max_{x:\, \|x\|_2 = 1} x^\top A x,$$

$$\lambda_n = \lambda_{\min}(A) \doteq \min_{x:\, \|x\|_2 = 1} x^\top A x.$$

Moreover, the optimum in the first problem is attained at the optimal solution $x^* = u^{(1)}$, and the optimum in the second problem is attained at the optimal solution $x^* = u^{(n)}$.

The first result in Corollary A.1 can be easily proved by considering that, by using the Spectral theorem and $\lambda_1 \geq \cdots \geq \lambda_n$,

$$x^\top A x = \sum_{i=1}^n \lambda_i (u^{(i)^\top} x)^2 \leq \lambda_1 \sum_{i=1}^n (u^{(i)^\top} x)^2 = \lambda_1 \|U^\top x\|_2^2 = \lambda_1 \|x\|_2^2 = \lambda_1.$$

Also, setting $x = u^{(1)}$ and recalling that $u^{(1)}, \ldots, u^{(n)}$ are orthonormal, we have that

$$x^\top A x = u^{(1)^\top} \left(\sum_{i=1}^n \lambda_i u^{(i)} u^{(i)^\top} \right) u^{(1)^\top} = \lambda_1,$$

which shows that the upper bound on $x^\top A x$ is indeed achieved for $x = u^{(1)}$. The second result follows similarly, upon changing A to $-A$.

A.1.3 Positive semidefinite matrices

The spectral theorem assures that the eigenvalues of a matrix $A \in \mathbb{S}^n$ are all real numbers, hence they can be ordered as $\lambda_{\max}(A) = \lambda_1 \geq \cdots \geq \lambda_n = \lambda_{\min}(A)$. If it happens that $\lambda_{\min}(A) \geq 0$, then we say that A is *positive semidefinite* (PSD), a condition that we shall indicate with the notation $A \succeq 0$. If strict inequality holds, that is, $\lambda_{\min}(A) > 0$, then we say that A is *positive definite* (PD), a condition that we shall indicate with the notation $A \succ 0$. Similarly, we say that A is *negative semidefinite* (NSD), denoted with $A \preceq 0$, if $-A$ is PSD, and we say that A is *negative definite* (ND), denoted with $A \prec 0$, if $-A$ is PD. Equivalently, $A \preceq 0$ if $\lambda_{\max}(A) \leq 0$, and $A \prec 0$ if $\lambda_{\max}(A) < 0$. Clearly, from these definitions follows that for a PSD (resp. NSD) matrix all the eigenvalues are nonnegative (resp. nonpositive), and for a PD (resp. ND) matrix all the eigenvalues are positive (resp. negative). We denote with $\mathbb{S}_+^n$ the set of all $n \times n$ PSD (and hence symmetric) matrices, and with $\mathbb{S}_{++}^n$ the set of all $n \times n$ PD matrices. Likewise, we denote with $\mathbb{S}_-^n$ and $\mathbb{S}_{--}^n$ the set of NSD and ND matrices, respectively.

An equivalent definition is that a matrix $A \in \mathbb{S}^n$ is PSD if the corresponding quadratic form is always nonnegative, that is, if $x^\top A x \geq 0$ for all $x \in \mathbb{R}^n$. Equivalent definitions of PD, NSD and ND matrices follow analogously. These equivalences with the previous definitions are easily proved by using the spectral decomposition of A.

If $A \in \mathbb{S}^n$ is PSD then every principal submatrix extracted from A is also PSD. As a consequence of this fact, all the diagonal entries A_{ii},

$i = 1, \ldots, n$, of A (which are one-dimensional principal submatrices) are real and nonnegative. Further, considering that all eigenvalues, and hence the determinant, of any 2×2 principal submatrix is nonnegative, we have that

$$|A_{ij}| \leq \sqrt{A_{ii} A_{jj}}, \quad \forall i \neq j.$$

Another important feature of PSD matrices is that they admit a *matrix square root*. In particular, for a symmetric matrix $A \in \mathbb{S}^n$:

- $A \succeq 0$ if and only if there exists some matrix B such that $A = B^\top B$;

- if $A \succeq 0$ then there exists a unique $B \succeq 0$ such that $A = BB$; we call such a matrix the matrix square root of A, and denote it by $B = A^{1/2}$.

A.2 Point estimates

We next derive the expressions given in Section 3.1.2 for the minimizers of the expected loss, under the posterior density, for three relevant special cases of loss functions.

Minimum mean square error (MMSE) estimator For the quadratic loss $\mathcal{L}(\theta - \hat{\theta}) = (\theta - \hat{\theta})^2$, we have that

$$\mathrm{E}\{\mathcal{L}(\theta - \hat{\theta}) | \mathcal{D}\} = \int (\theta - \hat{\theta})^2 p(\theta|\mathcal{D}) \mathrm{d}\theta.$$

To compute the minimum of the above expected loss with respect to $\hat{\theta}$, we compute the derivative of the objective:

$$\frac{\partial}{\partial \hat{\theta}} \int (\theta - \hat{\theta})^2 p(\theta|\mathcal{D}) \mathrm{d}\theta = \int \frac{\partial}{\partial \hat{\theta}} \left((\theta - \hat{\theta})^2 p(\theta|\mathcal{D}) \right) \mathrm{d}\theta$$

$$= \int -2(\theta - \hat{\theta}) p(\theta|\mathcal{D}) \mathrm{d}\theta.$$

Setting the previous derivative to zero, we obtain

$$\int -2(\theta - \hat{\theta}) p(\theta|\mathcal{D}) \mathrm{d}\theta = 0 \Leftrightarrow \int \hat{\theta} p(\theta|\mathcal{D}) \mathrm{d}\theta = \int \theta p(\theta|\mathcal{D}) \mathrm{d}\theta$$

$$\Leftrightarrow \hat{\theta} = \int \theta p(\theta|\mathcal{D}) \mathrm{d}\theta = \mathrm{E}\{\theta|\mathcal{D}\}.$$

Thus, we have that the point estimator which minimizes the quadratic loss coincides with the mean of the posterior distribution $p(\theta|\mathcal{D})$:

$$\hat{\theta}_{\mathrm{MMSE}} = \int \theta p(\theta|\mathcal{D}) \mathrm{d}\theta = \mathrm{E}\{\theta|\mathcal{D}\}.$$

This point estimator is called the *minimum mean square error* estimator, because it minimizes the average squared error.

Minimum mean absolute error (MMAE) estimator For the absolute value loss $\mathcal{L}(\theta - \hat{\theta}) = |\theta - \hat{\theta}|$, we have that

$$E\{\mathcal{L}(\theta - \hat{\theta})|\mathcal{D}\} = \int |\theta - \hat{\theta}| p(\theta|\mathcal{D}) d\theta.$$

It can be shown[21] that the minimum w.r.t. $\hat{\theta}$ is obtained when

$$\int_{-\infty}^{\hat{\theta}} p(\theta|\mathcal{D}) d\theta = \int_{\hat{\theta}}^{\infty} p(\theta|\mathcal{D}) d\theta.$$

[21] A well-known fact, see, for example, J. B. S. Haldane, "Note on the Median of a Multivariate Distribution," *Biometrika*, vol. 35, pp. 414–417, 1948.

In words, the estimate $\hat{\theta}$ is the value which divides the probability mass into equal proportions:

$$\int_{-\infty}^{\hat{\theta}} p(\theta|\mathcal{D}) d\theta = \frac{1}{2},$$

which is the definition of the *median* of the posterior distribution. The point estimate that minimizes the absolute value loss thus coincides with the median of the posterior distribution, and it is called the *minimum mean absolute error* estimator.

Maximum a-posteriori (MAP) estimator For the hit-or-miss loss we have

$$E\{\mathcal{L}(\theta - \hat{\theta})|\mathcal{D}\} = \int_{-\infty}^{\hat{\theta}-\delta} p(\theta|\mathcal{D}) d\theta + \int_{\hat{\theta}+\delta}^{\infty} p(\theta|\mathcal{D}) d\theta$$
$$= 1 - \int_{\hat{\theta}-\delta}^{\hat{\theta}+\delta} p(\theta|\mathcal{D}) d\theta.$$

This is minimized by maximizing $\int_{\hat{\theta}-\delta}^{\hat{\theta}+\delta} p(\theta|\mathcal{D}) d\theta$. For small $\delta > 0$ and suitable assumptions on $p(\theta|\mathcal{D})$ (e.g., smooth, quasi-concave) the maximum occurs at the maximum of $p(\theta|\mathcal{D})$. Therefore, the estimator is the *mode* (i.e., the peak value) of the posterior distribution $p(\theta|\mathcal{D})$, thus the name maximum a posteriori estimator:

$$\hat{\theta}_{\text{MAP}} = \arg\max_{\theta} p(\theta|\mathcal{D})$$
$$[\text{by Bayes' rule}] = \arg\max_{\theta} p(\mathcal{D}|\theta) p(\theta).$$

Since $p(\mathcal{D}|\theta)$ is the likelihood and $p(\theta)$ is the prior, we observe from the previous expression that if the prior is uniform, that is, $p(\theta) = \text{const.}$, then maximizing $p(\theta|\mathcal{D})$ is equivalent to maximizing the likelihood $p(\mathcal{D}|\theta)$. The so-called *maximum likelihood* estimator

$$\hat{\theta}_{\text{ML}} = \arg\max_{\theta} p(\mathcal{D}|\theta),$$

is therefore equivalent to the MAP estimator, under uniform prior.

A.3 Identity plus low-rank covariance approximation

We next provide a derivation for the optimal solution of problem (3.12):

$$(\hat{L}, \hat{\sigma}) = \arg \min_{L \in \mathbb{R}^{n,k}, \sigma \in \mathbb{R}} \|S - (LL^\top + \sigma^2 I_n)\|_F.$$

We let $S = U \Lambda U^\top$ be the eigenvalue decomposition of S, with $\Lambda = \mathrm{diag}(\lambda_1, \ldots, \lambda_n)$, $\lambda_1 \geq \cdots \geq \lambda_n \geq 0$, and $U = [u^{(1)}, \ldots, u^{(n)}]$. Noting that $LL^\top$ is PSD and has rank $\leq k$, our problem is related to finding a matrix $Q \succeq 0$ of rank at most k that minimizes the residual $\|S(\alpha) - Q\|_F$, where, for $\alpha \doteq \sigma^2 \geq 0$,

$$S(\alpha) \doteq S - \alpha I = U(\Lambda - \alpha I)U^\top = \sum_{i=1}^n (\lambda_i - \alpha) u^{(i)} u^{(i)\top}.$$

Let $n_+(\alpha)$ denote the number of elements in the above summation for which $\lambda_i - \alpha > 0$. Since the λ_i are ordered, we have $\lambda_i - \alpha > 0$ for $i = 1, \ldots, n_+(\alpha)$, and $\lambda_i - \alpha \leq 0$ for $i = n_+(\alpha) + 1, \ldots, n$. Now, an extension to the positive semidefinite matrix approximation of the Eckart–Young–Mirsky theorem[22] states that the optimal solution to

$$\min_{Q \succeq 0} \|S(\alpha) - Q\|_F, \quad \text{s.t.: } \mathrm{rank}\, Q \leq k$$

is given by

$$Q(\alpha) = \sum_{i=1}^{\min(k, n_+(\alpha))} (\lambda_i - \alpha) u^{(i)} u^{(i)\top}. \tag{A.1}$$

[22] See A. Dax, "Low-Rank Positive Approximants of Symmetric Matrices," *Advances in Linear Algebra and Matrix Theory*, vol. 4, pp. 172–185, 2014.

Substituting this $Q(\alpha)$ back into the objective, and observing that squaring a nonnegative objective does not change its minimizers, we have that problem (3.12) can be restated equivalently as

$$\min_{\alpha \geq 0} \|S(\alpha) - Q(\alpha)\|_F^2 = \left\| \sum_{i=\min(k, n_+(\alpha))+1}^{n} (\lambda_i - \alpha) u^{(i)} u^{(i)\top} \right\|_F^2$$

$$= \sum_{i=\min(k, n_+(\alpha))+1}^{n} (\lambda_i - \alpha)^2.$$

We now have two possible cases, depending on whether $n_+(\alpha) \geq k$ (i.e., $\alpha \leq \lambda_k$), or $n_+(\alpha) < k$ (i.e., $\alpha > \lambda_k$), hence we can write

$$\|S(\alpha) - Q(\alpha)\|_F^2 = \begin{cases} \sum_{i=k+1}^{n} (\lambda_i - \alpha)^2, & \text{if } n_+(\alpha) \geq k, \\ \sum_{i=n_+(\alpha)+1}^{k} (\lambda_i - \alpha)^2 + \sum_{i=k+1}^{n} (\lambda_i - \alpha)^2, & \text{if } n_+(\alpha) < k. \end{cases}$$

We then observe that $\sum_{i=k+1}^{n} (\lambda_i - \alpha)^2$ is increasing for $\alpha > \lambda_k$ and moreover the objective function has an additional nonnegative term in such range. It follows that the α that minimizes $\|S(\alpha) - Q(\alpha)\|_F^2$ will be contained in the range where $\alpha \leq \lambda_k$. For $0 \leq \alpha \leq \lambda_k$ the objective is given by $\sum_{i=k+1}^{n}(\lambda_i - \alpha)^2$, and we obtain the minimum by setting to zero the derivative with respect to α:

$$\sum_{i=k+1}^{n} \lambda_i = (n-k)\alpha \quad \Rightarrow \quad \hat{\alpha} = \frac{\sum_{i=k+1}^{n} \lambda_i}{n-k}.$$

We notice that indeed $\hat{\alpha} \geq 0$ and $\hat{\alpha} \leq \lambda_k$, as it should be. The previous formula gives us the optimal estimate $\hat{\sigma}^2 = \hat{\alpha}$ of the noise level, which is equal to the average value of the $n-k$ smallest eigenvalues of S. The optimal Q is then obtained by substituting $\hat{\alpha}$ back into (A.1), which results in

$$\hat{Q} = Q(\hat{\alpha}) = \sum_{i=1}^{k}(\lambda_i - \hat{\alpha})u^{(i)}u^{(i)\top}.$$

The optimal factor $\hat{L}$, such that $\hat{L}\hat{L}^\top = \hat{Q}$, is then found as

$$\hat{L} = \left[\sqrt{\lambda_1 - \hat{\alpha}}\, u^{(1)} \cdots \sqrt{\lambda_k - \hat{\alpha}}\, u^{(k)}\right].$$

4
Principal component analysis and low-rank approximation

UNDERSTANDING DATA STRUCTURE AND REDUCING the data dimensionality is a fundamental task in data analysis. In the previous chapter we observed that in many situations the dimension n of the sample vectors can be very large, in some cases even larger than the number of samples m. Such high-dimensional data can be very difficult to interpret, not to mention visualize. For example, Figure 4.1 shows the daily log-returns of 50 companies belonging to the Standard & Poor (S&P) 500 index over a one year period. We can observe that it is extremely difficult to extract valuable knowledge from this data representation. Perhaps the only information that we could easily obtain from this plot is that the returns are approximately zero, which is not very informative.

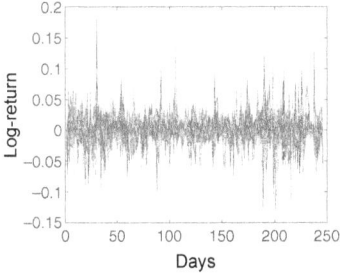

Figure 4.1 Daily market log-returns of 50 stocks from the S&P 500 basket in the period 25/01/2021–13/01/2022; see `Fig_SP500_2021.m` in the online resources.

A common practice to improve the interpretability of high-dimensional data is to reduce its dimensionality by transforming a large set of variables into a smaller one that still contains most of the information of the original set. Low-dimensional datasets are easier to visualize and process, making the analysis of data more straightforward. In this chapter, we introduce principal component analysis (PCA), which is perhaps the most popular technique for dimensionality reduction. We also discuss the link between PCA and low-rank approximations, showing that PCA implicitly forms a low-rank approximation of the empirical covariance matrix. At the end of the chapter we discuss some limitations of PCA and present some extensions that address these issues.

4.1 Principal component analysis

Principal component analysis was first introduced in psychometrics in the 1930s and rapidly became a fundamental tool for data analysis. Its application fields are countless, ranging from computer vision to medicine. In a financial context, for instance, PCA is useful for obtaining a general understanding of market data. It also underlies many theoretical models that are commonly used in finance, such as the capital asset pricing model (CAPM) or the models describing the behavior of the term structure of interest rates. In addition, PCA is frequently applied in risk analysis and portfolio immunization.

The main idea of PCA is to transform the data to a new coordinate system, where each dimension represents a *principal component* of the data. Principal components are defined as orthogonal directions where the data exhibits the maximal amount of variability. The reason why we are interested in maximizing the variance is that there is a strong relationship between variance and information: the larger the dispersion of the data along a direction, the more information it contains. In simple words, we can think of principal components as a new coordinate system that provides the best point of view to analyze the data, where the differences between the samples are more visible. This allows to compress most of the information into a few principal components and discard the components with low information, thus reducing the data dimensionality without losing too much information.

4.1.1 Iterated variance maximization

Given a data matrix $X = [x^{(1)} \cdots x^{(m)}] \in \mathbb{R}^{n,m}$, we next discuss how to compute the principal components of the data set. In the following we assume without loss of generality that the data points $x^{(i)}$, $i = 1, \ldots, m$, have already been centered around their centroid.

For a given direction described by a vector u, $\|u\|_2 = 1$, we know from Section 2.5 that the scores α_i of the projections of the data points along u are given by $\alpha_i = u^\top x^{(i)}$, $i = 1, \ldots, m$, and that the variance of these scores is given by $\sigma_u^2 = u^\top S u$, a quantity known as the *directional variance* of the data, where $S = XX^\top/m \in \mathbb{S}_+^n$ is the sample covariance matrix of X.

The directional variance is easy to compute by evaluating the quadratic form $u^\top S u$, for given direction u, and once the sample covariance matrix S has been constructed. The problem we want to address next, however, is not simply evaluating σ_u^2 for given u but rather *finding a direction u along which σ_u^2 is maximal*. Such direction is called the "first principal component" of the data matrix, and it is computed by solving the following *variance maximization problem*, see Figure 4.2:

$$\max_{u} u^\top S u, \quad \text{s.t.:} \; \|u\|_2 = 1. \tag{4.1}$$

From Corollary A.1, we have that the solution of problem (4.1) is $u^* = u^{(1)}$, where $u^{(1)}$ is an eigenvector of S corresponding to its largest eigenvalue λ_1, and that the optimal value of problem (4.1) is λ_1.

The first principal direction $u^{(1)}$ is thus a direction that captures the highest directional variability in the data. Once we have $u^{(1)}$, let us see how we can find a second direction $u^{(2)}$, orthogonal to $u^{(1)}$, capturing the second-largest directional variability in the data, and so on for the other directions. In Section 2.5.1, we already introduced the concept of *deflation*, which consists in removing from the original data the components along a given direction before projecting the data on a new direction. We can use this approach to find the other principal components, by applying the following steps:

1. Deflate the data from their variability along the directions of the principal components already found.

2. Find the direction of maximal variance for the projected data.

3. Iterate n times.

It is worth noting that, even though the process has to be iterated n times to find all the n principal components, it can also be stopped at an earlier iteration, finding only the first k components, where $k \ll n$.

Let us now analyze in detail the steps of the deflation method presented above. For the sake of simplicity, in the following we focus on the second principal component. The other components can be found in an analogous way. We recall from Section 2.5.1 that we can deflate the data points $x^{(i)}, i = 1, \ldots, m$, from their components along a direction $u^{(1)}$ as follows:

$$x^{(i)}_{\neg u^{(1)}} = x^{(i)} - x^{(i)}_{u^{(1)}} = x^{(i)} - (x^{(i)\top} u^{(1)}) u^{(1)} = P_1 x^{(i)}, \quad i = 1, \ldots, m,$$

where we defined the symmetric matrix $P_1 \doteq I - u^{(1)} u^{(1)\top}$. This operation projects the data points on a subspace orthogonal to $u^{(1)}$, as shown in Figure 4.3.

After having projected the data points on the subspace orthogonal to the first principal components, we have to find in that subspace the direction of maximal variance of the projected data. The sample covariance matrix $S^{(2)}$ of the projected data points $x^{(i)}_{\neg u^{(1)}}, i = 1, \ldots, m$, is

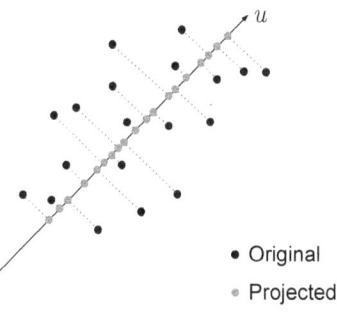

Figure 4.2 Variance maximization problem: we seek a direction u such that the variance of the projected points on the line generated by u is maximal.

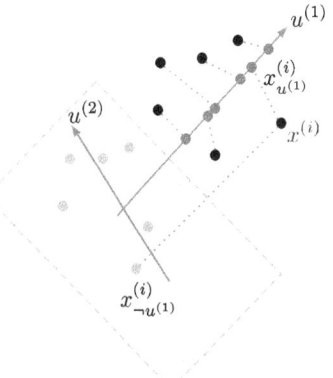

Figure 4.3 Deflation consists in projecting the data on a subspace orthogonal to the directions found before; a new maximum-variance direction contained in such subspace is then found, and the process is iterated.

$$S^{(2)} = \frac{1}{m}\sum_{i=1}^{m} x^{(i)}_{\neg u^{(1)}} x^{(i)\top}_{\neg u^{(1)}} = \frac{1}{m}\sum_{i=1}^{m} P_1 x^{(i)} x^{(i)\top} P_1 = P_1 S P_1.$$

By using the spectral factorization of S we have that

$$S^{(2)} = P_1 S P_1 = (I - u^{(1)} u^{(1)\top}) \left(\sum_{i=1}^{n} \lambda_i u^{(i)} u^{(i)\top}\right)(I - u^{(1)} u^{(1)\top})$$

$$= \sum_{i=2}^{n} \lambda_i u^{(i)} u^{(i)\top},$$

being $\lambda_1 \geq \cdots \geq \lambda_n$ the eigenvalues of S and $u^{(1)}, \ldots, u^{(n)}$ the corresponding eigenvectors. We can observe that the projected covariance $S^{(2)}$ is obtained by zeroing out the first eigenvalue of S. Therefore, the largest eigenvalue of $S^{(2)}$ is by construction λ_2 and a corresponding eigenvector is $u^{(2)}$. To find the second principal component, we consider the variance maximization problem:

$$\max_{u} u^\top S^{(2)} u, \quad \text{s.t.: } \|u\|_2 = 1. \tag{4.2}$$

Applying again Corollary A.1, we obtain that the solution of problem (4.2) is $u^{(2)}$. Iterating the deflation process k times we obtain that the first k principal components are given by the k eigenvectors $u^{(1)}, \ldots, u^{(k)}$ corresponding to the first k eigenvalues of S.

In practice, we can also compute all the principal components of X in one shot via the eigenvalue factorization of S. Letting $S = U\Lambda U^\top$ be such eigenvalue factorization, where $U \in \mathbb{R}^{n,n}$ is orthogonal and $\Lambda = \text{diag}(\lambda_1, \ldots, \lambda_n)$, we have that the first k principal components are given by the first k columns of U, which we collect in the matrix $U_k = [u^{(1)} \cdots u^{(k)}] \in \mathbb{R}^{n,k}$, and that the projections $x^{(1)}_{\mathcal{U}_k}, \ldots, x^{(m)}_{\mathcal{U}_k} \in \mathbb{R}^n$ of the points $x^{(1)}, \ldots, x^{(m)} \in \mathbb{R}^n$ onto the subspace $\mathcal{U}_k$ generated by $(u^{(1)}, \ldots, u^{(k)})$ are given by

$$x^{(i)}_{\mathcal{U}_k} = \sum_{j=1}^{k} x^{(i)}_{u^{(j)}} = \sum_{j=1}^{k} \left(u^{(j)\top} x^{(i)}\right) u^{(j)} = \sum_{j=1}^{k} \alpha_j(x^{(i)}) u^{(j)}, \quad i = 1, \ldots, m,$$

where $\alpha_j(x) \doteq u^{(j)\top} x$ is the *score* of x along the direction $u^{(j)}$. In matrix notation, we write

$$X_k = U_k A_k, \quad A_k \doteq U_k^\top X, \tag{4.3}$$

where $A_k \in \mathbb{R}^{k,m}$ contains by columns the scores of each original data point along the k principal directions, and $X_k \in \mathbb{R}^{n,m}$, $X_k = \left[x^{(1)}_{\mathcal{U}_k} \cdots x^{(m)}_{\mathcal{U}_k}\right]$ contains by columns the original data projected onto the subspace $\mathcal{U}_k$. Observe that the projected points $x^{(1)}_{\mathcal{U}_k}, \ldots, x^{(m)}_{\mathcal{U}_k}$ are still n dimensional, although they lie on the subspace $\mathcal{U}_k$ of dimension $k \leq n$. Actual dimensionality reduction is obtained by working with the score vectors $\alpha^{(1)}, \ldots, \alpha^{(m)} \in \mathbb{R}^k$ (the columns of A_k), which represent the coordinates of the original data points in the basis $u^{(1)}, \ldots, u^{(k)}$ of the subspace $\mathcal{U}_k$.

4.1.2 Approximation error and explained variance

We next analyze how well the data is approximated by its projection on the subspace $\mathcal{U}_k$ span by the first k principal directions. For $k = 1$, the approximation error is given by the sum over i of the distances between $x^{(i)}$ and its projection $x^{(i)}_{u^{(1)}}$, see Figure 4.4, that is,

$$e_1^2 = \frac{1}{m}\|X - X_1\|_F^2 = \frac{1}{m}\|X - u^{(1)}u^{(1)\top}X\|_F^2 = \frac{1}{m}\|P_1 X\|_F^2$$
$$= \operatorname{Tr}(P_1 S P_1) = \lambda_2 + \cdots + \lambda_n,$$

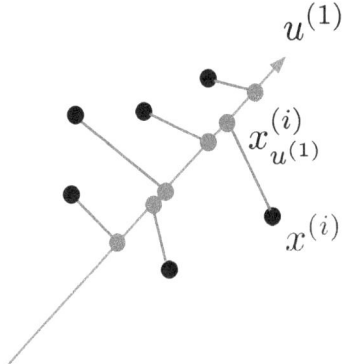

Figure 4.4 3D plot showing the distances between data points and their projections.

where $X_1 = u^{(1)}u^{(1)\top}X$ is the matrix of the projected points. More generally, after the kth deflation and projection, the matrix of the projected points is $X_k = U_k U_k^\top X$, and the approximation error is

$$e_k^2 \doteq \frac{1}{m}\|X - X_k\|_F^2 = \frac{1}{m}\|X - U_k U_k^\top X\|_F^2$$
$$= \frac{1}{m}\operatorname{Tr}(XX^\top) - \frac{2}{m}\operatorname{Tr}(U_k^\top XX^\top U_k) + \frac{1}{m}\operatorname{Tr}(U_k U_k^\top XX^\top U_k U_k^\top)$$
$$= \operatorname{Tr}(S) - \operatorname{Tr}(U_k^\top S U_k)$$
$$= \sum_{i=1}^n \lambda_i - \sum_{i=1}^k \lambda_i = \sum_{i=k+1}^n \lambda_i, \qquad (4.4)$$

where we used commutativity of the matrix product under the trace, the spectral factorization $S = U\Lambda U^\top$, and the fact that $U_k^\top U = [I_k \ 0_{k,n-k}]$. The approximation error e_k^2 is thus related to the complementary notion of *explained variance*, which compares the sum of variances contained in the first k principal directions (i.e., $\operatorname{Tr}(U_k^\top S U_k) = \sum_{i=1}^k \lambda_i$) with the total variance of the data set. We recall from Section 3.2 that the total variance σ^2 of X is defined as $\sigma^2 = \operatorname{Tr}(S) = \lambda_1 + \cdots + \lambda_n$, hence we define the relative explained variance as the ratio

$$\eta_k \doteq \frac{\lambda_1 + \cdots + \lambda_k}{\sigma^2} = \frac{\sigma^2 - e_k^2}{\sigma^2} = 1 - \frac{e_k}{\sigma}, \quad \eta \in [0,1]. \qquad (4.5)$$

Remark 4.1 (*PCA projection has minimum error*) We have seen how PCA computes k principal directions $U_k = [u^{(1)} \cdots u^{(k)}]$ so that the data projected on the subspace span by U_k has the maximal possible variance.

Then in (4.4) we evaluated the approximation error between the original data and its projection. An interesting fact is that this error is actually the minimum possible. In other words, the orthogonal matrix U_k found via PCA is also the optimal solution of the problem

$$\min_{Q\in\mathbb{R}^{n,k}} \frac{1}{m}\|X - QQ^\top X\|_F^2, \quad \text{s.t.:} \ Q^\top Q = I_k.$$

The reason for this is apparent by manipulating the objective of the above minimization problem as in (4.4), obtaining the equivalent formulation

$$\min_{Q\in\mathbb{R}^{n,k}} \operatorname{Tr}(S) - \operatorname{Tr}(Q^\top S Q), \quad \text{s.t.:} \ Q^\top Q = I_k.$$

The first term $\operatorname{Tr}(S)$ does not depend on Q, hence the problem reduces to

$$\max_{Q\in\mathbb{R}^{n,k}} \operatorname{Tr}(Q^\top S Q), \quad \text{s.t.:} \ Q^\top Q = I_k.$$

Letting $Q = [q^{(1)} \cdots q^{(k)}]$, since $\operatorname{Tr}(Q^\top S Q) = \sum_{i=1}^k {q^{(i)}}^\top S q^{(i)}$ represents the sum of the directional variances along the orthogonal directions $q^{(1)},\ldots,q^{(k)}$, we know from PCA that this quantity is maximized by choosing $q^{(i)} = u^{(i)}$, $i = 1,\ldots,k$, which proves the claim. Therefore, the directions defined by the principal components are also the ones that minimize the approximation error, see Figure 4.5.

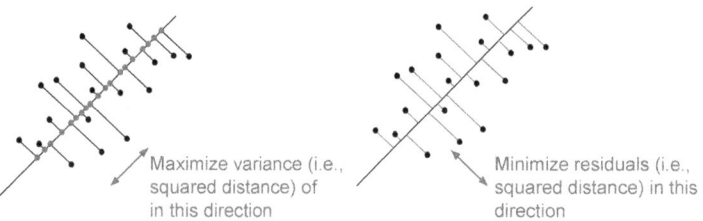

Figure 4.5 Two equivalent interpretations of PCA.

Example 4.1 (*Market data*) We considered the daily returns of the companies included in the Standard & Poor (S&P) 500 index over a period of more than five years, as shown in Figure 4.6, see the data file SP500_20150101_20200910.csv in the online resources.

We computed the principal components of this data set, and then evaluated the directional variance for each of the principal directions. Figure 4.7 shows the explained variance as a function of the number of principal components. We can observe that the first 80 components explain approximately 80% of the variance.

We then analyzed the first principal component. Table 4.1 lists the companies that correspond to the top 20 elements of the first principal component. We observe that its top elements correspond to various industries, such as insurance, cruise, oil, mining, and the values are all of the same sign. Hence we can think of the first component as a portfolio consisting of a weighted average of the market. This is confirmed by the

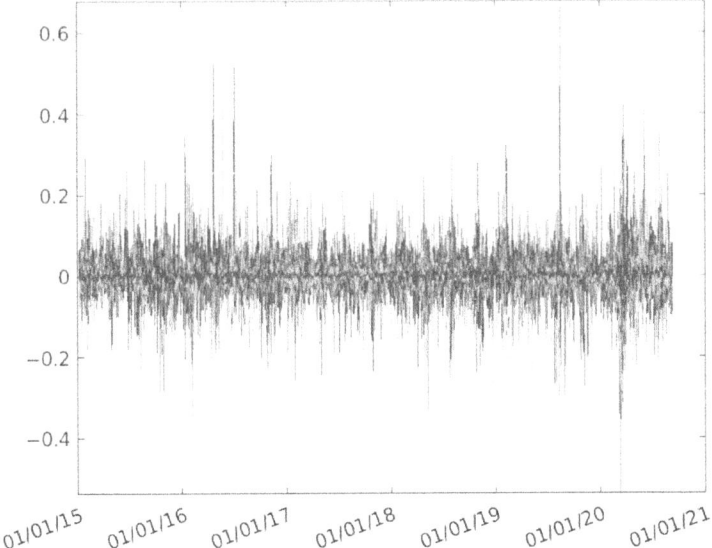

Figure 4.6 Standard & Poor 500 returns, 1/2015–4/2020.

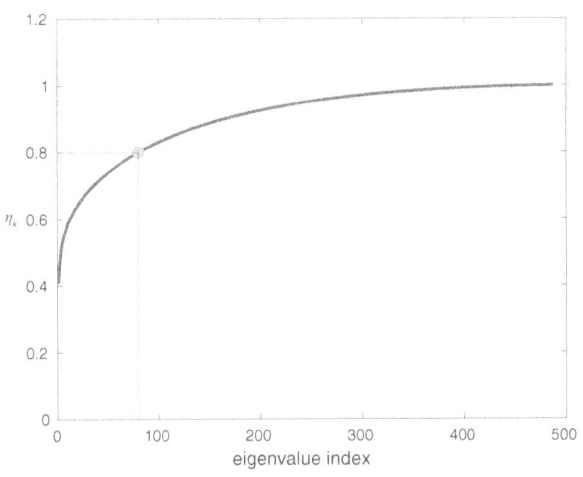

Figure 4.7 Explained variance η_k as function of the number of principal components. Data: S&P 500 log-returns, 1/2015–4/2020; see ex_PCA_sp500.m in the online resources.

plot shown in Figure 4.8, which shows a comparison between the first principal component and the S&P 500 index itself: the two curves exhibit similar behavior, and the first principal component alone is able to describe the main trends of the index.

4.2 Low-rank approximations

We next introduce the concept of low-rank approximation and analyze its link with PCA. Before this analysis, however, we briefly recall some notions of linear algebra that will be useful in the rest of this section.

Company	Symbol	Industry sector
Lincoln National	LNC	Multi-line insurance
Norwegian Cruise Line Holdings	NCLH	Hotels, Resorts and cruise lines
APA Corporation	APA	Oil and gas exploration and production
Freeport-McMoRan	FCX	Copper
Noble-Energy	NBL	Oil and gas drilling
Devon Energy	DVN	Oil and gas exploration and production
MGM Resorts International	MGM	Casinos and gaming
Royal Caribbean Group	RCL	Hotels, resorts and cruise lines
DXC Technology	DXC	IT consulting and other services
Carnival Corporation	CCL	Hotels, resorts and cruise lines
Halliburton	HAL	Oil and gas equipment and services
Marathon Oil	MRO	Oil and gas exploration and production
United Rentals	URI	Trading companies and distributors
Discovery Financial Services	DFS	Consumer finance
PVH	PVH	Apparel, accessories and luxury goods
Ameriprise Financial	AMP	Asset management and custody banks
United Airlines	UAL	Airlines
Wynn Resorts	WYNN	Casinos and gaming
SVB Financial	SIVB	Regional banks

Table 4.1 Companies associated with the top 20 elements of the first principal component.

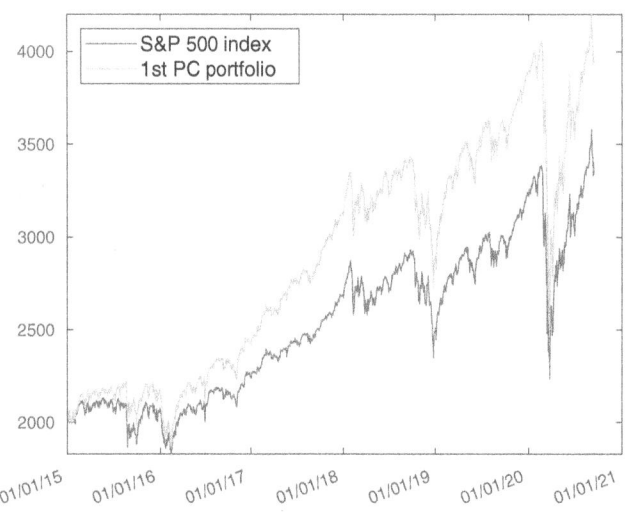

Figure 4.8 Comparison between the first principal component and the S&P 500 index; see ex_PCA_sp500.m in the online resources.

4.2.1 Singular value decomposition

The singular value decomposition (SVD) is a generalization of the eigenvalue decomposition discussed in the previous chapter. The Spectral theorem in Appendix A.1.2 provides a decomposition technique for symmetric matrices. The SVD extends this result, providing a general decomposition that can be applied to any nonzero $n \times m$ real matrix. The main result is presented in the following theorem.

Theorem 4.1 (SVD of general matrices) *We can decompose any non-zero $n \times m$ matrix X as*

$$X = U\Sigma V^\top = \sum_{i=1}^{r} \sigma_i u^{(i)} v^{(i)^\top},$$

with
$$\Sigma = \left[\begin{array}{c|c} \text{diag}(\sigma_1,\ldots,\sigma_r) & 0 \\ \hline 0 & 0 \end{array}\right] \in \mathbb{R}^{n,m},$$

where $\sigma_1 \geq \cdots \geq \sigma_r > 0$ are the singular values, and $U = [u^{(1)} \cdots u^{(n)}]$, $V = [v^{(1)} \cdots v^{(m)}]$ are square, orthogonal matrices (i.e., $U^\top U = I_n$, $V^\top V = I_m$). The integer $r \leq \min(m,n)$ (i.e., the number of nonzero singular values) is called the rank of X. The first r columns of U, V contain the left and right singular vectors of X, respectively, that is,

$$X^\top u^{(i)} = \sigma_i v^{(i)}, \quad X v^{(i)} = \sigma_i u^{(i)}, \quad i = 1,\ldots,r.$$

The SVD of an $n \times m$ matrix X is strongly linked to the eigenvalue decomposition of two positive semidefinite matrices related to X, as stated by the following corollary.

Corollary 4.1 *If $X = U\Sigma V^\top$ is the SVD of $X \in \mathbb{R}^{n,m}$, then the eigenvalue decomposition of $XX^\top$ is $U\Lambda U^\top$, with $\Lambda = \text{diag}(\sigma_1^2,\ldots,\sigma_r^2,0,\ldots,0) \in \mathbb{R}^{n,n}$. Also, the eigenvalue decomposition of $X^\top X$ is $V\Lambda V^\top$, with $\Lambda = \text{diag}(\sigma_1^2,\ldots,\sigma_r^2,0,\ldots,0) \in \mathbb{R}^{m,m}$.*

Hence, the left (resp. right) singular vectors of X are the eigenvectors of the PSD matrix $XX^\top$ (resp. $X^\top X$) and the singular values are the square roots of the nonzero eigenvalues of $X^\top X$ or of $XX^\top$ (which are the same).

Via the SVD we can define generalized inverses of a matrix. In particular, the Moore–Penrose pseudo inverse of $X = U\Sigma V^\top \in \mathbb{R}^{n,m}$ is defined as

$$X^\dagger = V\Sigma^\dagger U^\top \in \mathbb{R}^{m,n},$$

where
$$\Sigma^\dagger = \left[\begin{array}{c|c} \text{diag}(\sigma_1^{-1},\ldots,\sigma_r^{-1}) & 0 \\ \hline 0 & 0 \end{array}\right] \in \mathbb{R}^{m,n}.$$

The pseudo inverse has the property that $XX^\dagger X = X$, and $X^\dagger X X^\dagger = X^\dagger$. If X is invertible then $X^\dagger = X^{-1}$. If X has rank n (full row rank) then $X^\dagger = X^\top(XX^\top)^{-1}$ is a right-inverse of X, since $XX^\dagger = I_n$. If X has rank m (full column rank) then $X^\dagger = (X^\top X)^{-1}X^\top$ is a left-inverse of X, since $X^\dagger X = I_m$. The pseudo inverse is often used to represent a solution to least-squares problems of the form

$$\min_x \|Ax - y\|_2^2,$$

whereby it holds that an optimal solution is given by $x^* = A^\dagger y$, see Exercise 4.2.

4.2.2 SVD and the power iteration algorithm

From the numerical point of view, the factors U, V, Σ of the SVD factorization $X = U\Sigma V^\top$ can be computed using algorithms for the eigen-decomposition of the symmetric matrices $XX^\top$ and $X^\top X$. Singular value decomposition factorization is available in all the popular computational platforms, see, for example, the `svd` command in MATLAB®, R, and in Python/numpy.

When X is very large and/or structured, and only the first few singular values and singular vectors are needed, a fast iterative approach called the *power iteration* (PI) algorithm can be used. For the first singular vectors $u^{(1)}, v^{(1)}$, the PI algorithm works as follows:

1. Given $X \in \mathbb{R}^{m,n}$, initialize $u \in \mathbb{R}^m$ and $v \in \mathbb{R}^n$ to random values, then normalize $u \leftarrow u/\|u\|_2$, $v \leftarrow v/\|v\|_2$.

2. Compute $u = \frac{Xv}{\|Xv\|_2}$, $v = \frac{X^\top u}{\|X^\top u\|_2}$.

3. Iterate step 2 until convergence.

For any arbitrary initialization of u and v, this method converges to the first left and right singular vectors $u^{(1)}, v^{(1)}$, under mild conditions on X. The rate of convergence depends on the ratio $\rho = \frac{|\lambda_2|}{|\lambda_1|}$, where $|\lambda_1|$ is the eigenvalue of largest modulus of $XX^\top$ and $|\lambda_2|$ is the second largest modulus eigenvalue. We thus have a fast convergence for $\rho \ll 1$.

Once $u^{(1)}, v^{(1)}$ have been obtained, the corresponding singular value σ_1 can be computed as $\sigma_1 = u^{(1)\top} X v^{(1)}$. The PI iterations can be next used again on the deflated matrix $X_1 = X - \sigma_1 u^{(1)} v^{(1)\top}$ for finding the second left and right singular vectors $u^{(2)}, v^{(2)}$ and the corresponding $\sigma_2 = u^{(2)\top} X_1 v^{(2)}$, and so on for the other successive singular vectors.

The PI algorithm may require a high number of matrix-vector multiplications to reach convergence, which can be time-consuming in high dimensions. However, computation may be sped up significantly in the case of sparse matrices, for which the computational time required for matrix-vector multiplications is significantly lower. The PI algorithm has a wide range of applications. For instance, the PageRank algorithm used by Google to rank web pages in their search results is based on a version of the PI algorithm, see further discussion in Section 13.4.

Remark 4.2 (*SVD and the PCA*) The SVD is a general matrix factorization tool having a multitude of practical applications. Here, we remark that the SVD applied to a centered data matrix $X \in \mathbb{R}^{m,n}$ is a computational way for obtaining the PCA. Indeed, we have seen in Section 4.1.1

that the principal components of a centered data set X are given by the columns of the orthogonal matrix U, where U comes from the spectral factorization of the sample covariance matrix $S = \frac{1}{m}XX^\top = U\Lambda U^\top$. Suppose now that instead of working with the data covariance matrix S we work directly with the (centered) data matrix X, and compute its SVD $X = \tilde{U}\Sigma\tilde{V}^\top$. Then, substituting X in the definition of S we have

$$S = \frac{1}{m}XX^\top = \frac{1}{m}\tilde{U}\Sigma\tilde{V}^\top \tilde{V}\Sigma^\top \tilde{U}^\top = \frac{1}{m}\tilde{U}\Sigma\Sigma^\top \tilde{U}^\top = \tilde{U}\Lambda\tilde{U}^\top,$$

for $\Lambda = \frac{1}{m}\Sigma\Sigma^\top$, from which we conclude that the $\tilde{U}$ matrix coming from the SVD of X can be used to obtain a spectral factorization of S, and hence it provides by columns the desired principal directions.[1]

[1] Notice that in general we cannot say that $\tilde{U}$ coincides exactly with U, since both the SVD and the spectral factorization are not unique, in general. For example, given a U we can multiply some of its columns by -1 and still obtain a valid factorization, or, in the case of groups of eigenvalues with identical value, the columns of U corresponding to those groups can be rearranged in an arbitrary order.

4.2.3 Low-rank matrix approximation

A data matrix $X \in \mathbb{R}^{n,m}$ having low-rank $k \ll \min(n,m)$ has linear correlations among the rows and columns. These correlations could be exploited to substantially reduce the complexity of the data set. The following remark illustrates this fact on an example of time series data.

Remark 4.3 (*Interpretation for time-series data*) To provide an interpretation of the low-rank approximation of a data matrix, we first focus on the rank-one case. Let us assume that the data matrix $X \in \mathbb{R}^{n,m}$ represents time-series data, where each column is a time series of length n:

$$X = [x^{(1)} \cdots x^{(m)}], \quad x^{(i)} \in \mathbb{R}^n, \quad 1 \leq i \leq m.$$

Let us also assume that X is rank-one, that is, $X = \sigma_1 u v^\top$, where σ_1 is the first (and only) singular value of X and u, v are the corresponding left and right singular vectors, respectively. Then, we obtain that

$$x_t^{(j)} = X_{tj} = \sigma_1 u_t v_j, \quad 1 \leq j \leq m, \quad 1 \leq t \leq n.$$

Thus, each time series is a scaled copy of the time series represented by u, with scalings given by v. We can think of u as a "factor" that drives all the time series. Therefore, if a data matrix is rank-one, then all the data points are on a single line defined by u.

Let us now assume that the data matrix X is rank k, where $k \ll n$:

$$X = U\Sigma V^\top, \quad U \in \mathbb{R}^{n,k}, \quad \Sigma = \mathrm{diag}(\sigma_1, \ldots, \sigma_k) \in \mathbb{R}^{k,k}, \quad V \in \mathbb{R}^{m,k}.$$

We can thus express the jth column of X as

$$x_t^{(j)} = \sum_{i=1}^k \sigma_i u_t^{(i)} v_j^{(i)}, \quad 1 \leq t \leq n, \, 1 \leq j \leq m.$$

If X contains time-series data, this means that each time series is the *sum* of scaled copies of k time series represented by $u^{(1)}, \ldots, u^{(k)}$, with scalings given by $v^{(1)}, \ldots, v^{(k)}$. Therefore, we can think of vectors $u^{(i)}$, $i = 1, \ldots, k$, as the factors that drive all the time series.

Oftentimes although the data matrix X is not itself low rank, it can be closely approximated by a matrix of small rank and, if the approximation error is small, the approximated low-rank matrix can be used in place of X in the subsequent analysis. We thus next study the problem of low-rank matrix approximation, and discuss its relations with the SVD. For a given $n \times m$ matrix X, not necessarily centered, and integer $k \leq r \doteq \text{rank}(X)$, the rank-$k$ approximation problem is

$$\hat{X}^{(k)} \doteq \arg\min_{\hat{X}} \|X - \hat{X}\|_F, \quad \text{s.t.: rank } \hat{X} \leq k. \quad (4.6)$$

The Eckart–Young–Mirsky theorem states that the solution of problem (4.6) is

$$\hat{X}^{(k)} = \sum_{i=1}^{k} \sigma_i u^{(i)} v^{(i)\top} = U \Sigma^{(k)} V^\top,$$

where $X = \sum_{i=1}^{r} \sigma_i u^{(i)} v^{(i)\top}$ is an SVD of matrix X, and

$$\Sigma^{(k)} = \left[\begin{array}{c|c} \text{diag}(\sigma_1, \ldots, \sigma_k) & 0 \\ \hline 0 & 0 \end{array} \right] \in \mathbb{R}^{n,m}.$$

Hence, the SVD provides an exact solution for problem (4.6). The relative approximation error is defined as the ratio

$$\zeta_k \doteq \frac{\|X - \hat{X}^{(k)}\|_F}{\|X\|_F}.$$

Recalling that $\|X\|_F = \sqrt{\text{Tr}(XX^\top)} = \sqrt{\sum_{i=1}^{r} \sigma_i^2}$, we have that

$$\zeta_k^2 = \frac{\|X - \hat{X}^{(k)}\|_F^2}{\|X\|_F^2} = \frac{\|U(\Sigma - \Sigma^{(k)})V^\top\|_F^2}{\|X\|_F^2}$$

$$= 1 - \frac{\sigma_1^2 + \cdots + \sigma_k^2}{\sigma_1^2 + \cdots + \sigma_n^2}.$$

In the case when X is centered (i.e., $X\mathbf{1} = 0$) the above quantity is related to the relative explained variance η_k as

$$\frac{\|X - \hat{X}^{(k)}\|_F^2}{\|X\|_F^2} = \frac{\text{Tr}(S - \hat{S}^{(k)})}{\text{Tr}(S)} = 1 - \frac{\lambda_1 + \cdots + \lambda_k}{\lambda_1 + \cdots + \lambda_n} = 1 - \eta_k,$$

where $\hat{S}^{(k)}$ is the rank-k approximation of the data covariance S.

Remark 4.4 (*PCA and low-rank approximation*) We have shown in Remark 4.2 that PCA on the data covariance matrix S is equivalent and can be obtained by performing an SVD on the centered data matrix X. We can now observe that PCA also provides a rank-k approximation of the data covariance matrix, and vice versa. Indeed, given $S = U\Lambda U^\top$, we have that the best rank-k approximation of S is given by

$$\hat{S}^{(k)} = U_k \Lambda^{(k)} U_k^\top,$$

where $\Lambda^{(k)} = \mathrm{diag}(\lambda_1, \ldots, \lambda_k)$, and U_k contains the first k columns of U, which are also the k principal directions provided by the PCA. Similarly, solving the rank-k approximation problem $\min_{\hat{X}} \|X - \hat{X}\|_F^2$ subject to rank $\hat{X} \leq k$ via the SVD of X (centered) gives the first k principal directions. In particular, for $k = 1$, since rank-one matrices must have the form $X = \sigma u v^\top$, we have that the rank-one approximation problem can be written as

$$\min_{u \in \mathbb{R}^n, v \in \mathbb{R}^m, \sigma \geq 0} \|X - \sigma u v^\top\|_F^2, \quad \text{s.t.:} \ \|u\|_2 = 1, \ \|v\|_2 = 1, \quad (4.7)$$

which returns among its optimal variables the first principal direction $u^{(1)}$.

Alternating minimization for rank-one matrix approximation Consider problem (4.7), and expand the objective as

$$f(u, v, \sigma) = \|X - \sigma u v^\top\|_F^2 = \mathrm{Tr}\left((X - \sigma u v^\top)(X - \sigma u v^\top)^\top\right)$$
$$= \mathrm{Tr}(XX^\top) - 2\sigma u^\top X v + \sigma^2,$$

where we used the fact that $u^\top u = v^\top v = 1$. We examine an alternating minimization approach for $f(u, v, \sigma)$ which cyclically minimizes over $\{u \colon \|u\|_2 = 1\}$ for v, σ fixed, then on $\{v \colon \|v\|_2 = 1\}$ with u, σ fixed, hence on $\sigma \geq 0$ with u, v fixed, and then starts the cycle again until eventual convergence. The minimization over u is equivalent to maximizing the inner product $u^\top (Xv)$ over $\{u \colon \|u\|_2 = 1\}$, which has the optimal solution $u = (Xv)/\|Xv\|_2$. Similarly, the minimization over v is equivalent to maximizing $(u^\top X)v$ over $\{v \colon \|v\|_2 = 1\}$, which has the optimal solution $v = (X^\top u)/\|X^\top u\|_2$. Finally, computing the derivative of f with respect to σ and equating it to zero we have that the minimum over σ is attained for $\sigma = u^\top X v$.[2] The steps to be performed cyclically until convergence (starting from random initial guesses) are therefore:

$$u \leftarrow \frac{Xv}{\|Xv\|_2}, \quad v \leftarrow \frac{X^\top u}{\|X^\top u\|_2}, \quad \sigma \leftarrow u^\top X v. \quad (4.8)$$

[2] Notice that after the v step one has that $\sigma = u^\top X v = (u^\top X)(X^\top u)/\|X^\top u\|_2 = \|X^\top u\|_2 \geq 0$, hence σ is guaranteed to be nonnegative at each iteration.

This turns out to be precisely the power iteration algorithm that we discussed in Section 4.2.1 for the computation of the first singular vectors.

4.2.4 SVD-based auto-encoder

Auto-encoders are usually employed in machine learning to learn efficient data codings, that is, a compact data representation whose dimension is significantly lower than the one of the original data. An auto-encoder has two parts: an encoder that maps the input data into the code, and a decoder that maps the code to an output reconstruction of the data. Figure 4.9 shows the scheme of a basic auto-encoder.

We can think of SVD as a special case of auto-encoder, where encoder and decoder are linear mappings. Indeed, given an $n \times m$ data matrix X, we can consider the SVD-based rank-k approximation $\hat{X}^{(k)}$ to X:

$$X = U\Sigma V^\top \approx \hat{X}^{(k)} = U\Sigma^{(k)} V^\top.$$

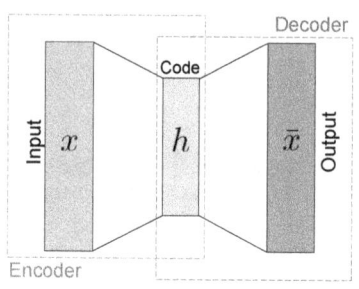

Figure 4.9 Architecture of a basic auto-encoder.

Thus, for each input point $x^{(j)} = Xe_j$, where $e_j \in \mathbb{R}^m$ is a vector with all zeros but a one in the jth position, we obtain the approximation

$$\hat{x}^{(j)} \doteq \hat{X}^{(k)} e_j = U\Sigma^{(k)} V^\top e_j = U_k h^{(j)},$$

where

$$h^{(j)} = \text{diag}(\sigma_1, \ldots, \sigma_k) V_k^\top e_j,$$

and $U_k = [u^{(1)} \cdots u^{(k)}]$ and $V_k = [v^{(1)} \cdots v^{(k)}]$ contain the first k left and right singular vectors, respectively. Vector $h^{(j)} \in \mathbb{R}^k$ is a low-dimensional representation of the data point $x^{(j)}$. Therefore, we can interpret SVD as a linear auto-encoder where any data point $x^{(j)}$ is encoded to a low-dimensional vector $h^{(j)}$. Then, given the code $h^{(j)}$, we can decode it via $\hat{x}^{(j)} = U_k h^{(j)}$ obtaining the output approximation. Note that both the input encoding and the decoding are defined via linear maps. Indeed, the output decoding is defined via the linear map $\hat{x}^{(j)} = U_k h^{(j)}$, and the input encoding is defined via the linear map $h^{(j)} = U_k^\top x^{(j)}$.[3] Since the error between the original data points and the reconstructed ones is minimum in the mean square error sense by construction, we can interpret SVD as an optimal one-layer linear auto-encoder.

[3] You can prove this latter fact as an exercise; as a hint, check algebraically that $\hat{X}^{(k)} = U_k U_k^\top X$.

4.2.5 PCA and factor models

Recalling the factor models presented in Section 3.5, we can introduce a stochastic interpretation of low-rank approximations. Let us assume that the observations $x^{(i)} \in \mathbb{R}^n$, with $i = 1, \ldots, m$, are generated by a stochastic model of the form

$$x = Lz, \qquad (4.9)$$

where $L \in \mathbb{R}^{n,k}$ is the factor loading matrix with $k \leq n$, and $z \in \mathbb{R}^k$ is a random vector with zero mean and identity covariance matrix. If we choose $k \ll n$, we assume that there are few factors that drive the observations. This behavior can also be observed from the covariance matrix of x:

$$\Sigma = \mathrm{E}(xx^\top) = \mathrm{E}(Lzz^\top L^\top) = LL^\top,$$

which is a matrix of rank $\leq k$. Therefore, when we approximate a sample covariance matrix with a low-rank matrix as done in Section 4.4, we are implicitly imposing a factor structure on the observations as the one described in (4.9).

More generally, we can assume that the observations are of the form
$$x = Lz + \sigma e,$$
where z and e are zero-mean random variables with identity covariance matrix, and $\sigma > 0$ is a parameter. The variable z contains the driving factors, while e is a noise term that affects each observation independently, that is, idiosyncratic noise. We can compute the covariance matrix of x as
$$\Sigma = LL^\top + \sigma^2 I.$$

If we collect samples $x^{(1)}, \ldots, x^{(m)}$ of x, and construct the sample covariance matrix S from these sample, we can estimate L and σ. The optimal estimate L that minimizes the approximation error $\|S - (LL^\top + \sigma^2 I)\|_F$ can be found via the method described in Section 3.5.1. More general factor models that allow for idiosyncratic noises with different variances can also be estimated via the technique described in Section 3.5.2.

4.3 Extensions

In this section, we first discuss some of the limitations of PCA and then present extensions that address these issues. In particular, we will introduce generalized low-rank models and show how they can be used to extend PCA in various directions. At the end of the section, we will also show how these models can be used for matrix completion.

4.3.1 Issues with PCA

One of the main drawbacks of PCA is its high sensitivity to outliers. As shown in Figure 4.10, adding a few outliers to the data can have a large effect in the computation of the principal components. This is due to the fact that PCA minimizes a criterion based on the sum of squared errors. Thus, errors from outliers can be significantly larger than errors from other data points and will dominate the overall approximation error, significantly affecting the principal components. This is the same behavior observed in the computation of the mean of a set of data values, where changing a few entries can have a large impact on the mean value. Later in this chapter we introduce an extension of PCA that is more robust to outliers.

Another limitation of PCA is that it does not handle some data sets well, in particular nonlinear data. Since PCA defines a linear transformation of the data, it can detect only linear relationships between features, failing to detect more complex correlations. In some cases, this issue can be overcome by introducing coordinate transformations that map the data samples in a higher-dimensional space where they can be linearly separated. This approach is named Kernel PCA.[4] The kernel trick, presented in Chapter 8 in the context of support vector machines, allows computations in the higher-dimensional space to be avoided, which would have a very high computational cost.

The use of PCA can also be limited by its lack of interpretability. Since the principal components are eigenvectors in data space, their interpretation is not straightforward, especially if one seeks to find features (i.e., dimensions) in the data that are important. To overcome this limitation, a usual practice is to find the large elements in the principal components, as done in the Example 4.1. However, this might not identify correctly the important features. This motivates the introduction of sparse principal component analysis, which will be discussed in Section 4.3.5.

Another restriction of standard PCA is that it cannot handle some data types, such as Boolean, categorical, and nonnegative.

In the next section, we introduce generalized low-rank models, which offer a flexible way to model data and can overcome some of the limitations that we have discussed. In contrast to classical low-rank models, however, the convergence for their solution methods is not guaranteed. At the end of the chapter, we will show that these models can be also extended to handle matrix completion problems.

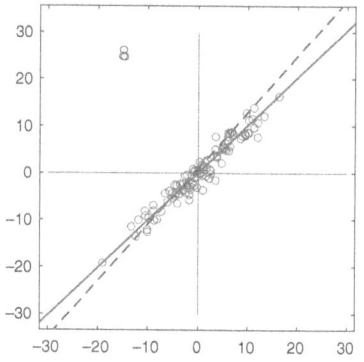

Figure 4.10 Solid line: first principal component of the data set without outliers. Dashed line: first principal component of the data set with outliers (the circles in the upper left); see ex_PCA_outliers.m in the online resources.

[4] See B. Schölkopf, A. Smola, and K.-R. Müller, "Nonlinear Component Analysis as a Kernel Eigenvalue Problem," *Neural Computation*, vol. 10(5), pp. 1299–1319, 1998.

4.3.2 Generalized low-rank models

In Section 4.2.3, we have introduced the low-rank matrix approximation problem:

$$X \approx X^{(k)} \doteq \arg\min_{\hat{X}} \|X - \hat{X}\|_F, \quad \text{s.t.: } \operatorname{rank} \hat{X} \leq k.$$

The rank-k approximation matrix can be rewritten as $X^{(k)} = LR^\top$, where $L \in \mathbb{R}^{n,k}$, $R \in \mathbb{R}^{m,k}$, with $k \ll m, n$. We can thus reformulate the low-rank matrix approximation problem as follows:

$$\min_{L,R} \|X - LR^\top\|_F : L \in \mathbb{R}^{n \times k},\ R \in \mathbb{R}^{m \times k}. \tag{4.10}$$

Observe that $(LR^\top)_{ij} = l_i^\top r_j$, where

$$L = \begin{bmatrix} l_1^\top \\ \vdots \\ l_n^\top \end{bmatrix}, \quad R = \begin{bmatrix} r_1^\top \\ \vdots \\ r_m^\top \end{bmatrix}.$$

Thus, we can rewrite problem (4.10) as

$$\min_{L,R} \sum_{i,j} \mathcal{L}(X_{ij}, l_i^\top r_j) : l_i \in \mathbb{R}^k, \ i = 1, \ldots, n, \ r_j \in \mathbb{R}^k, \ j = 1, \ldots, m,$$

with loss function $\mathcal{L}(a,b) = (a-b)^2$. We can now generalize the problem by considering more general convex losses and adding suitable penalties to the objective:

$$\min_{L,R} \sum_{i,j} \mathcal{L}(X_{ij}, l_i^\top r_j) + \sum_i p_i(l_i) + \sum_j q_j(r_j), \qquad (4.11)$$

where $\mathcal{L}$ is convex, and functions p_i, q_j are convex penalties. Problem (4.11) is not jointly convex, due to the presence of the cross terms $l_i^\top r_j$, but it is convex with respect to X, R (respectively X, L) when L (respectively R) is fixed. An effective heuristic can then be used for solving the problem approximately, based on iterative alternate minimization, with rounds repeatedly minimizing over L (with R fixed) and over R (with L fixed), until possible convergence. However, in most cases there is no guarantee of convergence to a global minimum.

An alternative approach to the alternating minimization method presented above is the following convex model:

$$\min_Z \sum_{i,j} \mathcal{L}(X_{ij}, Z_{ij}) + \lambda \|Z\|_*,$$

where $\|Z\|_*$ denotes the *nuclear norm* (i.e., the sum of the singular values of Z). In this case, we try to minimize the rank of a matrix by actually minimizing a tractable proxy of the rank, given by the sum of the singular values. The rationale behind this approach is that the nuclear norm is akin to the ℓ_1 norm of the vector of singular values. As seen in Section 3.7.1, the ℓ_1 norm encourages sparsity of its argument and thus minimizing the nuclear norm is similar to minimizing the number of nonzero singular values. The advantage of this approach is that we obtain a convex problem, which can, however, be still challenging to solve in high-dimensional instances. In practice, alternate minimization is a good heuristic when squared regularization is included (as described in the next section). For many finance applications, when the data size is not too big, the convex model based on the nuclear norm relaxation may be a reliable alternative.

4.3.3 Regularized PCA

By defining different losses $\mathcal{L}$ and penalties p_i and q_j of the generalized low-rank problem (4.11), we can model various extensions of the basic PCA problem. We first consider the ℓ_2-regularized PCA, which is defined by the following problem:

$$\min_{L,R} \|X - LR^\top\|_F^2 + \gamma \left(\|L\|_F^2 + \|R\|_F^2\right) : L \in \mathbb{R}^{n,k}, \ R \in \mathbb{R}^{m,k},$$

where $\gamma > 0$ is a regularization parameter. This problem has a closed-form solution: given the SVD of $X = U\Sigma V^\top$, we set

$$\tilde{\Sigma}_{ii} = \max(0, \Sigma_{ii} - \gamma), \ i = 1, \ldots, k,$$

and $L = U_k \tilde{\Sigma}^{1/2}, R = V_k \tilde{\Sigma}^{1/2}$, where U_k, V_k contain the first k columns of U, V, respectively. Note that the singular values below γ are truncated to zero. When $\gamma = 0$ the solution reduces to the solution of the standard PCA. Regularized PCA is used in Section 4.3.8 in the context of matrix completion.

4.3.4 Robust PCA

In Section 4.3.1 we observed that PCA may have high sensitivity to outliers, that is, the presence of a few outliers can significantly affect the computation of the principal components. To overcome this issue, a generalization of PCA, called *robust PCA*, can be considered. In robust PCA one assumes that X has a "low-rank plus sparse" structure:

$$X = LR^\top + N,$$

where $N \in \mathbb{R}^{n,m}$ is a sparse noise matrix. This assumption improves the robustness since it takes into account the fact that the data observations can be corrupted by the presence of outliers, represented by the sparse noise matrix N. Since $N = X - LR^\top$ must be sparse, we should minimize the cardinality[5] of $X - LR^\top$, that is, solve $\min_{L,R} \|X - LR^\top\|_0$. However, since the cardinality function $\|\cdot\|_0$ is nonconvex and difficult to treat numerically, a standard approach is to use the ℓ_1 norm instead, and hence solve the problem

$$\min_{L,R} \sum_{i,j} |X_{ij} - l_i^\top r_j| = \|X - LR^\top\|_1, \qquad (4.12)$$

[5] The cardinality function $\|Z\|_0$ returns the number of nonzero elements contained in matrix Z.

where $\|\cdot\|_1$ represents the sum of the absolute values of the entries of its matrix argument. This problem has the form of (4.11), with $\mathcal{L}(a,b) = |a-b|$. We can hence solve approximately problem (4.12) by alternate minimization over L, R, and each step of this approach amounts to solving a linear programming problem.

Alternatively, we can consider the following convex problem:[6]

$$\min_{N} \|X - N\|_* + \lambda \|N\|_1.$$

As discussed in Section 4.3.2, the nuclear norm is related to the ℓ_1 norm of the vector of singular values, and thus its minimization encourages low-rank matrices. At the optimum, therefore, $X - N$ is usually low-rank. The L, R factors can then be recovered a posteriori by imposing that $X - N = LR^\top$ and using the SVD of $X - N$. Note that while this approach tends to make $LR^\top$ of low rank, it is not possible to guarantee that the rank will be smaller than an a-priori given k. The workaround is to solve the problem for decreasing values of $\lambda \geq 0$, until the desired rank is possibly achieved.

Independently of the solution approach taken, however, one drawback of robust PCA is that it does not possess the nice orthogonality properties of regular PCA, since the resulting principal directions need not be orthogonal. In addition, robust PCA still lacks interpretability of the factors, just as regular PCA does.

[6] For further details see E. J. Candès et al., "Robust Principal Component Analysis?," *Journal of the ACM (JACM)*, vol. 58(3), pp. 1–37, 2011.

Example 4.2 (*Robust PCA of raw log-returns*) From the data contained in the file SP500_20150101_20200910.csv in the online resources, we consider the daily log-returns of 20 tech companies included in the S&P 500 index for the fourth quarter of 2015. The list of the companies considered in this example is: Adobe (ADBE), Analog Devices (ADI), ADP, Autodesk (ADSK), AMD (AMD), Broadcom (AVGO), Cadence Design Systems (CDNS), Salesforce (CRM), Cisco (CSCO), Facebook (FB), Fortinet (FTNT), HP (HPQ), IBM, Intel (INTC), Illinois Tool Works (ITW), Microsoft (MSFT), Nvidia (NVDA), Oracle (ORCL), Qualcomm (QCOM), and Texas Instruments (TXN). We first center the data and then apply robust PCA in order to compute the low-rank matrix $LR^\top$ and the sparse component N. Figure 4.11 shows the data matrix and the resulting decomposition in low-rank and sparse components, for $k = 10$.

4.3.5 Sparse PCA

As discussed in the previous section, robust PCA does not address the lack of interpretability of PCA: in both PCA and robust PCA, the principal directions are usually dense vectors that represent combinations of all the input features. In sparse PCA we aim at finding principal directions that are sparse. This provides a direct interpretability in terms of the features that are relevant in a principal direction (those corresponding to nonzero coefficients) and those that are not (corresponding to the zero coefficients). Irrelevant features may, for example, be removed altogether from the data set, thus providing direct data simplification.

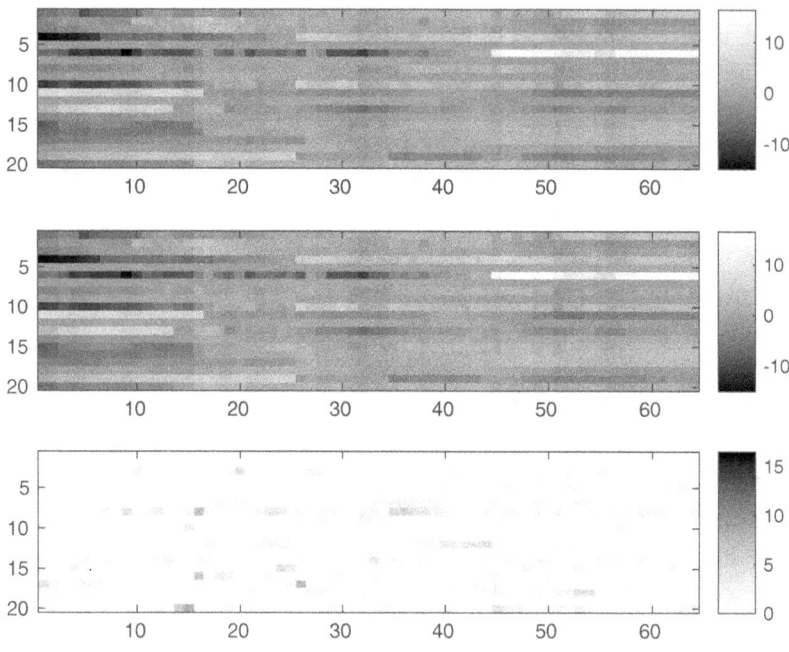

Figure 4.11 Robust PCA of 2015 Q4 raw log-returns for 20 tech companies. From top to bottom: original centered data matrix, low-rank component, absolute value of the sparse component; see ex_robustpca.m in the online resources.

One approach to gain sparsity consists in simply thresholding the principal components, selecting only the features with components larger than a given threshold and zeroing out the others. However, this method can often lead to a strong degradation of the explained variance. A more sophisticated approach, called *sparse PCA*, consists in modifying the basic variance maximization problem to

$$\max_{u} u^\top S u - \lambda \|u\|_0, \quad \text{s.t.:} \ \|u\|_2 = 1, \qquad (4.13)$$

where $\lambda \geq 0$ is a penalty parameter, $S \in \mathbb{R}^{n,n}$ is the sample covariance matrix of X, and $\|u\|_0$ is the number of nonzero elements in u. Adding the ℓ_0 norm term encourages sparsity of the principal components, hence improving their interpretability. The sparse PCA problem reduces to standard PCA when $\lambda = 0$.

Due to the presence of the ℓ_0 norm, (4.13) is a hard combinatorial problem, although it can be approximated via convex relaxation.[7] Further, in sparse PCA we introduce a *safe feature elimination* method which can applied before solving the problem to reduce the problem size. The following theorem holds.[8]

Theorem 4.2 (Safe feature elimination) *Given* $S = F^\top F$ *where* $F = [f^{(1)} \ldots f^{(n)}] \in \mathbb{R}^{m,n}$ *and each* $f^{(i)}$ *corresponds to one feature, we have*

$$\max_{u: \|u\|_2 = 1} u^\top S u - \lambda \|u\|_0 = \max_{u: \|u\|_2 = 1} \sum_{i=1}^{n} \max(0, (f^{(i)\top} u)^2 - \lambda).$$

[7] See A. d'Aspremont, L. El Ghaoui, M. I. Jordan, and G. R. G. Lanckriet, "A Direct Formulation for Sparse PCA Using Semidefinite Programming," *SIAM Review*, vol. 49(3), pp. 434–448, 2007.

[8] The proof is omitted, we refer the interested reader to A. d'Aspremont, F. Bach, and L. El Ghaoui, "Optimal Solutions for Sparse Principal Component analysis," *Journal of Machine Learning Research*, vol. 9(7), pp. 1269–1294, 2008.

An optimal nonzero pattern corresponds to indices i with $\lambda < (f^{(i)^\top} u)^2$ at optimum.

From the above result, we can derive the following corollary.

Corollary 4.2 *We can safely remove feature i, if $\lambda \geq \|f^{(i)}\|_2^2 = S_{ii}$.*

Proof From Theorem 4.2 we have that the ith feature is irrelevant at optimum if $(f^{(i)^\top} u)^2 \leq \lambda$ for all $u: \|u\|_2 = 1$. Since $\max_{\|u\|_2=1} (f^{(i)^\top} u)^2 = \|f_i\|_2^2$, we have that $\lambda \geq \|f^{(i)}\|_2^2$ implies that the contribution of the ith feature in the objective function is identically zero, hence that feature can be removed. □

The above result shows that the presence of the penalty parameter λ allows to prune out dimensions in the problem, without affecting the solution. The decision criterion for choosing the dimensions to prune is simply based on the variance of each feature (i.e., the directional variance along unit vectors). In practice, a higher λ results in a more important reduction of the problem size and may lead to substantial computational savings.

The iterative thresholded PI algorithm An unsophisticated alternative to the convex relaxation of (4.13) for computing a sparse principal component is given by a thresholded version of the classical PI algorithm. We start by posing the PCA problem in the form of a rank-one approximation problem as in (4.7). Then, we add to the problem cardinality constraints, obtaining

$$\min_{\|u\|_2=1, \|v\|_2=1, \sigma \geq 0} \|X - \sigma u v^\top\|_F^2, \quad \text{s.t.: } \|u\|_0 \leq k, \|v\|_0 \leq h. \quad (4.14)$$

This problem can be tackled via a simple heuristic, which consists in adapting the standard Power Iteration algorithm in (4.8) by adding thresholding to u, v at each iteration:

1. Randomly initialize $u \in \mathbb{R}^n$ and $v \in \mathbb{R}^m$, then normalize $u \leftarrow u/\|u\|_2$, $v \leftarrow v/\|v\|_2$.

2. Compute $u = P(T_k(Xv))$, $v = P(T_h(X^\top u))$.

3. Iterate step 2 until convergence.

Here X is the data matrix, P is the ℓ_2 normalization operator (i.e., for $z \neq 0$, $P(z) = z/\|z\|_2$), and T_k is the thresholding operator that removes all but the k largest-magnitude components of its input. The thresholded power iteration method reduces to the standard PI method when $k = n$ and $h = m$, in which case the optimal u and

v are the left and right singular vectors of X, respectively. Instead, when $k < n$ and/or $h < m$, the presence of the thresholding operators modifies these singular vectors to make them sparser, while keeping the corresponding rank-one approximation close to X. As discussed in Section 4.2.1, the power iteration method is particularly efficient with sparse data matrix. Thus, for sparse data this heuristic can be used in high dimensions, although convergence to the global optimum is not guaranteed.

Example 4.3 (*Sparse PCA of New York Times headlines*) We consider the New York Times text collection, which contains $m = 300,000$ articles and has a dictionary of $n = 102,660$ unique words.[9] We define the document encoding matrix $X \in \mathbb{R}^{n,m}$ where, for each word i and j, the element X_{ij} contains the number of occurrences of word i in document j. Thus, the jth column of X represents the jth document of the collection and the ith row of X represents the vector of expressions of word i of the dictionary over all documents. For each "probing document" (i.e., collection of words) $u \in \mathbb{R}^n$, the directional variance is proportional to $u^\top X_c X_c^\top u$, where X_c is the centered data matrix. Sparse probing documents of maximal variance can be found by solving the following problem:

$$\max_u u^\top X_c X_c^\top u - \lambda \|u\|_0, \quad \text{s.t.:} \ \|u\|_2 = 1.$$

[9] This dataset is taken from the UCI machine learning repository.

In order to reduce the computational complexity, we can reduce the problem size with safe feature elimination. Figure 4.12 shows the variance of the features (i.e., words). We can observe that the variance decreases very fast, which allows to drastically reduce the number of features: with a target number of words less than 10, the safe elimination method allows to reduce the number of features from $n \approx 100,000$ to $n = 500$. Table 4.2 shows the words associated with the top 5 sparse principal components

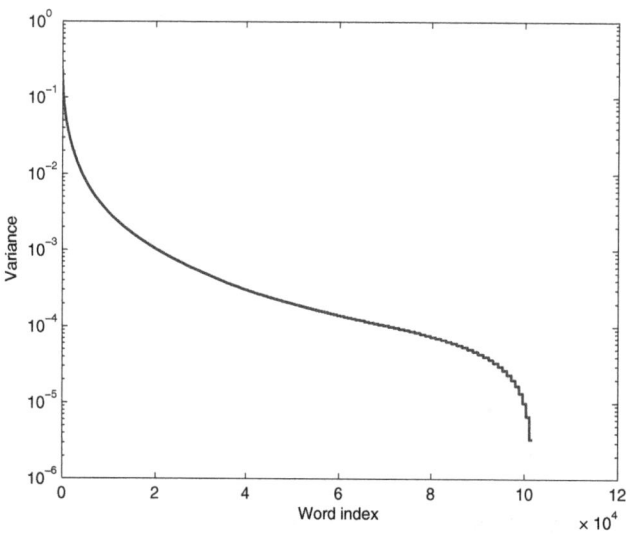

Figure 4.12 Sorted variances of 102,660 words in the New York Times dataset.

of the dataset. We can observe that the words of each principal component correspond to a specific argument or "topic" (e.g., finance, sports, politics, etc.). It is important to note that the algorithm found those terms without any information on the subject headings of the corresponding articles. We compare this result with the principal components found by a thresholded PCA, which involves simply thresholding the principal components. Table 4.3 shows the first principal component obtained with this method for various cardinality of k. Differently from the sparse PCA approach, we can observe that in this case the results contain many non-informative words.

1st PC (6 words)	2nd PC (5 words)	3rd PC (5 words)	4th PC (4 words)	5th PC (4 words)
million	point	official	president	school
percent	play	government	campaign	program
business	team	united_states	bush	children
company	season	u_s	administration	student
market	game	attack		
companies				

Table 4.2 Words associated with the top 5 sparse principal components in the New York Times.

$k=2$	$k=3$	$k=9$	$k=14$
even	even	even	would
like	like	we	new
	states	like	even
		now	we
		this	like
		will	now
		united	this
		states	will
		if	united
			states
			world
			so
			some
			if

Table 4.3 First PC from Thresholded PCA for various cardinality k.

4.3.6 Nonnegative matrix factorization

As discussed in Section 4.3.1, one of the limitations of PCA is that it cannot handle natively some data types, such as Boolean, categorical, and nonnegative. In this section, we will focus on the case where the data matrix is nonnegative and we discuss a generalization of PCA that is specific for this case.

Nonnegative matrix factorization (NNMF) is a variant of PCA where the factors are required to be nonnegative. That is, we seek to solve

$$\min_{L,R} \|X - LR^\top\|_F^2, \text{ s.t.: } L \geq 0, \ R \geq 0,$$

with inequalities understood component-wise and $L \in \mathbb{R}^{n,k}$, $R \in \mathbb{R}^{m,k}$ with $k \leq \min(m,n)$. This problem addresses the cases when the data matrix is itself nonnegative and can be modeled using the generalized low-rank model presented in (4.11) by defining $\mathcal{L}(a,b) = (a-b)^2$ and choosing all the penalties p_i, q_j to be equal to

$$p(z) = \begin{cases} 0 & \text{if } z \geq 0, \\ +\infty & \text{otherwise.} \end{cases}$$

4.3.7 Abstract data types

Principal component analysis cannot handle directly data types such as Boolean and ordinal data. These types of data are common in financial applications, where, for instance, they might be used for describing known characteristics of companies, asset classes, ratings, or analyst's recommendations such as "Strong Buy," "Buy," "Hold," "Underperform," and "Sell." In this section, we will see how we can extend PCA to handle these data types.

We first consider the case when the data matrix X contains Boolean entries, that is, $X_{ij} \in \{-1,1\}$. We can model these entries using the generalized low-rank model presented in (4.11) by defining a suitable loss function as

$$\mathcal{L}(a,b) = \max(0, 1 - ab) = (1 - ab)_+,$$

which is called the *hinge loss*. This definition means that if $a \in \{-1,1\}$, $\mathcal{L}(a,b) = 0$ implies b has the same sign as a. If the data matrix X is entirely Boolean, that is, $X \in \{-1,1\}^{n,m}$, we can define a variant of PCA as follows:

$$\min_{L,R} \sum_{i,j} (1 - X_{ij} l_i^\top r_j)_+.$$

The idea here is to find L, R so that $LR^\top$ match entry-wise as much as possible the signs of X.

Ordinal data is another data type that cannot be handled by standard PCA. We can encode ordinal data using consecutive integers representing the consecutive levels of the variable, for example, $\{1,\ldots,K\}$, where K is the number of categories. We can then use the generalized low-rank model presented in (4.11) by defining

$$\mathcal{L}(a,u) = \sum_{b=1}^{a-1}(1 - u + b)_+ + \sum_{b=a+1}^{K}(1 + u - b)_+. \quad (4.15)$$

This loss penalizes the entries which deviate by many levels from the encoded ordinal value, as shown in Figure 4.13. It is important to point out that this approach assumes that every increment of error is equally bad.[10] For example, if we use a ordinal variable to encode the analyst's recommendations of strong buy/buy/ hold/underperform/sell, this approach assumes that approximating "strong buy" by "buy" is just as bad as approximating "buy" by "hold."

[10] A more flexible approach is presented in M. Udell, C. Horn, R. Zadeh, and S. Boyd, "Generalized Low rank Models," *Foundations and Trends in Machine Learning*, vol. 9(1), pp. 1–118, 2016.

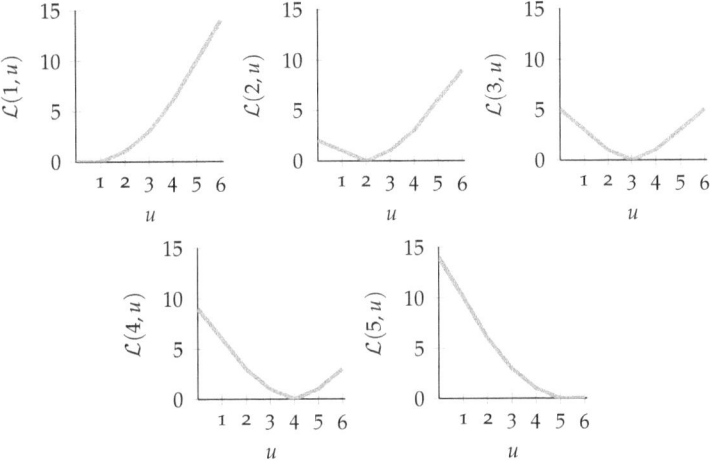

Figure 4.13 Loss (4.15) for a variable defined on five categories.

4.3.8 Matrix completion

In many applications, we do not have full knowledge of the data matrix: some entries are known and other are unknown. Matrix completion addresses this problem trying to fill the unknown entries. There are many heuristics that can be used for this scope. A very straightforward approach involves filling the missing entries with zeros. A slightly more sophisticated method computes averages across features or data points to fill the unknown entries. Alternatively, we may also fill the missing entries based on the assumption that the true data matrix is close to low-rank. Under this assumption, the completion is made so that the completed matrix has the lowest rank possible. This is obtained by using the regularized PCA introduced in Section 4.3.3:

$$\min_{L,R,X \in \mathcal{X}} \|X - LR^\top\|_F^2 + \gamma \left(\|L\|_F^2 + \|R\|_F^2 \right) : L \in \mathbb{R}^{n,k}, \ R \in \mathbb{R}^{m,k},$$

where it is important to highlight that X is a variable, differently to the standard problem introduced in Section 4.3.3, and $\mathcal{X}$ is the set of $n \times m$ matrices that have the required known entries pattern. This problem can be solved using alternating minimization as done in the previous sections, but in this case we add the missing entries in X as variables. Theoretical analysis shows that if the locations of missing entries are randomly distributed, then convergence to the global minimum is guaranteed.[11] In practice, this means that the approach is successful if the missing entries do not follow a clear pattern. For example, they should not all be located in a row or in a column of the data matrix.

[11] See R. Ge, J. D. Lee, and T. Ma, "Matrix Completion Has no Spurious Local Minimum," *Advances in Neural Information Processing Systems*, pp. 2973–2981, 2016.

In order to be able to perform matrix completion when the data matrix has some abstract data types, we can extend this approach to the generalized low-rank model presented in (4.11):

$$\min_{L,R,X \in \mathcal{X}} \sum_{i,j} \mathcal{L}(X_{ij}, l_i^\top r_j) + \sum_i p_i(l_i) + \sum_j q_j(r_j).$$

Again, this problem can be solved with alternate minimization optimizing with respect to L, X and R, X. Each sub-problem is relatively easy to solve, but convergence is not guaranteed. For example, we can consider the case where the matrix contains some Boolean entries, that is, $X_{ij} \in \{-1, 1\}$, defining for those entries:

$$\mathcal{L}(a, b) = \max(1 - ab, 0).$$

Hence, we obtain

$$\hat{X}_{ij} = \arg\min_{a \in \{-1,1\}} \max(0, 1 - al_i^\top r_j).$$

4.4 Exercises

Exercise 4.1 *Low-rank matrices*

1. Consider the matrix $X = pq^\top$, with $p \in \mathbb{R}^n$, $q \in \mathbb{R}^m$. Explain how to compute an SVD of X.

2. Consider the 2×2 matrix
$$X = \frac{1}{\sqrt{10}} \begin{pmatrix} 2 \\ 1 \end{pmatrix} \begin{pmatrix} 1 & -1 \end{pmatrix} + \frac{2}{\sqrt{10}} \begin{pmatrix} -1 \\ 2 \end{pmatrix} \begin{pmatrix} 1 & 1 \end{pmatrix}.$$

   What is an SVD of X? Make sure to check all the properties required.

3. Consider the image represented in Figure 4.14. We build a matrix X of 256×256 pixels based on this image: the upper-left block (corresponding to columns and rows from 1 to 199) is full of ones, as well as the lower-right block (columns and rows from 201 to 256). All the other elements in X are zero. Find an SVD for X.

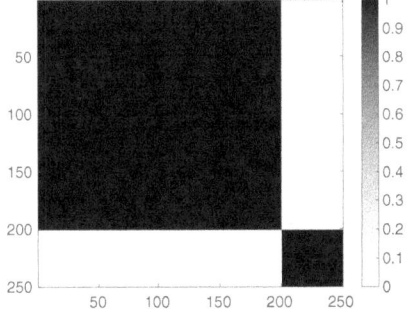

Figure 4.14 Block-diagonal image.

Exercise 4.2 *Least-squares solution via SVD*

Given $A \in \mathbb{R}^{m,n}$, $y \in \mathbb{R}^m$, consider the least-squares (LS) problem
$$\min_x \|Ax - y\|_2^2.$$

1. Prove that an optimal solution of this problem can be expressed via the pseudo inverse of A as $x^* = A^\dagger y$.

2. Derive a condition on the A matrix which guarantees that the LS solution is unique.

3. If the LS solution is *not* unique, describe the set of all optimal solutions.

Exercise 4.3 *On rank-k data approximation*

Consider a data matrix $X \in \mathbb{R}^{n,m}$ and assume that rank $X = n$ (i.e., X is full-row rank). For given $k \le n$, let X_k denote the optimal rank-k approximation of X (in the Frobenius norm sense), and let $x \in \mathbb{R}^n$ be any given vector. Since rank $X = n$, it follows that x can be expressed as $x = X\zeta$, for some $\zeta \in \mathbb{R}^m$. That is, x is a linear combination of the columns $x^{(i)}$ of X, with coefficients ζ_i, $i = 1, \dots, m$.

Suppose we want to approximate x with $\hat{x} = X_k z$, that is, via a linear combination of the columns of X_k. Show that $z^* = \zeta$ minimizes the approximation error $\|x - \hat{x}\|_2$. Is this still true if we take $\|x - \hat{x}\|_1$ as an error criterion?

Exercise 4.4 *Centered and non-centered PCA*
Given a data matrix $X \in \mathbb{R}^{n,m}$ the SVD factorization $X = U\Sigma V^\top$ and the related rank-k approximation problem in (4.6) can be performed and make sense both when X is centered (i.e., such that $X\mathbf{1} = 0$) and when it is not. The connection of these problems with PCA, however, is valid only when working with centered data matrices. That is, the SVD of X provides the maximum variance directions that are of interest in PCA only when X is centered.

What happens then if we perform SVD (or, equivalently, rank-k approximation) on an uncentered data matrix? What kind of directions do we find in this case in terms of directional data variance? Is there any relation you can find between the SVD of X and the SVD of $X - \hat{x}\mathbf{1}^\top$, where $\hat{x} = \frac{1}{m}X\mathbf{1}$ is the data center?

Exercise 4.5 *Low-rank factorization of stock investment matrix*
Given n individuals and m stocks, consider a matrix $X \in \mathbb{R}^{n,m}$ where $X_{ij} = 1$ if person i invested in stock j and $X_{ij} = 0$ if person i did not invest in stock j. For simplicity all decisions are assumed to be binary. What does it mean if $X \simeq uv^\top$ for vectors $u \in \mathbb{R}^n$, $v \in \mathbb{R}^m$? What does v represent in that case? What if we compute a rank-k approximation of X using the SVD $X \simeq \sum_{i=1}^k \sigma_i u^{(i)} v^{(i)\top}$? Interpret the σ_is, $u^{(i)}$s, and $v^{(i)}$s in this case.

Exercise 4.6 *Low-rank approximation of log-return data*
You are given an $n \times m$ market data matrix X, with $n = 100$ and $m = 250$, where each column represents a time series of length n of log-returns for a specific asset. Assume that the columns are centered, so that $X\mathbf{1} = 0$. You observe that:

- the Frobenius norm of X is $\|X\|_F = 5$;

- the top 4 singular values are $\sigma = (3, \sqrt{5}, \sqrt{2}, 0.05)$.

Answer the following questions:

1. How much total variance is present in the data?

2. How much percentage of total variance is explained by a rank-3 approximation to the data?

3. What is the largest variance a portfolio $w \in \mathbb{R}^n$ such that $w^\top w = 1$ can attain? Express your result as a fraction of the total variance.

4. Show that there exist a portfolio $w \in \mathbb{R}^n$ such that $w^\top w = 1$, with a variance less than 0.01% of the total variance.

Exercise 4.7 *Generalized low-rank models*
Consider the data given in the file INTLFXD_csv.zip in the online

resources, which contains the exchange rates against the USD for several currencies, from 1999 until 2018.

1. Use PCA to try to answer the following qualitative questions:

    (a) Prepare data sets that correspond to daily, weekly, monthly, or annual fluctuations using 1999_2018_complete.csv. Normalize the data so that the data points are all on the same scale.

    (b) Run PCA using the entire data set. Based on coefficients found in the eigenvectors, what are the main drivers (i.e., countries) of exchange rate fluctuations?

    (c) Re-do the previous part for every day, week, and month per year. How do coefficients in the eigenvectors shift over the years for each country, or a particular group of countries? Illustrate the findings with a few appropriate tables or line graphs.

    (d) Based on the answers from the above two parts, what are the main clusters of currencies? How do these compare to external knowledge, such as geography?

2. There are many variants of PCA, as described at the end of this chapter. Focus on the *matrix completion* PCA described in Section 4.3.8, that allows to run PCA on a data matrix with some elements unknown. For the rest of this problem, use the currency data only for the year 2018. For this problem, we will use the h2o package[12] in Python.

    (a) Remove between 5% to 40% of the currency data matrix randomly in 5% increments and test whether the algorithm successfully recovers the removed data. Discuss the appropriate metrics. Make a plot where the y-axis is your appropriate metric and the x-axis is the percentage of the data which was randomly removed (include error bars).
    Parameters: Set $k = 5$ (the rank of the approximation) and don't use any regularization. Set the number of max_iterations to 100.
    Note: In order to use the matrix completion functionality in h2o, your training frame must include NaN entries. You may then find the indices of the randomly places NaN entries, train your model, and then view the prediction on these entries.

    (b) In many practical instances, a whole chunk of data is missing; for example, some currency rates might not be available before a certain date. How do you expect matrix completion algorithms to behave in that case? Test this on the currency data, where the data for the first few months of a specific currency is removed.

[12] Use the function H2OGeneralizedLowRankEstimator to define a generalized low-rank model. See https://docs.h2o.ai/h2o/latest-stable/h2o-docs/data-science/glrm.html for further details on this function.

4 PRINCIPAL COMPONENT ANALYSIS AND LOW-RANK APPROXIMATION

Exercise 4.8 *PCA for financial data analysis*

We consider the daily log-return of a collection of 476 stocks chosen from the S&P 500 companies over the time period from January 1, 2010, until January 1, 2015. A 476 × 1258 matrix is used to represent the data, with each column corresponding to a day and each row to a stock. We will apply PCA on the data to help us visualize and understand it.

1. Load the data from 476Stocks.mat in the online resources. The desired matrix $X \in \mathbb{R}^{476,1258}$ is stored in variable LogReturn. You can also find other information such as each stock's ticker and sector. Plot the log-return versus time curve of all 476 stocks on the same figure. Can you interpret any information from this figure?

2. Compute the centered and normalized data matrix $\tilde{X}$. Each row of the normalized matrix $\tilde{X}$ is zero-centered and scaled to be length-1, as measured by the ℓ_2 norm.

3. Compute the SVD on $\tilde{X} = U\Sigma V^\top$ and plot the percentage of variance explained versus the number of principal components in "stock space." Notice that size of the columns of U is 476, which is the number of stocks. We can thus say that U represents the "stock space," and each column in U can be interpreted as a portfolio or mix of the different stocks. Similarly, the size of the columns of V is 1258, which is the number of days, so V represents the "time space" and each column in V represents a time profile.

4. We want to visualize stocks in a way that stocks with similar properties appear in clusters and can be separated from other stocks. We can project each row $\xi^{(i)\top}$ of $\tilde{X}$ (corresponding to a stock) onto $v^{(1)}$, the "time space" singular vector corresponding to the largest singular value as

$$\alpha_1(\xi^{(i)}) = \xi^{(i)\top} v^{(1)}, i = 1, \ldots, 476.$$

Each stock will be assigned a score $\alpha_1(\xi^{(i)})$. For stocks from the financial, health care, energy, and information technology sectors, plot each stock's score on a single line and color stocks from different sectors differently. Can you separate the above sectors in the figure? Now project $\tilde{X}$ onto "time space" singular vectors corresponding to the second and third largest singular values, to obtain $\alpha_2(\xi^{(i)})$ and $\alpha_3(\xi^{(i)})$ for each stock, respectively. Using $(\alpha_1(\xi^{(i)}), \alpha_2(\xi^{(i)}))$ and $(\alpha_1(\xi^{(i)}), \alpha_2(\xi^{(i)}), \alpha_3(\xi^{(i)}))$ as the coordinates of each stock, plot stocks from the above four sectors on a 2D plane and 3D space, respectively. Now can you separate the sectors in the figure? Briefly comment.

5
Clustering methods

CLUSTERING refers to a fundamental task of data analysis in which one seeks to discover similarities and community structures in data. Clustering is a typical *unsupervised* learning task, whereby data are grouped into clusters such that points in the same cluster have high similarity while points in different clusters have low similarity. Similarity between points is typically measured by standard distance measures such as the Euclidean or the ℓ_1 norm, by geodesic graph distance, or by location in the same contiguous region of high point density. In all cases, the data points are initially unlabeled, and the output of the clustering algorithm is the assignment to each point of a label representing the cluster to which the point is likely to belong. The number k of clusters into which the data is to be partitioned can be an input for the clustering algorithm, as is the case for the k-means algorithm, or it can be determined by the algorithm itself, as in the DBSCAN algorithm, or still it can be selected a posteriori on the basis of the algorithm's output, as happens in the hierarchical clustering approach. The applications of clustering in economics and finance are many and range, for instance, from marketing applications such as customers segmentation by purchase history, profiling based on activity monitoring, inventory categorization by sales activity or by manufacturing metrics, to finance applications in analysis, and categorization of time series and of investment opportunities. In this chapter we briefly present some of the most popular approaches to clustering, and discuss general techniques for evaluating and validating the quality of a data partition.

5.1 k-means

The *k*-means algorithm takes as input a set of feature vectors $x^{(1)}, \ldots, x^{(m)} \in \mathbb{R}^n$ and an integer $k \leq m$, and returns k vectors $c^{(1)}, \ldots, c^{(k)} \in \mathbb{R}^n$ representing the cluster centers. Once we have the cluster centers, the membership of points to clusters is obtained by assigning each point to its closest center, where the distance is the standard Euclidean norm. That is, for each $i = 1, \ldots, n$,

$$I(x^{(i)}) = \arg\min_{j=1,\ldots,k} \|x^{(i)} - c^{(j)}\|_2^2 \tag{5.1}$$

denotes the cluster to whom $x^{(i)}$ is assigned. The clusters centers, therefore, define a Voronoi partition of $\mathbb{R}^n$ into k cells $V_1, \ldots, V_k$, and $x^{(i)}$ is assigned to cluster j if $x^{(i)} \in V_j$.

In turn, the centers $c^{(1)}, \ldots, c^{(k)}$ should be determined so that the average (squared) distance of the points from their respective cluster center is minimized overall, that is, one should solve

$$J_k = \min_{c^{(1)},\ldots,c^{(k)}} \sum_{i=1}^{m} \min_{j=1,\ldots,k} \|x^{(i)} - c^{(j)}\|_2^2, \tag{5.2}$$

where $\min_{j=1,\ldots,k} \|x^{(i)} - c^{(j)}\|_2^2$ is the squared distance from $x^{(i)}$ from its cluster's center. Problem (5.2), unfortunately, is nonconvex and hard to solve exactly. As a matter of fact, it can be cast as a Boolean optimization problem by defining a matrix variable containing the centers $C = [c^{(1)} \cdots c^{(k)}] \in \mathbb{R}^{n,k}$ and a Boolean matrix variable $U = (u_{ij}) \in \{0,1\}^{k,m}$ such that

$$u_{ij} = \begin{cases} 1 & \text{if } x^{(j)} \text{ belongs to cluster } i, \\ 0 & \text{otherwise,} \end{cases}$$

subject to the constraint that $\mathbf{1}^\top U = \mathbf{1}^\top$, meaning that each $x^{(j)}$ belongs to one and only one cluster. Now, CU is an $n \times m$ matrix such that its jth column represents the center of the cluster for point $x^{(j)}$. Letting the data matrix be $X = [x^{(1)} \cdots x^{(m)}]$, we can rewrite problem (5.2) compactly as

$$J_k = \min_{C \in \mathbb{R}^{n,k}, U \in \{0,1\}^{k,m}} \|X - CU\|_F^2, \quad \text{s.t.: } \mathbf{1}^\top U = \mathbf{1}^\top, \tag{5.3}$$

or, equivalently, as

$$J_k = \min_{C \in \mathbb{R}^{n,k}, U \in \{0,1\}^{k,m}} \sum_{j=1,\ldots,m} \|x^{(j)} - Cu^{(j)}\|_2^2, \quad \text{s.t.: } \mathbf{1}^\top U = \mathbf{1}^\top, \tag{5.4}$$

where $u^{(j)}$ denotes the jth column of U. The *k*-means algorithm can now be interpreted simply as an alternate minimization method for solving approximately (5.3): given an initial assignment of the center matrix C, for example, chosen at random or selected according to prior knowledge, k-means iterates the following two steps:

1. *The cluster assignment step,* in which for fixed C we solve (5.3) with respect to the cluster assignment matrix U.

2. *The center update step,* in which for fixed cluster assignments U we solve (5.3) with respect to the center matrix C.

It turns out that each of the above steps has an easily computable solution. Indeed, if the centers in C are fixed, then problem (5.4) to be solved in the cluster assignment step reduces to

$$\min_{u^{(1)},\ldots,u^{(m)} \in \{0,1\}^k} \sum_{j=1,\ldots,m} \|x^{(j)} - Cu^{(j)}\|_2^2, \quad \text{s.t.: } \mathbf{1}^\top u^{(j)} = 1, \; j = 1,\ldots,m.$$

The above problem is separable in the $u^{(j)}$ variables, which can then be optimized independently by solving

$$\min_{u^{(j)} \in \{0,1\}^k} \|x^{(j)} - Cu^{(j)}\|_2^2 = \min_{u_i^{(j)} \in \{0,1\}, i=1,\ldots,k} \|x^{(j)} - \sum_i c^{(i)} u_i^{(j)}\|_2^2,$$

under the constraint $\sum_i u_i^{(j)} = 1$. This problem is clearly solved by assigning $u_i^{(j)} = 1$ to the $c^{(i)}$ at minimum distance from $x^{(j)}$, and zero to all other centers; this is indeed equivalent to the cluster assignment prescribed by (5.1).

For the center update step we have instead fixed U and need to solve (5.4) with respect to the centers $c^{(1)},\ldots,c^{(k)}$. Since the partition is assigned, the summation in (5.4) can be regrouped so to consider first the points in the first cluster, then those in the second cluster, and so on. That is, we solve

$$\min_{c^{(1)},\ldots,c^{(k)}} \sum_{i=1,\ldots,k} \sum_{j:I(x^{(j)})=i} \|x^{(j)} - c^{(i)}\|_2^2 = \sum_{i=1,\ldots,k} \min_{c^{(i)}} \sum_{j:I(x^{(j)})=i} \|x^{(j)} - c^{(i)}\|_2^2.$$

This problem is separable in the $c^{(i)}$s, so we solve

$$\min_{c^{(i)}} \sum_{j:I(x^{(j)})=i} \|x^{(j)} - c^{(i)}\|_2^2$$

by setting the gradient to zero, that is,

$$\sum_{j:I(x^{(j)})=i} 2c^{(i)} - 2x^{(j)} = 0 \iff c^{(i)} = \frac{1}{m_i} \sum_{j:I(x^{(j)})=i} x^{(j)},$$

where m_i is the number of data points contained in the ith cluster. The centers computed in the second step of the algorithm are thus the centroids of the points in each cluster.

The two steps of k-means are repeated iteratively until one sees that the clusters' centers do not change from one iteration to the next, in which case the algorithm stops. Provided some deterministic tie-break rule is inserted in the algorithm in cases when a point has the same distance from two or more centers, the k-means algorithm stops in a finite number of iterations. The centers obtained upon exit, however, are not guaranteed to yield the global optimum of the criterion in (5.3). Indeed, k-means generally converges only to a local optimum of the cost function, and the returned solution depends on and it is sensitive to the algorithm's initialization.

The underlying criterion in k-means is minimization of the within-cluster variance (sum or mean of squared distances from the centroids), and as such it suffers from the usual drawbacks of variance-based criteria. First, it is sensitive to outliers in the data, since the squared distance amplifies the weight of even a single far-form-center point and may skew the center estimates. Second, it is a method geared towards point collections that have clusters with essentially spherical shape, since the variance criterion gives the same weight to any spatial direction. Third, it suffers from the curse of dimensionality problem: as the dimension n of the feature space increases, and for certain data distributions, the Euclidean distance may become more and more inefficient in distinguishing data among clusters.[1] Dimensionality problems can be alleviated by performing feature selection prior to applying k-means, or by dimensionality reduction via PCA.

[1] For more in-depth discussion on clustering for high-dimensional data, see, for example, M. Steinböz, L. Ertöz, and V. Kumar, "The Challenges of Clustering High Dimensional Data," in L. T. Wille, (ed.) *New Directions in Statistical Physics*, Berlin, Heidelberg: Springer, 2004.

5.1.1 Measuring the cluster quality

The number k of clusters is a given input datum in the k-means algorithm. Typically, however, this number is not known in advance, and the usual approach is to solve k-means for several increasing values of k, and then to seek for the solution that yields the best clustering according to some appropriate metric. For instance, one may consider the mean distance between data points and their respective centroids as a clustering score, although such score will decrease with k, to the extreme of reaching zero when k is the same as the number of data points. Rather than seeking to minimize such score with respect to k, we shall hence plot it as a function of k and look for the "elbow point" where the rate of decrease sharply shifts; the corresponding k value is a rough estimate for the correct number of clusters.

An alternative is to use a performance measure that takes into account not only the within-cluster similarity of a point to its own cluster (cohesion), but also the level of dissimilarity with other clusters (separation). One such measure is the so-called *silhouette* value $s_i \in [-1, 1]$, which is computed for each data point $x^{(i)}$ as[2]

$$s_i \doteq \frac{b_i - a_i}{\max(a_i, b_i)}, \tag{5.5}$$

where a_i (cohesion) is the mean distance of $x^{(i)}$ from the other points in its cluster C_i:

$$a_i = \frac{1}{|C_i| - 1} \sum_{j \in C_i} \|x^{(i)} - x^{(j)}\|_2,$$

and b_i (separation) is the smallest dissimilarity from $x^{(i)}$ to other clusters, that is,

$$b_i \doteq \min_{C_j \neq C_i} d(i, C_j).$$

The dissimilarity $d(i, C_j)$ is the mean distance of $x^{(i)}$ to points in another cluster $C_j \neq C_i$:

$$d(i, C_j) \doteq \frac{1}{|C_j|} \sum_{j \in C_j} \|x^{(i)} - x^{(j)}\|_2.$$

A well-clustered point has small cohesion and large separation, that is, $b_i \gg a_i$, and hence its s_i value is close to one. A point which is in the wrong cluster is nearer to some other cluster than to its own, that is, $b_i \ll a_i$, and hence its s_i value is close to minus one. Silhouette values near zero correspond to ambiguous points that are near the boundary of two clusters. Once the silhouette values have been computed for all points, their overall mean $\bar{s}$ can be taken as a synthetic indicator of the clustering quality. Therefore, plotting $\bar{s}$ as a function of k and looking for the maximum of the silhouette score can be a good way to determine the correct number of clusters.

For a given k, the quality of the clustering can also be validated by leaving aside part of the data points as a test set (e.g., by splitting the data 70% for clustering and 30% for testing), and evaluating the clustering objective on such set. This operation can actually be repeated for several splits, so that statistics of the performance score can be computed.

[2] The stated formula for s_i is valid if $x^{(i)}$ belongs to a cluster containing at least two points (including $x^{(i)}$); otherwise we let $s_i = 0$.

Example 5.1 (*Clustering assets by risk/return profile*) We considered price data for the $m = 500$ stocks included in the S&P 500 basket, during the period from Jan. 25, 2021 to Jan. 13, 2022 (246 data points), see data file `PriceDataFrame_SP500_2021.csv` in the online resources. From price data we computed daily return data, and then for each stock we calculated the sample mean $\hat{r}_i$ of its return sequence and its sample standard deviation $\hat{\sigma}_i$, so that each asset is described by the pair $(\hat{\sigma}_i, \hat{r}_i)$, $i = 1, \ldots, m$, and can be represented graphically as a point in a risk/return plane. To adjust the scale of the data, we actually normalized the $\hat{r}_i$s and $\hat{\sigma}_i$s by dividing them by their respective standard deviations. To this data we applied the *k*-means algorithm for increasing values of k, and for each k we computed the average of the square distances from points to their assigned centers. A plot of the discrete derivative of this function, suitably smoothed in order to highlight the main trend, is displayed in Figure 5.1. It can be seen from the plot that rate of decrease of the cost essentially stabilizes for $k \geq 15$, so we can take $k = 15$ as a good guess for the cluster cardinality.

The result of the *k*-means clustering with $k = 15$ is shown in Figure 5.2. In this case, *k*-means provides a quick tool for finding groups of stocks that had similar performance in the risk/return sense, in the considered time frame. Notice, however, that *k*-means always returns *some* clustering of the data, also in cases in which the data does not have any real cluster structure. In the present case, the clustering has a mean silhouette value of 0.51, which reflects the fact that the clustering in the data is informative but not very strong.

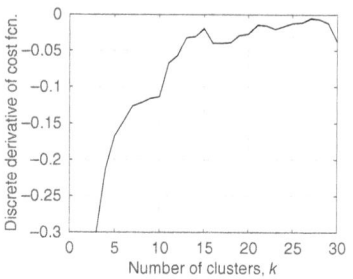

Figure 5.1 Plot of the smoothed difference of the average of the square distances criterion, as a function of k.

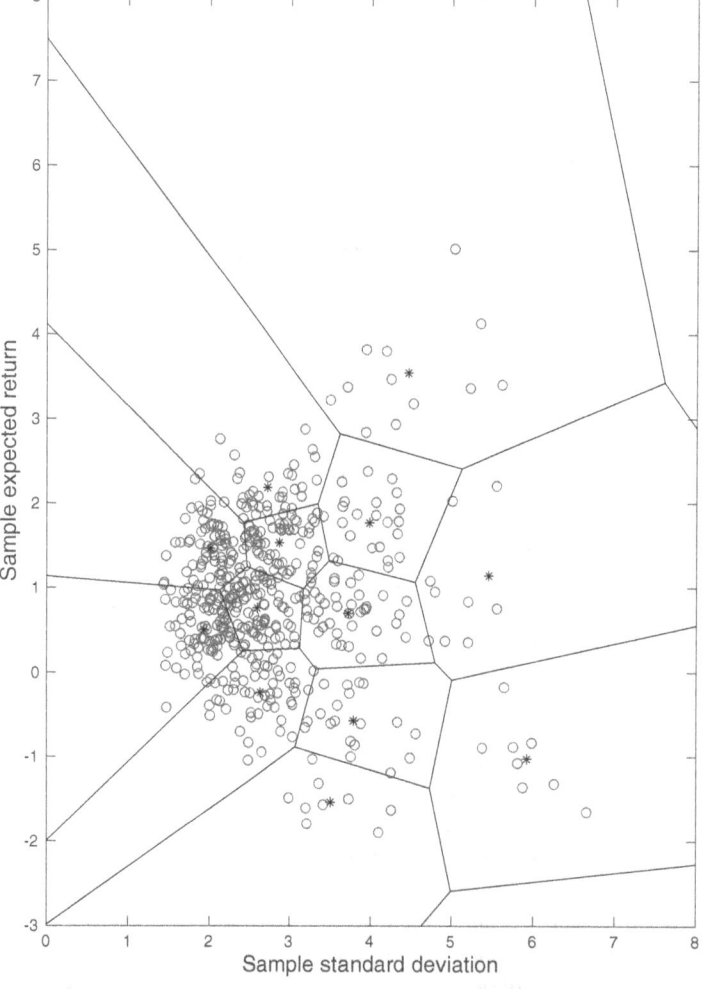

Figure 5.2 Circles represent stocks $(\hat{\sigma}_i, \hat{r}_i)$, $i = 1, \ldots, m$. Asterisks represent the cluster centers as determined by the *k*-means algorithm with $k = 15$. Solid lines show the Voronoi regions relative to each cluster center, see `clustering_SP500_retvar.m` in the online resources.

5.1.2 Variants of k-means

The basic *k*-means algorithm is based on the minimization of squared Euclidean distances of the data points from their corresponding centers. Several variants of the method can thus be obtained by considering different types of distance measures. A first variant arises if one considers plain Euclidean distances, instead of squared ones, that is, the J_k criterion in (5.2) is replaced by

$$J_k = \min_{c^{(1)},\ldots,c^{(k)}} \sum_{i=1}^{m} \min_{j=1,\ldots,k} \|x^{(i)} - c^{(j)}\|_2. \quad (5.6)$$

Considering plain Euclidean distances instead of squared ones improves the robustness of the method to the presence of outliers. However, the center update step in this case does not have a simple closed-form solution and it requires finding the so-called *geometric median* of a set of points, which can be done via a recursive algorithm.[3]

A further variant consists in replacing the Euclidean norm altogether with some other norm. A popular choice is to substitute it with the ℓ_1 norm, thus obtaining a problem of the form

$$J_k = \min_{c^{(1)},\ldots,c^{(k)}} \sum_{i=1}^{m} \min_{j=1,\ldots,k} \|x^{(i)} - c^{(j)}\|_1. \quad (5.7)$$

In this case, it can be proved that the center update step amounts to finding the coordinate-wise median of the data in each cluster, that is, the *i*th coordinate $c_i^{(j)}$ of the *j*th center is computed as the median of the *i*th coordinates of the points belonging to the *j*th cluster. For this reason, this version of the algorithm is also known as the *k-medians* algorithm.

Note that in all the previous declinations of the algorithm the cluster centers are in general new points, different from any of the given data points. In the *k-medoids* algorithm, instead, one of the data points in a cluster is selected as that cluster representative. Such point, called a *medoid*, is defined as the object in the cluster whose average dissimilarity to all the objects in the cluster is minimal, that is, it is a most centrally located point in the cluster. Given a cluster with ν elements, for finding the medoid we need to compute the average distance from each point in the cluster to all other points in the same cluster, based on the similarity matrix (d_{ij}), where d_{ij} is the distance from point *i* to point $j \neq i$ (we let $d_{ii} = 0$). Constructing the similarity matrix for a cluster requires $\nu(\nu-1)/2$ distance computations. One advantage is that the medoid can be computed with an identical approach, irrespective of the distance metric used. A classical two-step iterations of

---

[3] In the standard *k*-means algorithm the center update step required solving problems of the type $\min_c \sum_{i \in Q} \|x^{(i)} - c\|_2^2$, where Q represents the index set of some cluster, and we have seen that the solution is simply given by the centroid of the cluster: $c = \frac{1}{|Q|} \sum_{i \in Q} x^{(i)}$. In the version with plain distance, instead, we need to solve $\min_c \sum_{i \in Q} \|x^{(i)} - c\|_2$. This problem has no closed-form solution. Nevertheless, it can be solved efficiently via second-order cone-type convex programming.

cluster assignment and medoid computation can be used for the k-medoid algorithm. However, such an approach may perform poorly in this case, and superior algorithms exist, such as the partitioning around medoids (PAM) algorithm of Kaufman and Rousseeuw.[4]

[4] For additional details on this algorithm, see L. Kaufman and P. J. Rousseeuw, *Partitioning Around Medoids (Program PAM)*, Wiley Series in Probability and Statistics, Hoboken, NJ: John Wiley & Sons, 1990.

5.2 Spectral clustering

Spectral clustering methods exploit a graph representation of the data set. A simple way to construct a graph from data is to consider each point as a node in a graph[5] and create an edge from each node to its k-nearest neighbors. The notion of nearness can be taken to be simply the Euclidean distance in the original data space, or some other more general and problem-dependent measure of *affinity* or *similarity* between data points. In general, we construct from the given data points a weighted adjacency matrix A whose elements $a_{ij} \geq 0$ expresses how similar points i and j are to each other; if i,j are very dissimilar then $a_{ij} \simeq 0$, while if the points are identical, then $a_{ij} = 1$. This weighted adjacency matrix A defines a *similarity graph* G for the input data set. From the weighted adjacency matrix A we compute the weighted degree matrix $D = \text{diag}(A\mathbf{1})$ and then the graph Laplacian $L = D - A$, which is a symmetric and positive semidefinite matrix, see Section 13.1. Note that by construction $\lambda_{\min}(L) = 0$ is a minimum eigenvalue of L, and we let

[5] See Section 13.1 for an introduction to graph formalism and definitions.

$$0 = \lambda_1 \leq \lambda_2 \leq \cdots \leq \lambda_m$$

be the eigenvalues of L in increasing order. It is a well-known result in algebraic graph theory that the second-smallest eigenvalue λ_2 (known as the *Fiedler eigenvalue*) is positive if and only if the graph is connected. Actually, we define the *algebraic connectivity* $\alpha(G) = \lambda_2(L_G)$ of a graph G as a quantitative measure of connectivity, where L_G is the Laplacian matrix of G.[6]

The problem of clustering into $k \geq 1$ groups the original data points $X = [x^{(1)} \cdots x^{(m)}] \in \mathbb{R}^{n,m}$ can now be reformulated using the similarity graph: we want to find a partition of the graph such that the edges between different groups have very low weights (so that different clusters are dissimilar) while the edges within the same group have high weights (so that points in the same cluster are similar). This can be achieved by considering the orthogonal matrix U formed by k eigenvectors of the Laplacian matrix $L \in \mathbb{S}_+^m$ of the similarity graph (which is assumed to be connected):

[6] Notice that it holds that $L\mathbf{1} = 0$, which means that $u = \mathbf{1}$ is an eigenvector of L associated with the eigenvalue $\lambda_1(L) = 0$. If such eigenvalue has unit multiplicity, then the graph associated with L has only one connected component. If instead the algebraic multiplicity of $\lambda_1(L) = 0$ is $k \geq 1$, then the corresponding graph has k connected components, and there are k orthogonal eigenvectors $u^{(1)}, \ldots, u^{(k)}$ associated to $\lambda_1 = 0$, where $u_i^{(j)}$ is one if node i belongs to the jth component, and it is zero otherwise (note that $\sum_{j=1}^k u^{(j)} = \mathbf{1}$).

$$U = [u^{(1)} \cdots u^{(k)}] \in \mathbb{R}^{m,k},$$

where $u^{(i)}$ is the eigenvector of L corresponding to its eigenvalue λ_i, for $i = 1, \ldots, k$. The images of the original points $x^{(1)} \cdots x^{(m)}$ under the embedding into the lower-dimensional space $\mathbb{R}^k$ is given by the m columns $y^{(1)}, \ldots, y^{(m)} \in \mathbb{R}^k$ of $U^\top$. The spectral clustering algorithm works by clustering $y^{(1)}, \ldots, y^{(m)}$ using k-means and then returning for each $i = 1, \ldots, m$ the corresponding cluster in $\{1, \ldots, k\}$. A full explanation of the functioning of spectral clustering requires delving into graph embeddings and approximation approaches for graph cuts, which are out of the scope of our exposition; the interested reader may refer, for instance, to the tutorial paper by von Luxburg.[7]

[7] U. von Luxburg, "A Tutorial on Spectral Clustering," *Statistics and Computing*, vol. 17, pp. 395–416, 2007.

5.3 Density-based spatial clustering

Density-based clustering refers to a class of unsupervised learning methods that aim at identifying clusters in the data based on the notion that a cluster is a contiguous region of high point density, separated from other clusters by contiguous regions of low point density. Density-based spatial clustering of applications with noise (DBSCAN) is a base algorithm for density-based clustering, which can discover clusters of different shapes and sizes from data. In DBSCAN the number k of clusters need not be specified a priori. The key idea is that, for each point of a cluster, the neighborhood of a given radius ϵ has to contain at least a minimum number of points $m_{\min}$, that is, the density in the neighborhood has to exceed some threshold. The shape of a neighborhood is determined by the choice of a distance function $d(p,q)$ between points p and q; for instance, by choosing the Euclidean norm we obtain spherical neighborhoods, and choosing the ℓ_∞ norm we obtain hyper-cubical neighborhoods. In DBSCAN the notion of *density-reachable* points is defined: a point p is density-reachable from a point q (given ϵ and $m_{\min}$) if there exist a chain of points $p_1, \ldots, p_t$, with $p_1 = q$, $p_t = p$ such that (a) $d(p_i, p_{i+1}) \leq \epsilon$, and (b) the number of points in $\{x : d(p_i, x) \leq \epsilon\}$ is at least $m_{\min}$ (core point condition). To find a cluster, DBSCAN starts with an arbitrary point p and retrieves all points that are density reachable from p. If p is a core point, this procedure yields a cluster, otherwise no points are density-reachable from p and DBSCAN tries with the next point of the data set. Further details on the implementation of DBSCAN can be found in the original paper.[8]

[8] See M. Ester, H.-P. Kriegel, J. Sander, and X. Xu, "A Density-Based Algorithm for Discovering Clusters in Large Spatial Databases with Noise," *KDD'96: Proceedings of the Second International Conference on Knowledge Discovery and Data Mining*, pp. 226–231, 1996.

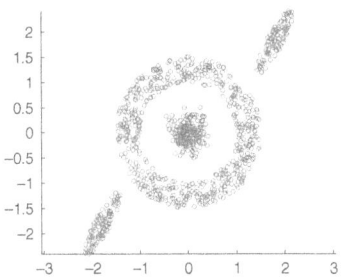

Figure 5.3 A two-dimensional data set, see data file ex_data_cluster.mat in the online resources.

DBSCAN is generally more successful than *k*-means in identifying clusters with non-spherical shapes. Figure 5.3, for example, shows a two-dimensional data set in which a human can easily recognize four separate clusters. The *k*-means algorithm, even when given in input the correct number of clusters $k = 4$, typically fails in detecting the correct clusters, see Figure 5.4(a). On the other hand, the DBSCAN algorithm, with $\epsilon = 0.2$ and $m_{\min} = 5$, correctly recognises the four clusters, see Figure 5.4(b).

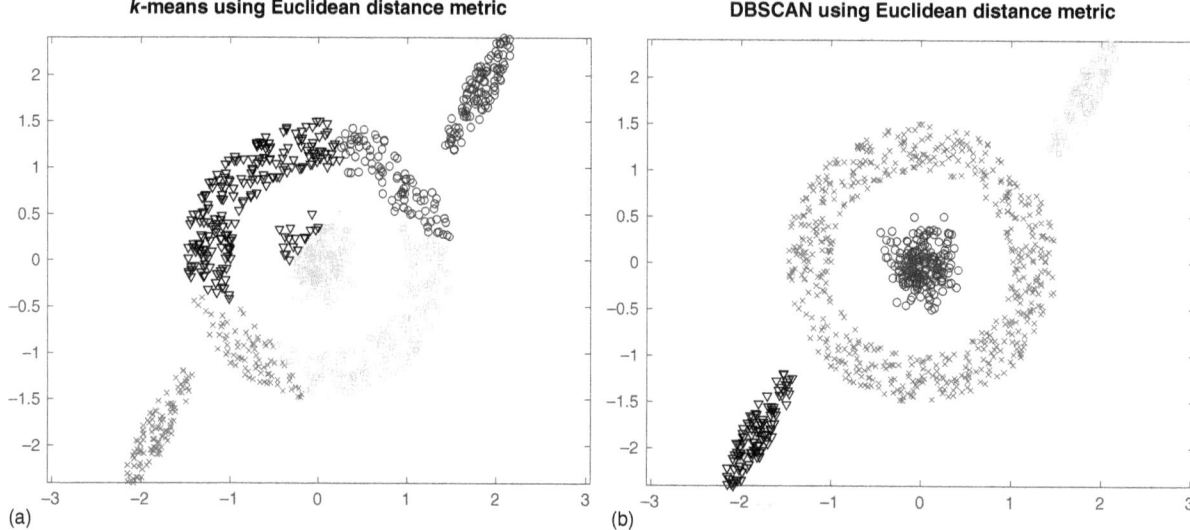

Figure 5.4 A two-dimensional data set clustered with *k*-means (a) and with DBSCAN (b), see file dbscan_example.m in the online resources.

5.4 Clusterpath

The *clusterpath* approach derives from a relaxation of the Boolean representation (5.3) of the *k*-means objective. The matrix $Y = CU$ appearing in (5.3) has as columns the k centers $c^{(1)}, \ldots, c^{(k)}$ repeated several times, that is, for instance, $c^{(1)}$ appears in Y in all the columns associated with the first cluster, $c^{(2)}$ appears in Y in all the columns associated with the second cluster, and so on. Given the data matrix $X \in \mathbb{R}^{n,m}$, the *k*-means problem can hence be rewritten in the form of a matrix approximation problem, in which we seek to minimize $\|X - Y\|_F^2$ with respect to $Y \in \mathbb{R}^{n,m}$, under the constraint that Y contains at most k unique columns (which correspond to the cluster centers). The first approximation is to use instead of this constraint an upper bound on the number of columns that are pair-wise different, that is,

$$\min_Y \|X - Y\|_F^2, \quad \text{s.t.:} \sum_{i=1}^{m-1} \sum_{j=i+1}^{m} I(\|y^{(i)} - y^{(j)}\|_2) \leq t, \quad (5.8)$$

where $I(x)$ is an indicator function, such that $I(x) = 1$ when $x \neq 0$ and $I(x) = 0$ otherwise. The double summation in the above problem runs over the $m(m-1)/2$ pairs of columns of Y and counts how many of these pairs are different. Clearly, if $t \geq m(m-1)/2$ the problem is in fact unconstrained, and the optimal solution is $Y = X$. If $t = m(m-1)/2 - 1$ then at least two columns in Y must coalesce to the same value; if $t = m(m-1)/2 - 2$ then at least two pairs of columns coalesce to the same value, and so on. In general, as t decreases, the columns of X tend to coalesce to few repeated column values. The formulation in (5.8) is still not tractable, hence we relax it by eliminating the indicator function and introducing weights $w_{ij} \geq 0$, obtaining

$$\min_Y \|X - Y\|_F^2, \quad \text{s.t.:} \sum_{i=1}^{m-1} \sum_{j=i+1}^{m} w_{ij} \|y^{(i)} - y^{(j)}\|_2 \leq t. \quad (5.9)$$

This problem is now convex and, in particular, it can be cast as a second-order cone program. For each t value we obtain a cluster, and the continuous regularization path of optimal solutions formed by varying t is what we call the clusterpath. The same clusterpath can be computed via a Lagrangian penalized version of the problem, that is, we solve

$$\min_Y \|X - Y\|_F^2 + \lambda \sum_{i=1}^{m-1} \sum_{j=i+1}^{m} w_{ij} \|y^{(i)} - y^{(j)}\|_2, \quad (5.10)$$

for increasing values of $\lambda \geq 0$. The clusterpath represents an *agglomerative* approach to clustering, in which we start with the full data set for $\lambda = 0$ and then the points (columns in Y) progressively aggregate and coalesce as λ increases, to the limit of converging all to the same overall mean of the columns of X. Notice that other norms can be used in (5.10) in place of the ℓ_2 norm; notably, the ℓ_1 norm and the ℓ_∞ norm are also considered in the original reference for clusterpath.[9]

5.5 Hierarchical clustering

We next illustrate a popular nonparametric method called *hierarchical agglomerative clustering* (HAC). In this approach we start from the individual data points $x^{(1)}, \ldots, x^{(m)}$, which represent the leaves of a tree and can be interpreted as trivial clusters containing only one element. Then, we define a *linkage function* which measures the distance between clusters. The linkage function may use different types of norms, and different criteria for assessing the distance. For instance,

[9] See T. D. Hocking, A. Joulin, F. Bach and J.-P. Vert, "Clusterpath: An Algorithm for Clustering using Convex Fusion Penalties," *Proceedings of the 28th International Conference on Machine Learning*, 2011.

the *single linkage* criterion computes the distance $d(I, J)$ between two groups of points with indices I, J based on the shortest distance over all possible pairs, that is,

$$d(I, J) = \min_{i \in I, j \in J} \|x^{(i)} - x^{(j)}\|,$$

for some norm of choice. There exist many alternative linkage criteria, such as, for instance, the *complete linkage*, which considers the maximum distance between points in the two sets:

$$d(I, J) = \max_{i \in I, j \in J} \|x^{(i)} - x^{(j)}\|,$$

or the *average linkage*, which considers the average distance between points in the two sets.

Equipped with the linkage function, we can determine which pair of clusters in the current level of the tree are the closest (starting from the individual data points in the leaves of the tree), and we fuse them into a new cluster. We then recompute the linkage function for all the new clusters, and repeat the process until all clusters are fused in a single cluster which sits at the root node of the tree. This approach can be conveniently represented graphically via a *dendrogram*, which is nothing but a representation of the tree in which the "legs" joining the clusters to be fused are represented with a length proportional to the linkage distance of the clusters. Once the dendrogram is complete, we have a full hierarchy of clusters, from single cluster at the root to single-element clusters at the leaves, and we can choose a cluster of desired cardinality by cutting the dendrogram at a given level.

Example 5.2 (*German credit data set*) We consider a data set regarding credit data of $m = 1,000$ individuals. The data contains $n = 24$ numerical features, such as status of existing checking account, credit history, purpose for loan request, credit amount, personal status, and so on.[10] Running HAC on this data set with average linkage we obtain the dendrogram shown in Figure 5.5.

Figure 5.5 shows the tree up to the lower level of 30 clusters (the full tree would go down to $m = 1,000$ leaf clusters). By cutting the dendrogram with an horizontal line at level γ we can obtain different clusters cardinalities. For instance, $\gamma = 70$ would yield two clusters, $\gamma = 50$ would yield four clusters, $\gamma = 40$ would yield eight clusters, and so on.

[10] This data set is found in the UCI Machine Learning Repository under the name of Statlog (German Credit Data) Data Set, see german.data-numeric.csv.

Example 5.3 (*Fundamentals of S&P 500 companies*) In this example we considered $m = 224$ companies in the S&P 500 basket corresponding to the top 50% quantile in that basket for market capitalization. For these companies, we collected $n = 5$ fundamental economic indicators: Number

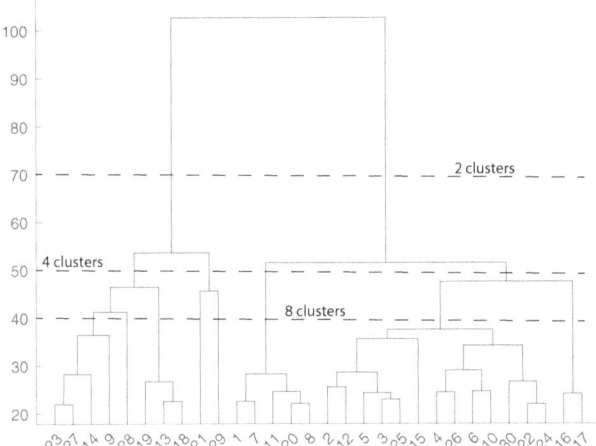

Figure 5.5 Dendrogram obtained by hierarchical agglomerative clustering on the German credit data set, see german_data_clustering.m in the online resources.

of employees [nemply], Market cap [mktcap], Revenue [ttmrev], Earnings Before Interest, Taxes, Depreciation (EBITD) [ttmebitd], and Net Income accruing to common shares for dividends and retained earnings [ttmniac], as of October 2020. The objective is to cluster these companies to discover similar ones in terms of the considered features; see the MATLAB® file clustering_SP500_fundamental.m in the online resources.[11]

An approach via k-means, for k in the range 2–30 yielded the average silhouette scores depicted in Figure 5.6. For $k = 10$ clusters we obtain an average silhouette score of 0.53, which denotes informative but not strong clustering performance.

Using an HAC approach with complete linkage and ℓ_1 norm distance criterion we obtain the dendrogram shown in Figure 5.7.

Cutting the dendrogram at a level that contains $k = 10$ clusters, we obtained one large cluster of 191 companies, and nine other clusters composed as detailed in Figure 5.8. The average silhouette score for this clustering is 0.75.

[11] A Python implementation of k-means clustering for this data is available in file SP500fundm_clustering_kmeans.py in the online resources. A hierarchical clustering approach is also available in SP500fundm_clustering_hier.py.

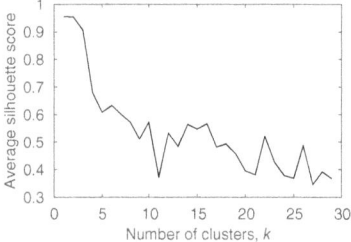

Figure 5.6 Average silhouette score from k-means clustering on the S&P 500 fundamental data.

5.6 Dimensionality reduction for clustering

As we discussed in previous sections, most of the clustering techniques rely in one way or another on some distance measure and, unfortunately, distance measures such as the Euclidean measure can be quite sensitive to data dimensionality and work progressively worse as the data dimension increases. Think for instance of the DBSCAN method, which needs to find neighboring points at a given distance ϵ from a current point; take 0 as the current point and suppose the data is distributed uniformly in a hypercube centered at zero and of unit side length. The volume of such hypercube is equal to $1^n = 1$,

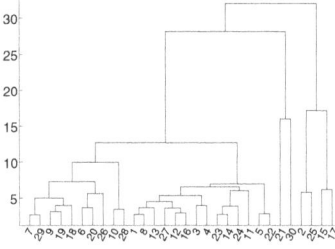

Figure 5.7 Dendrogram for hierarchical clustering on the S&P 500 fundamental data, stopped at the level of 30 clusters.

Cluster 2:

Anthem Inc.
CIGNA Corp.
Costco Wholesale Corp.
Chevron Corp.
Ford Motor
General Motors
McKesson Corp.
Walgreens Boots Alliance
Exxon Mobil Corp.

Cluster 3:

Accenture plc
Kroger Co.
Target Corp.
United Parcel Service

Cluster 4:

Citigroup Inc.
Comcast Corp.
Home Depot
International Business Machines
Johnson & Johnson
Procter & Gamble
Wells Fargo

Figure 5.8 Cluster composition resulting from HAC on the S&P 500 fundamental data.

Cluster 5:

Bank of America Corp
Intel Corp.
JPMorgan Chase & Co.
AT&T Inc
Verizon Communications

Cluster 6:

CVS Health
United Health Group Inc.

Cluster 7:

Amazon.com Inc

Cluster 8:

Wal-Mart Stores

Cluster 9:

Apple Inc.
Microsoft Corp.

Cluster 10:

Facebook, Inc.
Alphabet Inc Class A

irrespective of the data dimension n. A hyper-sphere centered at zero with radius ϵ has instead a volume equal to

$$v_\epsilon(n) = \frac{\pi^{n/2}}{\Gamma(n/2+1)} \epsilon^n,$$

where Γ is Euler's gamma function. The ratio between the volume of the ϵ-ball and the volume of the unit-side-length hypercube (whose volume is 1) gives the average fraction of uniformly distributed points in the hypercube that will be found in the ball. So, this fraction is simply $v_\epsilon(n)$.

Take $\epsilon = 0.5$, so that the ball is tightly inscribed in the hypercube, see Figure 5.9. We have for $n = 2$ that $v_\epsilon(2) = \pi\epsilon^2 \simeq 0.785$, so in dimension $n = 2$ about 78% of the points in the cube are also in the ball. In dimension $n = 3$ we have $v_\epsilon(3) \simeq 0.52$, and increasing the dimension we have $v_\epsilon(10) \simeq 0.0025$, $v_\epsilon(20) \simeq 2.46 \times 10^{-8}$, and so on, see Figure 5.10. Thus, by increasing n the volume ratio goes to zero at geometric rate, which means in practice that given a fixed number of points in the hypercube, these points become more and more rarefied as the dimension increases. For instance, given $m = 1,000$ uniformly distributed points in the hypercube $[-.5, .5]^n$, only 2.5 of these points, on average, are found inside a ball of radius 0.5 in dimension $n = 10$, and essentially no point is found in such ball for dimension $n = 20$. This is one manifestation of the "curse of dimensionality" principle.

In addition to the curse of dimensionality, there may be other reasons why it could be advantageous to work with lower-dimensional

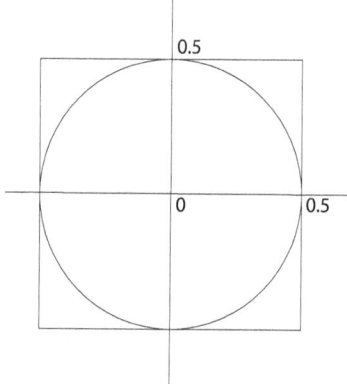

Figure 5.9 Ball of radius $\epsilon = 0.5$ inscribed in a unit-side-length hypercube, for $n = 2$.

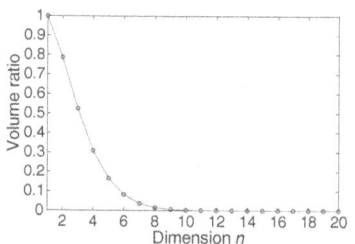

Figure 5.10 Plot of $v_\epsilon(n)$ as a function of n, for $\epsilon = 0.5$.

data. For instance, some features may be representative and important for the clustering purpose, while others may actually hinder the clustering task and add noise. Lowering the dimension of the clustering task in general helps finding the clusters more efficiently and understanding the data better. There are two main approaches for dimensionality reduction: one is based on projection of the data in a meaningful subspace, for example, by means of PCA. In this approach, one first computes the score vectors $\alpha^{(i)} \in \mathbb{R}^k$, $i = 1, \ldots, m$, where $k \ll n$, representing the data points in the principal axes, and then applies the clustering method of choice to these lower-dimensional points. The principal axes are found by constructing a centered data matrix $X \in \mathbb{R}^{n,m}$, computing its SVD factorization $X = U\Sigma V^\top$, and then computing the score vectors as

$$\alpha^{(i)} = U_k^\top x^{(i)}, \quad i = 1, \ldots, m,$$

where $U_k = [u^{(1)} \cdots u^{(k)}] \in \mathbb{R}^{n,k}$.

A second main approach is to reduce the dimensionality directly by removing some of the features (feature selection). While feature selection is well studied for supervised learning problems, for example, via the Lasso, sparse naïve Bayes, and so on, it is less developed in unsupervised situations such as clustering, due to the absence of a natural criterion for measuring the outcome quality of the algorithm, since there is no ground truth (labels) against which to compare. Nevertheless, one could use a measure of clustering efficiency, such as the mean silhouette value, to create a *wrapper* method that somehow optimizes over the feature subsets. These approaches are combinatorial in nature (choose the one subset of k features over the total of n features that yields the optimal clustering performance), but can be implemented in a greedy way, so to provide suboptimal solutions. For instance, one solves the clustering problem n times while removing one specific feature at each time, and then discards the feature whose removal yielded the best score improvement overall; then the process is repeated over the $n-1$ remaining features, and so on until one remains with a subset of $k \ll n$ selected features. Such an approach is subtractive, in the sense that one starts with the full set of features and then removes features one by one, but also additive approaches can be devised, whereby one finds the one single feature that yields the best clustering performance and puts it in the selected features set, then searches over the remaining features to find a second feature to add in the selected features set, and so on until the selected features set contains k elements. An approach to sparse clustering in which data points are clustered based on an adaptively chosen subset of the features has also been proposed in a paper by Witten and Tibshirani.[12] This approach uses a Lasso-type penalty to select the features, and it is applied to develop simple methods for sparse k-means and sparse hierarchical clustering.

[12] See D. M. Witten and R. Tibshirani, "A Framework for Feature Selection in Clustering," *Journal of the American Statistical Association*, vol. 105(490), pp. 713–726, 2010.

5.7 Exercises

Exercise 5.1 *Customer segmentation*
We consider a data set containing information about customers of a supermarket. We are interested in segmenting the customers in different clusters in order to be able to develop effective marketing strategies for each cluster of customers.

1. Load the data set contained in the file Mall_Customers.csv[13] in the online resources. This data set contains $m = 200$ customers. Each customer is described by $n = 4$ features: Gender, Age, Annual Income, and Spending Score.

    [13] This data set is also available on Kaggle at www.kaggle.com/datasets/vjchoudhary7/customer-segmentation-tutorial-in-python.

2. Cluster the customers using the *k*-means algorithm (for this task you should ignore the categorical feature Gender). In order to find the optimal value of *k*, try all values between 1 and 11. For each value of *k*, compute the average of the square distances between data points and their assigned centroid. Plot this score as a function of *k* and identify the value of *k* corresponding to the elbow point. This gives us a rough estimate of the optimal number of clusters.

3. Using the *k* value selected at the previous point, draw a scatterplot of the annual income and the spending score, using different colors to identify the *k* clusters. At visual inspection, does *k*-means correctly recognize the clusters present in the data?

Exercise 5.2 *Spectral clustering*
In this exercise we will focus on spectral clustering and perform a comparison between this method and *k*-means.

1. Load the data set contained in the file sine_dataset.csv in the online resources. This file contains a two-dimensional dataset. Visualize the datapoints using a scatterplot.

2. Center and normalize the data.

3. Cluster the data points using spectral clustering[14] with $k = 2$ clusters. Use a 15-nearest neighbor graph to build the similarity graph. Visualize the results using a scatterplot.

    [14] In Python, use the function SpectralClustering from the package sklearn.cluster.

4. Cluster the data points using *k*-means with $k = 2$ clusters. Use a scatterplot to visualize the results. Compare the obtained results with the ones of the previous point.

Exercise 5.3 *Clustering time series*

In this exercise, the focus is on clustering time-series data, and on reproducing the results of a research paper,[15] which argues that, for time series, the popular *k*-means algorithm may be meaningless when the subsequences are extracted using moving windows.

[15] E. Keogh and J. Lin, "Clustering of Time-Series Subsequences is Meaningless: Implications for Previous and Future Research," *Knowledge and Information Systems*, vol. 8(2), pp. 154–177, 2005.

1. Load the daily close log-return data from the Dow Jones Industrial Average data in the file dji_5yr_data.csv in the online resources. Generate subsequences based on a 5-year data span, and month-length subsequences (that is, each subsequence has 20 data points). Center and normalize the obtained data.

2. Implement and run a time-series clustering method based on a hierarchical clustering approach. Make sure you show the resulting tree, and comment; in particular, how many clusters appear to be reasonable? Discuss.

3. Cluster the subsequences using the *k*-means method with $k = 3$. Try to validate the claim that the cluster centroids found by using the moving window approach resemble sine waves, as shown in Figure 9 in the paper.

4. Load the cylinder-bell-funnel (CBF) data in the files 30_128_X.csv and 30_y.csv in the online resources, where 30_128_X.csv contains 30 sequences of length 128 and 30_y.csv contains 30 corresponding labels, where the labels 0, 1, and 2 represent cylinder, bell, and funnel respectively, which are three different classes of sequences. Visualize the three different classes of sequence. What have you observed?

5. Repeat parts 2 and 3 on the CBF data. You do not need to center and normalize the dataset.

6. Concatenate the 30 sequences from 30_128_X.csv into a single long sequence and perform the moving window approach with window size 128 to extract subsequences. Center and normalize your data. Then repeat parts 2 and 3 on the extracted subsequences. Contrary to part 5, you should observe sinusoidal waves like in part 3.

Exercise 5.4 *Banknote authentication*

Consider the data set contained in the file Banknote_Authentication.csv[16] in the online resources, which is composed of data extracted from images that were taken from genuine and forged banknote-like specimens. Each image is described by four features obtained by applying a wavelet transform, namely entropy of the image, variance,

[16] This data set is also available at https://archive.ics.uci.edu/ml/datasets/banknote+authentication.

skewness, and curtosis of the wavelet coefficients. For each image the class label is also indicated (i.e., forged/not forged).

1. Load the data set. Cluster the data points using k-means with $k = 2$ (do not use the class label).

2. Compare the clusters obtained using k-means with the true class labels to see if k-means was able to detect the true classes.

3. Cluster the data points using DBSCAN[17] with $\epsilon = 2$ (do not use the class label). As discussed in Section 5.3, DBSCAN does not require to specify a priori the number of clusters. Check the number of clusters obtained when running this algorithm.

[17] In Python, use the function DBSCAN from the package sklearn.cluster.

4. We now focus only on the two main clusters obtained with DBSCAN, the other clusters should be of small size and they can be considered as noise. Compare the two main clusters obtained using DBSCAN with the true class labels to see if DBSCAN was able to detect the true classes.

5. Compare the results obtained in point 2 with the ones obtained at the previous point. Use scatterplots to visualize the results and compare the clusters identified by the two clustering methods with the true classes.

Exercise 5.5 *Customer profiling*
The data set contained in the file marketing_campaign_clean.csv[18] in the online resources consists in a detailed analysis of $m = 2{,}212$ customers of a retail company. Each customer is represented by $n = 23$ features, which describe personal information of the customers and their shopping habits, such as the amount spent on different types of product and the sales channels used by the customer.

[18] This data set is a cleaned version of the data set available at https://www.kaggle.com/datasets/imakash3011/customer-personality-analysis.

1. Load the data set. Center and normalize the data set.

2. This data set contains a high number of features. As discussed in Section 5.6, clustering methods are quite sensitive to data dimensionality and work progressively worse as the data dimension increases. For this reason, we first have to perform dimensionality reduction before running the clustering algorithm. Use PCA to reduce the dimension of our problem to 3.

3. Cluster the projected data points using k-means. Try all values of k from 2 to 10. Select the value of k corresponding to the elbow point.

4. Analyze the clusters obtained at the previous point and try to define a customer profile for each cluster.

6
Linear regression models

LINEAR REGRESSION is the workhorse of statistics and (supervised) machine learning. A linear regression model relates an input vector of features $x \in \mathbb{R}^m$ to an output *response* $y \in \mathbb{R}$ by means of a linear combination of fixed functions of the input, called basis functions. The parameters of such model are given by the coefficients $w \in \mathbb{R}^d$ of the linear combination of the basis functions. The learning, or training, phase of a regression model entails using a labeled data set formed of m input–output pairs $\mathcal{D} = \{(x^{(1)}, y_1), \ldots, (x^{(m)}, y_m)\}$ for estimating the parameters w of the model, on the basis of a suitable *loss* function. The prediction phase amounts instead to computing the model output $\hat{y}$ for a given new input x. It has to be observed that the term "linear" referred to regression models hints to the fact that the model output is obtained as a linear combination of the basis functions. The actual input–output function realized by a linear regression model, however, is in general nonlinear, unless the basis functions are themselves linear functions of the input.

6.1 Linear regression

Similar to many statistical inference models, regression models are based on prior general assumptions on the mechanism that generates the output data. In particular, we assume that there exist an unknown function $f \colon \mathbb{R}^n \to \mathbb{R}$ such that

$$y = f(x) + \epsilon, \tag{6.1}$$

where ϵ represents a noise term, which is characterized by some given probability distribution, for instance a normal $\mathcal{N}(0, \sigma^2)$ distribution. When this type of input–output relation is probed via m input vectors $x^{(1)}, \ldots, x^{(m)}$, it produces outputs

$$y_i = f(x^{(i)}) + \epsilon_i, \quad i = 1, \ldots, m, \qquad (6.2)$$

where ϵ_i are i.i.d. random noise terms. The mechanism just described is the DGM that is assumed to generate the data set $\mathcal{D} = \{(x^{(1)}, y_1), \ldots, (x^{(m)}, y_m)\}$.

A *regression model* is built by postulating some proxy for the unknown function f, having the structure

$$\hat{y} = \sum_{j=1}^{d} w_j \phi_j(x), \qquad (6.3)$$

where $w = (w_1, \ldots, w_d)$ is a vector of model parameters, and $\phi_j \colon \mathbb{R}^n \to \mathbb{R}$, $j = 1, \ldots, d$, are given and fixed basis functions. When the input vectors $x^{(1)}, \ldots, x^{(m)}$ are fed to this model, it produces outputs

$$\hat{y}_i = \sum_{j=1}^{d} w_j \phi_j(x^{(i)}), \quad i = 1, \ldots, m. \qquad (6.4)$$

Clearly, there is in general a difference between the actual observed outputs y_i and the outputs generated by the synthetic regression model $\hat{y}_i$. These prediction errors

$$e_i \doteq y_i - \hat{y}_i, \quad i = 1, \ldots, m, \qquad (6.5)$$

are due to three main sources:

1. The first source of error is due to the fact that the assumed structural form $\sum_{j=1}^{d} w_j \phi_j(x)$ of the regression function may be a poor proxy of the actual unknown function $f(x)$ that we wish to approximate. Say, for instance, function f contains oscillating terms such as sinusoids, but we are trying to approximate it with basis functions ϕ_j who are merely affine. This contribution to the error is called structural error, or *bias*.

2. The second source of error is due to the fact that, even if f had exactly the structure supposed in the regression model, we still need to estimate the model parameters w from limited and noisy input–output data, and this induces uncertainty in the estimated value $\hat{w}$ of the parameter. This type of error is called predictor *variance* error.

3. The third source of error is due to the exogenous noise terms ϵ_i in (6.2), which produce an irreducible noise that would remain present in the prediction error even if the true model was precisely known.

Bias error can be reduced by accurate choice of the basis functions and of the model complexity d. Such choice is usually driven by prior knowledge of the phenomenon under study and/or experimentation. Bias error reduces by increasing d, since this increases the richness of the family of functions that can be generated via linear combinations of the basis functions. Variance error, instead, can be reduced by increasing the number m of observations. In theory, if there was no bias error, and appropriate estimation techniques were used, the variance error would go to zero as m goes to infinity. The interesting issue, however, is that bias and variance errors are connected: if one tries to reduce the bias by increasing d, then the number of parameters to be estimated in w increases and, for fixed m, their estimation error increases, thus increasing the variance error. The correct choice of basis functions and complexity d, in relation to the available number m of observations, is part of the "art" of modeling and it is subject to tradeoffs that need be assessed via proper calibrations and cross validations of the model.

Defining the feature map $\phi \colon \mathbb{R}^n \to \mathbb{R}^d$ taking vector values

$$\phi(x) \doteq [\phi_1(x) \cdots \phi_d(x)]^\top,$$

the regression model (6.3) is rewritten compactly as

$$\hat{y} = w^\top \phi(x). \tag{6.6}$$

In some cases, regression models are presented in a form that highlights the presence of a constant offset term $b \in \mathbb{R}$, that is,

$$\hat{y} = w^\top \phi(x) + b. \tag{6.7}$$

This form, however, can be obtained as a special case of (6.6), in which we take an augmented vector w of dimension $d+1$, with $w_{d+1} = b$ and $\phi(x) \doteq [\phi_1(x) \cdots \phi_d(x)\, 1]^\top$. For notational compactness we can thus use the model in the form (6.6), without loss of generality. We shall instead use the model in the form (6.7) when we wish to highlight the role of the offset term.

Notice that when the feature map is the identity, that is, $\phi(x) = x$, then we obtain the standard linear regression model, which is linear in both w and x:

$$\hat{y} = w^\top x.$$

Other choices of basis functions are possible. For instance, assuming for simplicity that $n = 1$, that is, x is scalar, one may take

$$\phi_j(x) = x^{j-1}, \quad j = 1, \ldots, d,$$

which yields a polynomial model of degree $d-1$:

$$\hat{y} = w_1 + w_2 x + w_3 x^2 + \cdots + w_d x^{d-1}.$$

Sigmoidal models are instead obtained via

$$\phi_j(x) = \sigma\left(\frac{x - \mu_j}{s}\right),$$

where σ is the logistic sigmoidal function $\sigma(z) = (1 - e^{-z})^{-1}$, and Gaussian basis function models are obtained via the choice of

$$\phi_j(x) = \exp\left(-\frac{(x - \mu_j)^2}{2s^2}\right),$$

where the μ_js govern the location of the basis functions, and s governs their spatial scale. Multi-dimensional generalizations of the basis functions exemplified above exist and are commonly used in practice. An alternative to selecting a specific family of basis functions is to elicit instead a *kernel*, which refers implicitly to basis functions, without ever needing to make them explicit.[1]

[1] For an account of kernel methods see, for example, J. Shawe-Taylor and N. Cristianini, *Kernel Methods for Pattern Analysis*, Cambridge: Cambridge University Press, 2004.

6.1.1 Bias and variance error

Consider a data generation model of the form (6.1), and assume we want to find a function $\hat{f}(x, \mathcal{D})$ of the form (6.3) that approximates $f(x)$, where $\hat{f}$ is estimated on the basis of the given data $\mathcal{D}$. Here, we are interested in considering the expected squared error between $y = f(x) + \epsilon$ and $\hat{f}(x, \mathcal{D})$ on a new, unseen, input sample x, that is,

$$e^2 \doteq \mathrm{E}\{(y - \hat{f}(x, \mathcal{D}))^2\},$$

where the expectation runs over the possible realizations of data sets $\mathcal{D}$ and noise term ϵ, which is here assumed to have zero mean and to be independent of $\mathcal{D}$. Simplifying the notation with $f = f(x)$, $\hat{f} = \hat{f}(x, \mathcal{D})$, we have that

$$\begin{aligned} e^2 &\doteq \mathrm{E}\{(y - \hat{f})^2\} = \mathrm{E}\{(f + \epsilon - \hat{f})^2\} \\ &= \mathrm{E}\{((f - \mathrm{E}\{\hat{f}\}) - (\hat{f} - \mathrm{E}\{\hat{f}\}))^2 + \epsilon\} \\ &= \mathrm{E}\{(f - \mathrm{E}\{\hat{f}\})^2\} + \mathrm{E}\{(\hat{f} - \mathrm{E}\{\hat{f}\})^2\} + \mathrm{E}\{\epsilon^2\} \\ &= (f - \mathrm{E}\{\hat{f}\})^2 + \mathrm{var}\{\hat{f}\} + \sigma^2, \end{aligned}$$

where we have used the fact that ϵ is independent of both f and $\hat{f}$, and that f is deterministic, whence $\mathrm{E}\{f\} = f$. In the expression above, $(f - \mathrm{E}\{\hat{f}\})^2$ denotes the squared *bias* error, describing how much the expected $\hat{f}$ differs from the true f, $\mathrm{var}\{\hat{f}\}$ is the *variance* error, describing how much $\hat{f}$ fluctuates around its mean on different realization of the data, and σ^2 is the irreducible error, due to additive noise on f. Summarizing, we have

$$e^2 \doteq \mathrm{E}\{(y - \hat{f})^2\} = \mathrm{bias}^2(\hat{f}) + \mathrm{var}\{\hat{f}\} + \mathrm{var}\{\epsilon\}. \tag{6.8}$$

We see that, in order to make the estimation error e^2 small, an estimation procedure should aim at making the sum of bias and variance small. *Unbiased* estimators, in particular, produce estimates that (under ideal conditions) have zero bias, so the the only error to control is the variance error. However, in practice, one may not know whether the regression model class is rich enough to provide zero bias error, hence an unbiased estimator may not be the best choice, since the smallest variance error achieved under zero bias may still be larger than the total error (bias plus variance) obtained via a *biased* estimator.

6.2 Training regression models

Training a regression model amounts to determining the appropriate value of the model parameters $w \in \mathbb{R}^d$ and $b \in \mathbb{R}$ on the basis of a set of observed input–output data $\mathcal{D} = \{(x^{(1)}, y_1), \ldots, (x^{(m)}, y_m)\}$. We shall assume that the *order d* of the model and the feature map ϕ have been selected and fixed in advance, so that w is the only unknown parameter in the model. We use an output prediction rule of the form

$$\hat{y} = w^\top \phi(x) + b, \tag{6.9}$$

where $w \in \mathbb{R}^d$ is the regression coefficient vector and $b \in \mathbb{R}$ is the offset term (which, as we already discussed, may also be incorporated in w, by setting one of the basis function to a constant). For each observation i we have a corresponding prediction error $e_i = y_i - \hat{y}_i$, $i = 1, \ldots, m$, which, using the expression in (6.9), is a function of w and b: $e_i = y_i - (w^\top \phi(x^{(i)}) + b)$. The key idea is to build a cost function $\mathcal{J}$ to be minimized over w, b, which is composed of a *loss* component $\mathcal{L}$ which accounts for the total in-sample error over the observations, plus a penalty component whose purpose is to improve the out-of-sample performance of the model. The loss component usually takes the additive form

$$\mathcal{L}(w, b) = \frac{1}{m} \sum_{i=1}^{m} \mathcal{L}_i(y_i - (w^\top \phi(x^{(i)}) + b)), \tag{6.10}$$

and the cost function takes the form

$$\mathcal{J}(w, b) = \mathcal{L}(w, b) + \lambda r(w), \tag{6.11}$$

where $r(w)$ is a regularization function, and $\lambda \geq 0$ is a penalty parameter, whose value is to be set via cross validation, as discussed in Section 6.4. The basic model training step thus amounts to solving an optimization problem of the form

$$\min_{w,b} \mathcal{L}(w,b) + \lambda r(w). \tag{6.12}$$

Many different choices are possible for the loss function and for the regularization function. By far, the most common choice for the loss function is the sum of squared prediction errors or, equivalently, the mean squared error (MSE) criterion, that is,

$$\mathcal{L}(w,b) = \mathrm{MSE}(w,b) \doteq \frac{1}{m}\sum_{i=1}^{m} e_i^2 = \frac{1}{m}\sum_{i=1}^{m}(y_i - (w^\top \phi(x^{(i)}) + b))^2$$

$$= \frac{1}{m}\|y - \Phi^\top w - b\mathbf{1}\|_2^2, \tag{6.13}$$

where

$$\Phi \doteq \begin{bmatrix} \phi(x^{(1)}) & \cdots & \phi(x^{(m)}) \end{bmatrix}, \quad y \doteq \begin{bmatrix} y_1 & \cdots & y_m \end{bmatrix}^\top. \tag{6.14}$$

Minimization of the MSE loss (6.13) with respect to w, b corresponds to the well-known least-squares (LS) problem.[2] Another frequently used alternative loss criterion is the mean of absolute errors criterion (MAE):

[2] See, for example, Chapter 6 of G. Calafiore and L. El Ghaoui, *Optimization Models*, Cambridge: Cambridge University Press, 2014.

$$\mathcal{L}(w,b) = \mathrm{MAE}(w,b) \doteq \frac{1}{m}\sum_{i=1}^{m}|e_i| = \frac{1}{m}\sum_{i=1}^{m}|y_i - (w^\top \phi(x^{(i)}) + b)|$$

$$= \frac{1}{m}\|y - \Phi^\top w - b\mathbf{1}\|_1. \tag{6.15}$$

Also, the so-called Huber loss is sometimes used, which blends together the squared errors and the absolute errors criteria, by using squared terms in case of small residuals and absolute terms in case of large residuals, that is,

$$\mathcal{L}(w,b) = \mathrm{HUB}(w,b) \doteq \frac{1}{m}\sum_{i=1}^{m} H_\delta(y_i - w^\top \phi(x^{(i)}) - b), \tag{6.16}$$

where, for given $\delta \geq 0$,

$$H_\delta(z) = \begin{cases} \frac{1}{2}z^2 & \text{if } |z| \leq \delta \\ \delta|z| - \frac{1}{2}\delta^2 & \text{if } |z| > \delta, \end{cases}$$

is the *Huber function*. The Huber function is smooth and convex, and penalizes small residuals quadratically, and large residuals linearly, see Figure 6.1.

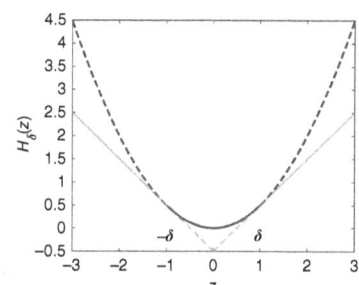

Figure 6.1 The Huber function.

Several choices are possible also for the penalty function, the most common being $r(w) = \|w\|_2^2$ and $r(w) = \|w\|_1$. Different choices of the loss and penalty function lead to models with different practical characteristics, as discussed in the following sections. Different models can also be viewed as the consequence of different noise distributions and different prior assumptions in a Bayesian learning framework; some methodological details in this direction are given in the Appendix B.1 for the interested readers. We next examine in some detail the specific types of optimization problems than arise from the most common combinations of loss and penalty functions.

6.3 Ridge regression, Lasso, Elastic Net, etc.

6.3.1 Ridge regression

Ridge regression refers to a training problem that arises from the combination of the MSE loss function and the squared ℓ_2 norm penalty, that is,

$$\min_{w,b} \mathcal{J}(w,b), \tag{6.17}$$

with

$$\mathcal{J}(w,b) \doteq \|y - \Phi^\top w - b\mathbf{1}\|_2^2 + \lambda \|w\|_2^2, \tag{6.18}$$

with $\Phi \in \mathbb{R}^{d,m}$, $y \in \mathbb{R}^m$. For $\lambda = 0$ we recover the standard Least-Squares regression problem, while values of $\lambda > 0$ progressively give more weight to the penalty term which promotes *shrinkage* in w, that is, preference towards w values with small Euclidean norm. Observe that in (6.18), for simplifying the notation, we removed the $1/m$ factor in front of the MSE loss function and we renamed $\lambda \leftarrow m\lambda$: this can be done without loss of generality, since the objective of a minimization problem can always be divided by a positive constant without changing the optimal solutions of the problem.

Problem (6.17)–(6.18) is convex in both w and b, and its solution can be expressed in closed form by equating the gradient of the cost function to zero. Expanding $\mathcal{J}(w,b)$ we have that

$$\mathcal{J}(w,b) = w^\top (\Phi\Phi^\top + \lambda I)w - 2y^\top \Phi^\top w \\ + 2bm\bar{\phi}^\top w - 2bm\bar{y} + mb^2 + \|y\|_2^2,$$

where we let

$$\bar{y} \doteq \frac{1}{m}\sum_{i=1}^m y_i, \quad \text{and} \quad \bar{\phi} \doteq \frac{1}{m}\sum_{i=1}^m \phi(x^{(i)})$$

denote the average observed output and the average input feature vector, respectively. Now, computing the gradient of $\mathcal{J}(w,b)$ with respect to b and equating it to zero we have

$$\nabla_b \mathcal{J}(w,b) = 2m\bar{\phi}^\top w - 2m\bar{y} + 2mb = 0,$$

from which we obtain that the optimal offset should satisfy

$$b = \bar{y} - \bar{\phi}^\top w. \tag{6.19}$$

Substituting this expression of the optimal b into $\mathcal{J}(w,b)$ we eliminate b and obtain an expression in terms of w only:

$$\begin{aligned} \mathcal{J}(w) &= \|\tilde{y} - \tilde{\Phi}^\top w\|_2^2 + \lambda \|w\|_2^2 \\ &= w^\top (\tilde{\Phi}\tilde{\Phi}^\top + \lambda I)w - 2\tilde{y}^\top \tilde{\Phi}^\top w + \|\tilde{y}\|_2^2, \end{aligned} \tag{6.20}$$

where

$$\tilde{y} \doteq y - \bar{y}\mathbf{1}, \quad \tilde{\Phi} \doteq \Phi - \bar{\phi}\mathbf{1}^\top. \tag{6.21}$$

Equating now to zero the gradient of $\mathcal{J}(w)$ with respect to w, we obtain

$$\nabla_w \mathcal{J}(w) = 2(\tilde{\Phi}\tilde{\Phi}^\top + \lambda I)w - 2\tilde{\Phi}\tilde{y} = 0.$$

The optimality conditions for w are thus expressed by a set of linear equations called the (regularized) *normal equations*

$$(\tilde{\Phi}\tilde{\Phi}^\top + \lambda I)w = \tilde{\Phi}\tilde{y}. \tag{6.22}$$

The matrix appearing on the left in (6.22) is invertible for any $\lambda > 0$, hence there exist a unique solution

$$w = (\tilde{\Phi}\tilde{\Phi}^\top + \lambda I)^{-1}\tilde{\Phi}\tilde{y}. \tag{6.23}$$

Expression (6.23) provides the optimal w parameter of the model, for any $\lambda > 0$ and, substituting this w into (6.19) we also obtain the optimal offset b.

6.3.2 Lasso regression

Lasso regression[3] refers to a training problem that arises from the combination of the MSE loss function and the ℓ_1 norm penalty, that is,

$$\min_{w,b} \mathcal{J}(w,b) \doteq \|y - \Phi^\top w - b\mathbf{1}\|_2^2 + \lambda \|w\|_1. \tag{6.24}$$

Solution of the Lasso problem requires solving a convex quadratic problem, which can be done via standard convex solvers such as CVX, for small-sized instances. Ad-hoc algorithms have been instead developed for solving the Lasso with high efficiency on large-size problem instances.[4] A key and well-known property of (6.24) is that

[3] Lasso stands for "least absolute shrinkage and selection operator." The original reference for the Lasso is R. Tibshirani, "Regression Shrinkage and Selection via the Lasso," *Journal of the Royal Statistical Society*, series B, vol. 58(1), pp. 267–288, 1996.

[4] See, for instance, the LARS method developed in B. Efron, T. Hastie, I. Johnstone and R. Tibshirani, "Least Angle Regression," *The Annals of Statistics*, vol. 32(2), pp. 407–499, 2004, and the coordinate-descent based methods discussed in J. Friedman, T. Hastie and R. Tibshirani, "Regularization Paths for Generalized Linear Models via Coordinate Descent," *Journal of Statistical Software*, vol. 33(1), pp. 1–22, 2010.

the regularization term $\lambda\|w\|_1$ biases the optimal solution towards zero (shrinkage) by promoting *sparsity* in the solution. The level of sparsity increases as λ increases, and for values of λ larger than a certain threshold $\bar\lambda$ the optimal w becomes all zeros.

The optimal offset b can be obtained by following the same approach as in the Ridge regression case, and it is given by

$$b = \bar y - \bar\phi w.$$

For this optimal b the Lasso objective becomes a function of w only:

$$\mathcal{J}(w) = \|\tilde y - \tilde\Phi^\top w\|_2^2 + \lambda\|w\|_1, \qquad (6.25)$$

with the centered data $\tilde y$ and $\tilde\Phi$ defined in (6.21). Contrary to the Ridge regression case, a minimizer of the cost function in (6.25) cannot be found in closed form, in general. However, $\mathcal{J}(w)$ can be efficiently minimized with respect to w numerically, by solving a convex quadratic problem of the form

$$\begin{aligned}\min_{w,v}\quad & \|\tilde y - \tilde\Phi^\top w\|_2^2 + \lambda\sum_{i=1}^d v_i,\\ \text{s.t.:}\quad & -v_i \le w_i \le v_i, \quad i=1,\ldots,d.\end{aligned}$$

In particular, it can be proved that that the optimal solution to this problem is $w \equiv 0$ if $\|\tilde\Phi\tilde y\|_\infty \le 2\lambda$, whence for all λ larger than the threshold $\bar\lambda = \|\tilde\Phi\tilde y\|_\infty/2$ the Lasso solution w is identically zero, while for $\lambda = 0$ it coincides with the plain least-squares solution. Intermediate values of $\lambda \in [0,\bar\lambda]$ yield optimal solutions with an increasing number of zeros as λ increases.

Remark 6.1 (*Offset term, centering, and scaling*) We observe that the Lasso problem (6.25) with *centered* data (6.25) is equivalent to the Lasso problem (6.24) with *uncentered* data, but in which an offset variable b is present. The same observation holds for the Ridge regression problem (6.20) with centered data and with uncentered data (6.18) with offset. In both problems we see that we can avoid pre-centering the data, and we can let the offset term take care of the centering instead. Similarly, we observe that a Ridge regression problem with centered and row-wise normalized data

$$\min_w \|\tilde y - \tilde\Phi^\top L^{-1} w\|_2^2 + \lambda\|w\|_2^2,$$

where L is a diagonal matrix of scalings, is equivalent to a Ridge regression problem with uncentered and unnormalized data but with an offset term and a penalty weight

$$\min_z \|y - \Phi^\top z - b\mathbf{1}\|_2^2 + \lambda\|Lz\|_2^2,$$

where $z = L^{-1}w$. The same holds for the Lasso problem. For both problems, therefore, data centering can be taken into account by means of the introduction of the offset term and data normalization (row-wise) can be taken into account via introduction of a weight matrix in the penalty term.

6.3.3 Elastic Net regression

The Elastic Net model[5] blends together the features of the Lasso and of the Ridge regression model, by proposing a training cost function in which $\mathcal{L}$ is the standard MSE loss, and the regularization function is of the form

$$r(w) = \alpha \|w\|_1 + (1-\alpha)\|w\|_2^2,$$

where $\alpha \in [0,1]$ is a parameter that blends the ℓ_1 penalty with the ℓ_2 penalty; for $\alpha = 1$ we have the Lasso, for $\alpha = 0$ we have the Ridge regression, and for the values of $\alpha \in (0,1)$ we have the actual Elastic Net criterion. One problem with the Lasso is its potential instability, meaning that the solution may "jump" as a consequence of small perturbations in the data, especially if groups of features are correlated. The Elastic Net addresses this problem by adding stability via the ℓ_2 term, which also provides an additional effect of grouping the correlated features together, so that groups of correlated features tend to enter or exit the model together as the level of sparsity is decreased or increased, respectively. Computationally, training an Elastic Net model amounts to solving a convex optimization problem of the form

$$\min_{w,b} \ \|y - \Phi^\top w - b\mathbf{1}\|_2^2 + \lambda(\alpha\|w\|_1 + (1-\alpha)\|w\|_2^2), \quad (6.26)$$

which may be easily restated in the explicit form of a convex quadratic problem. Numerical methods for solving the Lasso can be easily adapted for the solution of the Elastic Net problem (6.26); among these a simple and effective method is based on cyclic coordinate minimization.

[5] See H. Zou and T. Hastie, "Regularization and Variable Selection via the Elastic Net," *Journal of the Royal Statistical Society*, Series B. vol. 67(2), pp. 301–320, 2005.

6.3.4 ℓ_1 regression

One main problem with the sum or mean of squared errors loss is that it is extremely sensitive to *outliers* in the data, that is, to the presence in the data set $\mathcal{D}$ of observations that deviate in practice from the assumed theoretical noise model, which is implicitly Gaussian, see Appendix B.1.1. Intuitively, the presence of the square in the error terms is such that large residuals give substantially greater weight in the criterion, hence the presence of even one outlier in the data can substantially "skew" the estimate. It has been observed in the literature[6] that the use of a sum or mean of absolute errors criterion,

$$\mathrm{MAE}(w,b) = \frac{1}{m}\sum_{i=1}^{m} |y_i - w^\top \phi(x^{(i)}) - b|, \quad (6.27)$$

is much more resilient, or *robust*, to outliers in the data, thus making the trained model largely insensitive the the presence of bad data. In ℓ_1 regression (also named *robust* regression in the statistical literature) we thus solve a problem of the form

$$\min_{w,b} \|y_i - w^\top \phi(x^{(i)}) - b\|_1 + \lambda r(w),$$

which is readily cast as a linear program if $r(w) = \|w\|_1$, or as a convex quadratic program if $r(w) = \|w\|_2^2$.

[6] See, for example, P. Rousseeuw and A. M. Leroy, *Robust Regression and Outlier Detection*, Hoboken, NJ: Wiley, 1987.

6.3.5 Huber regression

The idea behind the MAE criterion was to avoid quadratic growth for large residuals. However, for small residuals the quadratic criterion has the advantage of being smooth. It make therefore sense to consider a "mixed" criterion that keeps the favorable aspects of both the quadratic error and the absolute error criteria. This is possible by considering a loss criterion of the form of the Huber function shown in Figure 6.1. In Huber regression we use the Huber function (6.16) as the loss function, possibly in conjunction with a regularizer. It can be proved that the Huber loss can be written in the equivalent form

$$H_\delta(z) = \inf_s \, \delta|s| + \frac{1}{2}(s - z)^2.$$

This expression follows (after nontrivial derivations) from the fact that $|x| = \sup_{v \in [-1,1]} vx$ and application of a min–max duality theorem. The Huber regression problem

$$\min_{w \in \mathbb{R}^d, b} \sum_{i=1}^{m} H_\delta(y_i - w^\top \phi(x^{(i)}) - b),$$

can then be written in the equivalent form

$$\min_{w \in \mathbb{R}^d, b, s_1, \ldots, s_m, t_1, \ldots, t_m} \sum_{i=1}^{m} \delta t_i + \frac{1}{2}(s_i - y_i + w^\top \phi(x^{(i)}) + b)^2,$$
$$\text{s.t.:} \quad -t_i \leq s_i \leq t_i, \; i = 1, \ldots, m,$$

which is a convex quadratic program that can be solved efficiently via suitable numerical methods.

6.3.6 Chebyshev regression

In Chebyshev regression the loss function only depends on the largest residuals, rather than on their sum or average. In Chebyshev regression one therefore minimizes the combined cost

$$\|y - \Phi^\top w - b\mathbf{1}\|_\infty + \lambda r(w),$$

where $\|y - \Phi^\top w - b\mathbf{1}\|_\infty = \max_{i=1,\ldots,m} |y_i - w^\top \phi(x^{(i)}) - b|$. The minimization of such cost can be cast as a linear program if $r(w) = \|w\|_1$, or as a convex quadratic program if $r(w)$ has the Ridge or Elastic Net format.

6.3.7 Quantile regression

For $p \in [0, 1]$, the pth quantile $Q(p)$ of a cumulative distribution function (cdf) $F(z)$ is defined as the minimum of the set of values z such that $F(z) \geq p$. For a data sample $\{z_1, \ldots, z_m\}$, we define the *empirical cdf* at z (or, the *sample distribution function*) as the proportion of sample values that are less than or equal to z, that is,

$$\hat{F}(z) = \frac{\text{number of } z_i \text{ such that } z_i \leq z}{m}.$$

The pth *sample quantile* $\hat{Q}(p)$ is then defined as the pth quantile of the empirical cdf of the data sample. If we denote the ordered data as $z_{[1]} \leq z_{[2]} \leq \cdots \leq z_{[m]}$, then for $k \in \{1, \ldots, m\}$ the $p = k/m$ sample quantile simply corresponds to $z_{[k]}$.

It is an interesting fact that the pth sample quantile $\hat{Q}(p)$ can be obtained as the minimizer of a particular asymmetric dispersion function. More precisely, define

$$h_p(u) \doteq \begin{cases} pu & \text{if } u > 0, \\ (p-1)u & \text{if } u \leq 0, \end{cases}$$

that is, $h_p(u) = \max(pu, (p-1)u)$ (see Figure 6.2), and consider a loss function of the form

$$\varphi_p(c) = \frac{1}{m} \sum_{i=1}^{m} h_p(z_i - c). \tag{6.28}$$

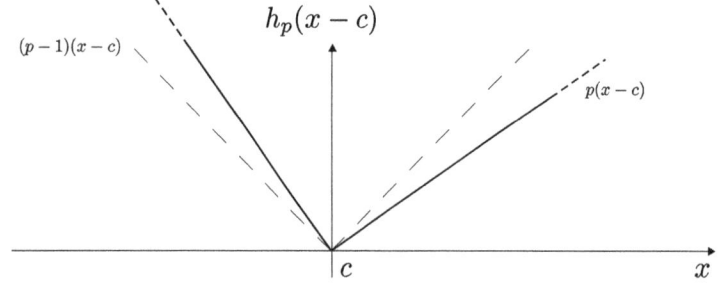

Figure 6.2 h_p dispersion function, $p \in [0, 1]$.

Then, it can be proved that the pth sample quantile is a minimizer of $\varphi_p(c)$, that is,

$$\hat{Q}(p) \in \arg\min_c \varphi_p(c).$$

Further, if $p = k/m$, then $\hat{Q}(p) = z_{[k]}$, that is, the *k*th smallest of the $\{z_1, \ldots, z_m\}$ is a minimizer of $\varphi_p(c)$.

Notice that for $p = 0.5$, we have $h_p(u) = 0.5|u|$, which is symmetric around zero. In this case, the minimizer of $\varphi_{0.5}(c)$ corresponds to the sample *median* of the data. For $p > 0.5$ the function $h_p(u)$ penalizes more the values u that are larger than 0, while for $p < 0.5$ it penalizes more the values u that are smaller than 0.

Quantile regression[7] deals with asymmetric error losses, namely we solve a problem of the form

$$\min_{w,b} \frac{1}{m} \sum_{i=1}^{m} h_p(y_i - w^\top \phi(x^{(i)}) - b) + \lambda r(w), \quad (6.29)$$

where, depending on p, the function h_p may penalize asymmetrically the positive and the negative residuals. At the optimum, the value of b will coincide with the p-quantile of the residuals $\{w^\top \phi(x^{(i)}) - y_i\}_{i=1,\ldots,m}$, and $m^{-1} \sum_{i=1}^{m} h_p(w^\top \phi(x^{(i)}) - b - y_i)$ will measure the corresponding average asymmetric dispersion of the residuals around the p-quantile. For $p = 0.5$, quantile regression simply becomes equivalent to a regularized version of the the mean absolute error regression discussed in Section 6.3.4.

With typical regularization functions $r(w)$, the quantile regression problem (6.29) can be cast and solved by means of linear programming (when $r(w) = \|w\|_1$), or by convex quadratic programming (when $r(w) = \|w\|_2^2$, or it has the Elastic Net structure). Quantile regression learning algorithms are included in most numerical machine learning packages, such as R or `scikit-learn`.

[7] See R. Koenker, *Quantile Regression*, Cambridge: Cambridge University Press, 2005, for a methodological introduction to quantile regression, and R. N. Rodriguez and Y. Yao, "Five Things You Should Know about Quantile Regression," *SAS Global Forum*, 2017, for an expository account and practical examples.

6.4 Cross validation and selection of the penalty parameter

A key aspect to consider when constructing a regression model is that the model should be good at predicting unseen data, rather than only good at explaining (fitting) the observed data. When we train a regression model on the basis of an observed data set $\mathcal{D} = \{(x^{(1)}, y_1), \ldots, (x^{(m)}, y_m)\}$, we obtain an estimated parameter $\hat{w}$ and offset $\hat{b}$ which depend on $\mathcal{D}$, and construct the model as $\hat{y} = \hat{w}^\top \phi(x) + \hat{b}$. For this model we have an *in-sample loss* or error

$$\mathcal{L}_{\text{is}} \doteq \mathcal{L}(\Phi^\top \hat{w} + \hat{b}\mathbf{1}, y),$$

where Φ, y contain the observed data, as defined in Appendix B.1.1. However, what we are really interested in is the error the model will provide when predicting the output for a new unseen input, that is, in the out-of-sample performance of the model. If we let x_{new} be a new unseen input and y_{new} the corresponding output, the model prediction would be $\hat{y}_{\text{new}} = \hat{w}^\top \phi(x_{\text{new}}) + \hat{b}$ and the *out-of-sample loss* would be

$$\mathcal{L}_{\text{oos}} \doteq \mathcal{L}(\hat{w}^\top \phi(x_{\text{new}}) + \hat{b}, y_{\text{new}}).$$

There is clearly a fundamental difficulty in minimizing $\mathcal{L}_{\text{oos}}$, since $x_{\text{new}}, y_{\text{new}}$ are typically not available at the time of model training. What we can do, however, is to build computational schemes that permit to estimate the expected value of the out-of-sample error. Such schemes essentially require splitting the available observed data set into a *training set* and a *validation set*. The training set is used to actually train the model and to find the parameter estimates $\hat{w}, \hat{b}$. The validation set is instead used as an artificial out-of-sample data set on which we can evaluate the out-of-sample performance of the trained model. The important point is that the validation set and the training set must not have common data points; in this way we guarantee that the estimate $\hat{w}, \hat{b}$ constructed in the training phase does not contain information that should be "unseen" at the time of training. This basic approach is known as the *hold-out method*, and it is described in further detail in the next paragraph.

The hold-out method Given the available data $\mathcal{D} = \{(x^{(i)}, y_i)\}_{i=1,\dots,m}$ we select $m' < m$ and let $\mathcal{D}_t, \mathcal{D}_v$ denote a disjoint partition of $\mathcal{D}$, where $\mathcal{D}_t$ contains m' data points from $\mathcal{D}$, and $\mathcal{D}_v$ contains the remaining $m - m'$ points. A typical choice is to take $m' \simeq 0.7m$. Then, we train the regression model on the basis of $\mathcal{D}_t$, obtaining the estimated parameter $\hat{w}$. With this parameter, we estimate the out-of-sample error as the average of the loss function values on the data in $\mathcal{D}_v$, that is,

$$\hat{\mathcal{L}}_{\text{oos}} = \frac{1}{m-m'} \sum_{i=1}^{m-m'} \mathcal{L}_i(\hat{w}^\top \phi(x[i]) + \hat{b}, y[i]),$$

where we denoted by $x[i]$ and $y[i]$, respectively, the ith input point and output value in the validation set $\mathcal{D}_v$. A schematic representation of the hold-out method is given in Figure 6.3.

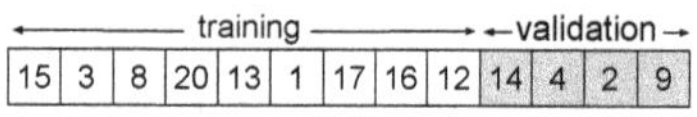

The k-fold cross-validation method The hold-out method works by partitioning the data set into one training set and one validation set. However, nothing prevents us to repeat the same process many times, using each time a different split of the original data, and in the end average over the obtained estimates of the out-of-sample error. This is indeed the idea behind k-fold cross-validation: the original data set is partitioned (randomly or according to a deterministic scheme) into

Figure 6.3 Scheme of the hold-out validation method: example with $m = 13$, $m' = 9$.

k equal-sized subsets. We then iterate for $i = 1, \ldots, k$ the following procedure: the ith subset is hold out as a validation subset, and the remaining $k - 1$ subsets are used as training data for the model. At each iterate i we obtain an estimated parameter $\hat{w}^{(i)}$, which we test on the ith validation subset, obtaining an estimate $\hat{\mathcal{L}}_{\text{oos}}^{(i)}$ of the out-of-sample error. At the end of the iterations, we compute the overall estimate as the average of all $\hat{\mathcal{L}}_{\text{oos}}^{(i)}$, that is,

$$\hat{\mathcal{L}}_{\text{oos}} = \frac{1}{k} \sum_{i=1}^{k} \hat{\mathcal{L}}_{\text{oos}}^{(i)}.$$

In this way, all observations are used for both training and validation, and each observation is used for validation exactly once. A schematic representation of the k-fold cross-validation method is given in Figure 6.4.

Also k-fold cross-validation can be implemented in a repeated randomized way, in which the data is randomly split into k partitions several times. In this way, the performance of the model can be averaged over several runs, or empirical distributions of the error can be obtained.

Figure 6.4 Scheme of the k-fold cross-validation method: example with $m = 12$, $k = 4$.

The random subsampling method In this method, also known as Monte Carlo cross-validation, we create multiple random splits of the dataset into training and validation data. For each such split, the model is fit to the training data, and its predictive accuracy is assessed using the validation data. The validation errors are then averaged over the splits, or their empirical distributions can be estimated. One advantage of this approach is that we can set the cardinality of the validation set and the desired number of splits independently. A drawback is that the assessments resulting from this method may vary due to their randomness. A schematic representation of the random subsampling validation method is given in Figure 6.5.

Figure 6.5 Scheme of the random subsampling validation method.

Selection of the penalty parameter The choice of the regularization parameter $\lambda \geq 0$ in (6.10) is typically done by gridding N values $\lambda_1 \geq \cdots \geq \lambda_N$ of $\lambda \in [\lambda_{\min}, \lambda_{\max}]$ and solving problem (6.10) repeatedly for each of these values, thus obtaining a *path of solutions* $\hat{w}(\lambda_k), \hat{b}(\lambda_k), k = 1, \ldots, N$. For numerical efficiency, the problems are typically solved going backwards from the largest λ_1, to the smaller values, each time initializing the solution algorithm with the previous solution (warm start). Each solution $\hat{w}(\lambda_k), \hat{b}(\lambda_k), k = 1, \ldots, N$, is evaluated in terms of its out-of-sample performance via cross validation, possibly also obtaining intervals of confidence for each estimated performance value. Then, a curve is obtained that shows the estimated out-of-sample performance versus λ, and the best λ value is selected, either simply as the value at which the average out-of-sample performance is minimum or, more safely, as the value at which the average out-of-sample performance is one standard error higher than the minimum. Figure 6.6 shows schematically a typical behavior of the training error (in-sample error) and of the validation error (out-of-sample error) as a function of the inverse of the regularization parameter λ. For small λ, all the focus in the training is on minimizing the in-sample loss, and this typically results in low training error but possibly larger validation error (over-fitting). For large λ, the training objective is much defocused from the in-sample loss, and this typically results in high training error and high validation error (under-fitting). The desirable results are found in some middle range of λ, typically at a level where the validation error is near a minimum.

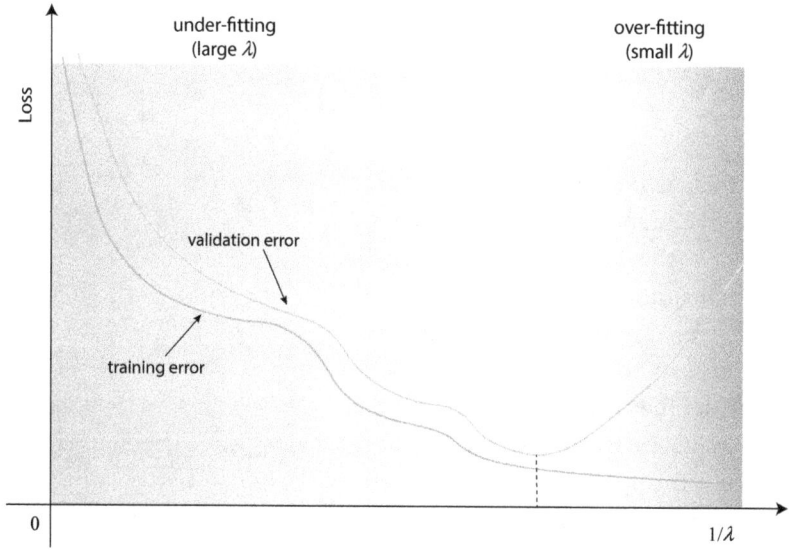

Figure 6.6 Typical behavior of the training and validation errors in function of the regularization parameter.

Example 6.1 (*US gasoline market*) We adapt an example[8] of a regression model describing the per-capita gasoline expenditure as a function of several variables, namely:

- Year: Year, in the range 1953–2004 (52 data points);
- GasExp: Total US gasoline expenditure;
- Pop: US total population in thousands;
- GasP: Price index for gasoline;
- Income: Per capita disposable income;
- Pnc: Price index for new cars;

[8] This model is taken from Chapter 2 of W. Greene, *Econometric Analysis*, 7th ed., Boston, MA: Pearson, 2012. The corresponding data table is available at https://pages.stern.nyu.edu/~wgreene/Text/Edition7/tablelist8new.htm; see also file USgasoline.csv in this book's online resources.

- Puc: Price index for used cars;
- Ppt: Price index for public transportation;
- Pd: Aggregate price index for consumer durables;
- Pn: Aggregate price index for consumer nondurables;
- Ps: Aggregate price index for consumer services.

The output y of the model is taken to be the logarithm of pro-capita gasoline expenditure, that is, $y = \log(\text{GasExp}/\text{Pop})$. The model we consider is of the form (6.9) where, for each year, x contains the $n = 8$ values of GasP, Income, Pnc, Puc, Ppt, Pd, Pn and Ps. The basis functions are taken to be equal to the variables, and to the logarithms of the variables, that is, $\phi_j(x) = \log x_j$ and $\phi_{j+8}(x) = x_j$, for $j = 1, \ldots, 8$, hence the dimension of the $\phi(x)$ vector is $d = 16$. We chose the mean square error as the loss criterion, and ℓ_1 penalty for the regularization. The training phase then results in the Lasso problem

$$\min_{w,b} \sum_{i=1}^{m} (y_i - w^\top \phi(x^{(i)}) - b)^2 + \lambda \|w\|_1,$$

where a suitable value of λ is determined via cross validation. For this example, a curve of the MSE performance as a function of λ, obtained via 10-fold cross validation, is given in Figure 6.7. Selecting the λ value for which the MSE is one standard error higher than the minimum, we obtained $\lambda = 3.06 \times 10^{-5}$ and estimated parameters

$\hat{w} = (1.14, 1.12, 0.28, -0.15, 0.24, 0.0, 0.0, 0.0,$
$\phantom{\hat{w} = (}0.0, 0.0, 0.0, 0.0, 0.0, 0.0, 0.0, 0.0),$

$\hat{b} = -24.21.$

We observe that, to two digit precision, many of the coefficients in $\hat{w}$ are numerically zero. In particular, some of the coefficients of the logarithmic terms are retained in the model, while all coefficients of the linear terms are neglected by the model. The resulting model has therefore the log-linear form

$$\begin{aligned}
y &= \log(\text{GasExp}/\text{Pop}) \\
&= -24.21 + 1.14 \log(\text{GasP}) + 1.12 \log(\text{Income}) \\
&\quad + 0.28 \log(\text{Pnc}) - 0.15 \log(\text{Puc}) + 0.24 \log(\text{Ppt}).
\end{aligned}$$

We also observe that the validation error curve in Figure 6.7, contrary to the schematic behavior of Figure 6.6, is monotonic decreasing. This suggest that the model does not overfit with the given data, even for very small λ values. In this case, detuning the model with an ℓ_1 regularization term does not give a significant out-of-sample performance improvement. However, ℓ_1 regularization, even with very small λ parameter, still yields a sparse coefficient vector, with negligible increase of the loss value.

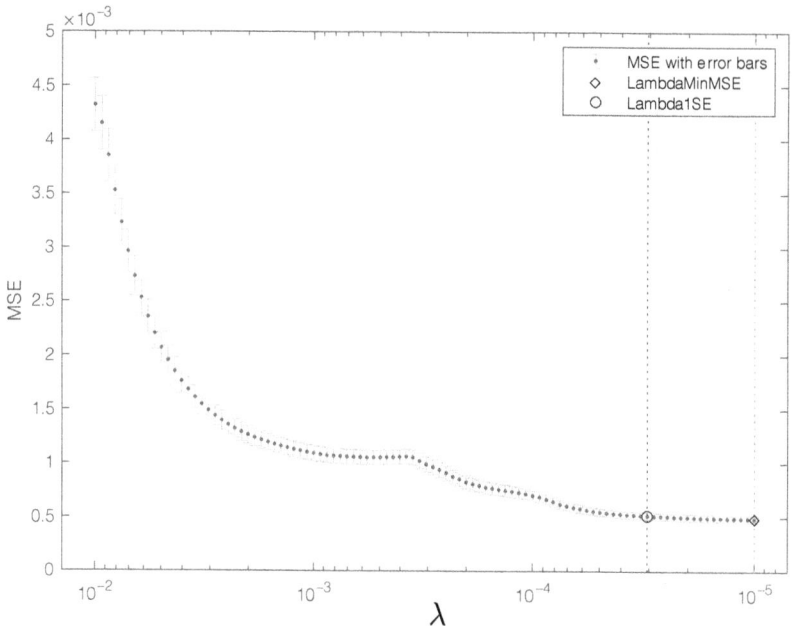

Figure 6.7 US gasoline market model. Cross-validated MSE performance as a function of λ, see file USgasoline.m in the online resources.

Example 6.2 (*Car fuel consumption*) Another classic regression dataset, known as the mtcars dataset,[9] contains the value y of the fuel consumption in miles per gallon for $m = 32$ different car models, together with $n = 10$ feature values for each car model, such as number of cylinders, horsepower, and so on; see the file mtcars.csv in the online resources. Considering a plain linear regression model of the form $\hat{y} = w^\top x + b$ for this data (i.e., $\phi(x) = x$), and choosing again a Lasso type of regularized loss function, we obtained the cross-validated MSE curve shown in Figure 6.8.

We see that in this case the model does overfit for small λ, and that a good choice for the regularization parameter is $\lambda = 1.097$. For this level of regularization the Lasso indeed selects as relevant only three features, namely the number of cylinders, the horsepower, and the weight of the car.

[9] The original source of the data is H. V. Henderson and P. F. Velleman, "Building multiple regression models interactively," *Biometrics*, vol. 37(2), pp. 391–411, 1981.

Example 6.3 (*Time-series prediction via autoregressive models*) A discrete-time signal y_t, $t = 1, \ldots,$ evolves according to an autoregressive (AR) model if the assumed mechanism that generates the data is of the form

$$y_t = \sum_{k=1}^{n} w_k y_{t-k} + b + \epsilon_t, \quad t = 1, \ldots, \qquad (6.30)$$

where ϵ_t is a zero-mean white noise process, see Section 3.9.2. In words, each time value y_t depends on n past values of the same signal, plus an offset b, plus an uncorrelated zero mean noise term ϵ_t. If we suppose such mechanism, and have available a stream of observed data

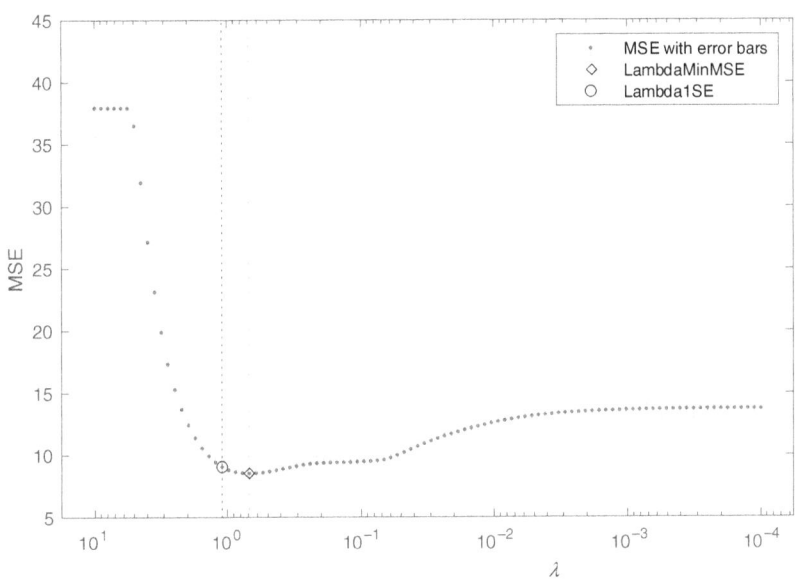

Figure 6.8 Car fuel consumption model. Cross-validated MSE performance as a function of λ; see `mtcars_ex.m` in the online resources.

$y_{1-n}, \ldots, y_0, y_1, \ldots, y_m$, $m > n$, then we can define vectors of regressors of the form

$$x^{(t)} \doteq [y_{t-1}\ y_{t-2} \cdots y_{t-n}]^\top, \quad t = 1, \ldots, m,$$

and consider a linear regression model for y_t of the form $\hat{y}_t = w^\top x^{(t)} + b$. Letting $y = (y_1, \ldots, y_m)$, and $X \doteq [x^{(1)} \cdots x^{(m)}]$, we can learn an AR model by solving a least-squares regression problem of the form

$$\min_{w,b} \|y - X^\top w - b\mathbf{1}\|_2^2.$$

As an example, we considered the log of the closing value of the Nasdaq composite index in the period from March 16, 2020 to March 12, 2021, for a total of $m = 251$ data points, see Figure 6.9.

For validation, we used a simple hold-out method, taking the first $m' = 172$ samples for training and the remaining ones for assessing the out-of-sample performance of the prediction. Taking $n = 5$ for the "memory" of the AR model, we obtained the predictor

$$\hat{y}_t = 0.7408 y_{t-1} + 0.3239 y_{t-2}$$
$$- 0.2145 y_{t-3} - 0.1778 y_{t-4} + 0.2935 y_{t-5} + 0.3184.$$

For such predictor, we have in-sample MSE of 0.0003 and out-of-sample MSE of 0.0002, which corresponds to a root mean squared (RMS) error of 0.0141. Suppose we intend to use such model for predicting the closing (log) value of the Nasdaq composite index for the next day, based on the information we have about the closing value in the current day and in the preceding four days. Will this be an effective prediction? Let's compare the percent relative prediction error $e_t \doteq (y_t - \hat{y}_t)/\hat{y}_t \times 100$ with the relative daily variation of y_t, that is, $\delta_t \doteq (y_t - y_{t-1})/y_{t-1} \times 100$, observing

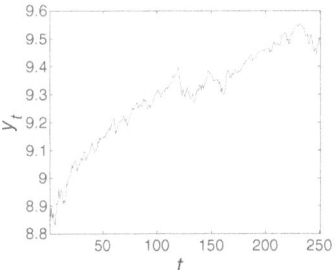

Figure 6.9 Logarithm of the closing value of the Nasdaq composite index, from March 16, 2020 to March 12, 2021, see `Nasdaq_comp_YD.csv` in the online resources.

that δ_t can also be interpreted as the relative prediction error we would commit if we used a trivial predictor which predicts that tomorrow's value y_t is simply equal to today's value y_{t-1}. On training data, we have that the average e_t is equal to 8.9×10^{-6}, with sample standard deviation 0.1853, while the average δ_t is equal to 0.0355, with sample standard deviation 0.2044, so indeed, as expected, the AR predictor works better than the naive predictor, on the training data. However, if we look at the validation data, the situation is that the average of e_t is 0.0637 and the sample standard deviation of this error is 0.1426, while the average of δ_t is 0.0148 and its standard deviation is 0.1402. Therefore, the AR predictor is actually performing worse than the trivial predictor on the validation data, which is where it counts. This fact confirms the common wisdom that predicting the stock market isn't easy.

Example 6.4 (*Long-term factors of economic growth*) According to empirical literature on economic growth there are variables that are partially correlated with a country's rate of economic growth. For instance, the initial level of income, the investment rate, various measures of education, some policy indicators, geographic position, and many other variables have been found to be significantly correlated with growth in regressions models. In this example, we use a subset of the data available in the literature[10] for constructing a Lasso regression model with the purpose of identifying the most significant features for the prediction of the average growth rate of the GDP (gross domestic product) per capita in the period 1960–1996. Our reduced data set is composed of $n = 67$ features relative to $m = 88$ countries; the regression output y is the growth rate, as described by the variable GR6096 of the data set, see the data file BACE_data_redux.csv in the online resources. We used a plain linear regression model of the form $\hat{y} = w^\top x + b$, where x is an 67-dimensional vector containing the countries' features. We computed the mean and the sample standard deviations of the features, and hence centered and standardized the data. We then run an Elastic Net fitting algorithm on the data, using 16 Monte Carlo repetitions of a 10-fold cross validation for each of the gridded values of the λ regularization parameter, with α being fixed to 0.8. Figure 6.10 shows the behavior of the Elastic Net fitting in function of the regularization parameter λ.

We chose the lambda value for which the cross-validated MSE was one standard error above the minimum, which resulted in $\lambda = 0.0033$. For this level of λ, the Elastic Net model selects 12 meaningful features.

[10] See X. Sala-i-Martin, G. Doppelhofer, and R. I. Miller, "Determinants of Long-Term Growth: A Bayesian Averaging of Classical Estimates (BACE) Approach," *American Economic Review*, vol. 94, pp. 813–835, 2004.

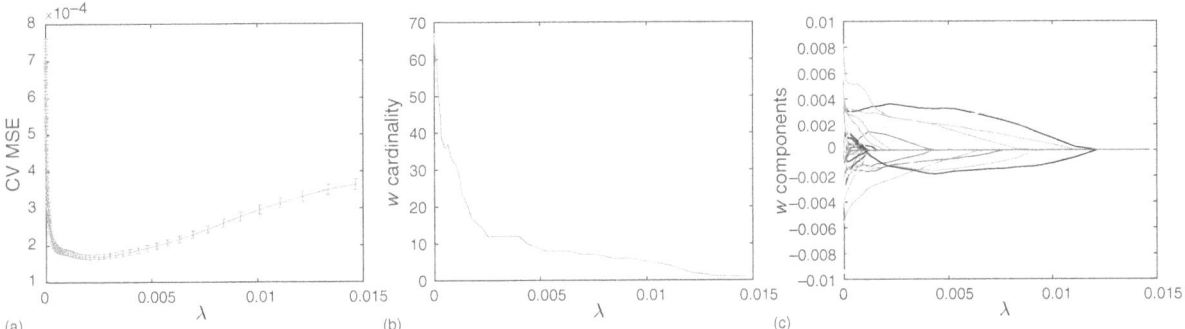

6.5 Exercises

Exercise 6.1 *Subgradient of the ℓ_1 norm*

The *subgradient* is a generalization of the gradient at points where the function is not differentiable. Let us consider the absolute value function $f\colon \mathbb{R} \to \mathbb{R}$, $f(x) = |x|$. Its graph is shown below by the solid line, in addition to the dashed lines $y = 0$, $y = \frac{1}{2}x$, $y = -\frac{1}{2}x$:

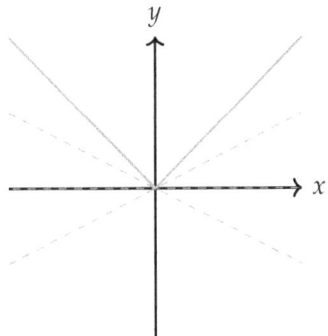

f is differentiable for all $x \neq 0$, with $\frac{df(x)}{dx} = \text{sign}(x)$. At 0 we can define the subgradient $\partial f(0) = [-1, 1]$ (i.e., all slopes with absolute value bounded by 1). Visually, we can see that the dashed lines correspond to subgradients.

1. The ℓ_1 norm $\|\cdot\|_1 \colon \mathbb{R}^n \to \mathbb{R}$ is differentiable for all x such that $x_i \neq 0$, with $0 \leq i \leq n$. Instead, if at least one component of x is equal to 0, we can define the subgradient of $\|x\|_1$ as:

$$\partial \|x\|_1 = \left\{ g \in \mathbb{R}^n \colon \forall 1 \leq i \leq n, \begin{array}{ll} g_i = \text{sgn}(x) & \text{if } x_i \neq 0 \\ g_i \in [-1, 1] & \text{if } x_i = 0 \end{array} \right\}.$$

Express the subgradient at w, $\partial \mathcal{J}(w)$, of the Lasso objective[11]

$$\mathcal{J}(w) = \|\mathbf{y} - \Phi^\top w\|_2^2 + \lambda \|w\|_1.$$

Figure 6.10 Economic growth model. (a) Shows the cross-validated MSE performance as a function of λ; (b) shows the cardinality (number of nonzero values) of the solution w; and (c) shows the value of the individual components of w as a function of λ; see code in BACE_lasso_ex.m in the online resources.

[11] In this exercise we suppose that the data is centered and normalized, and can thus neglect the offset term b.

To do so, use the additivity rule of the subgradient for two convex functions $f\colon \mathbb{R}^n \to \mathbb{R}$ and $g\colon \mathbb{R}^n \to \mathbb{R}$:

$$\partial(f+g)(x) = \partial f(x) + \partial g(x),$$

where the sum of two sets $A, B \subset \mathbb{R}^n$ is defined as $A + B = \{a + b\colon a \in A,\, b \in B\}$.

2. For a convex function f, x^* is a minimizer of f if and only if $0 \in \mathbb{R}^n$ is a subgradient; that is,

$$x^* \in \operatorname*{argmin}_{x} f(x) \iff 0 \in \partial f(x^*).$$

In light of this fact, give an informal explanation as to why the ℓ_1 norm encourages sparsity.

3. Compute the minimizer of

$$\mathcal{J}(w) = \|y - \Phi^\top w\|_2^2 + \lambda \|w\|_1$$

in the following two special cases:

(a) if $\Phi\Phi^\top = I$;

(b) if $\Phi \in \mathbb{R}^{1,m}$.

Exercise 6.2 *Least squares and regularization*

This exercise focuses on time-series modeling for a log-return data set. In the file AAPL_log_returns.mat in the online resources, you are provided with log-return time-series data for the Apple stock from 08-06-1997 to 08-24-2011 (the period when Steve Jobs returned to Apple as CEO), comprising 3,536 returns. First, form a training set by taking the first 2,500 log-returns. The remaining samples form the validation set. The training set will be used to train the model, and the validation set to test and evaluate the performance of the model.

We model the time series as an autoregressive process of order n. As discussed in Section 3.9.2, the AR(n) model is described as

$$y_t = \epsilon_t + \sum_{k=1}^{n} w_k y_{t-k}, \quad \forall t,$$

where ϵ_t is the noise term, and $w = (w_1, \ldots, w_n)$ are the model parameters that we need to fit.

1. As shown in Example 6.3, we can use ordinary least squares to fit model parameters based on a stream of observed data $y_{1-n}, \ldots, y_0, y_1, \ldots, y_m$, $m > n$. The problem is formulated as

$$\min_{w \in \mathbb{R}^n} \|y - \Phi^\top w\|_2^2,$$

where $y = (y_1, \ldots, y_m)^\top$, $\Phi \in \mathbb{R}^{n,m}$ and the tth column of Φ is $(y_{t-1}, \ldots, y_{t-n})^\top$.

Sweep n from 1 to 100 with increments of 1. In each step:

(a) Use ordinary least squares to determine the parameter vector w of the AR(n) process. Compute the average absolute training error.

(b) Use w from the previous training step to compute the average absolute prediction error on the validation set.

Then plot the average absolute training error and the average absolute validation error as a function of the model order n. Comment on the results.

2. We now use Ridge regression to determine the parameter vector of the AR process. As discussed in Section 6.3.1, in Ridge regression we add an ℓ_2 norm penalty to the objective function. The regularized problem can be written as

$$\min_{w \in \mathbb{R}^n} \|y - \Phi^\top w\|_2^2 + \lambda \|w\|_2^2,$$

where λ is a nonnegative parameter.

(a) In this model, we have two parameters to sweep, the order n and the penalty parameter λ. For $n \in \{15, 20\}$:

- sweep parameter λ from 0 to 100 (with reasonable intervals), fit w for each λ;
- plot the average absolute training error as a function of λ;
- plot the average absolute validation error as a function of λ.

Comment on the quality of predictions on the validation set as a function of λ.

(b) Let $\lambda = 30$, sweep parameter n from 1 to 100 again, and plot the average absolute training error and validation error versus n. As a comparison, also plot in the same figure the error versus n of the ordinary least squares from the previous part of the exercise. Briefly comment on the effect of regularization.

3. Replace the ℓ_2 norm penalty with an ℓ_1 norm penalty in the regularized objective as

$$\min_{w \in \mathbb{R}^n} \|y - \Phi^\top w\|_2^2 + \lambda \|w\|_1.$$

Train an AR model of order $n = 5$, sweep λ from 0 to 0.5 with increments of 0.05.

(a) For each λ, fit w by solving the above ℓ_1-regularized least-squares problem.

(b) Plot the average absolute training and validation error, as functions of λ. Comment on the quality of predictions on the validation data in function of λ.

(c) For each element in w, plot the absolute value of w_i as a function of λ (Plot all the $|w_i|-\lambda$ curves in the same figure). Comment on the result.

Exercise 6.3 *Constraints and penalties in regression problems*

In this exercise, we are interested in comparing the solutions of penalized and constrained regression problems. In fact, we are going to show that they are always equivalent, in the sense that, with appropriate penalization strength, the penalized problem can generate the same minimizer of the constrained problem, and vice versa. In this exercise, cvxpy[12] might be a good tool to use.

[12] See www.cvxpy.org/. You can install cvxpy in Python using "pip install cvxpy."

We try to predict stock price change based on market prices in the previous month. We use the five-year daily close log-return data from the Dow Jones Industrial Average data. Month-long subsequences are generated (i.e., each subsequence has 20 data points). The generated subsequences with length 20 are stacked as columns of the input data matrix X and the difference of log price between the 21st day and 20th day for the subsequences populates the entries of y (i.e., we are using the previous 20 days to predict the difference in log price of the next day). Each subsequence of X is re-scaled to have zero mean. Only the first 1,000 subsequeces are used in the following questions. The X and y matrices are found in file `DJ5y_logret.mat` in the online resources.

1. Solve the following constrained optimization problem with cvxpy using $\delta = 0.1$. Report your optimal value for w.

$$\min_{w} \|y - X^\top w\|_2 : \|w\|_2 \leq \delta.$$

2. Consider the following Ridge regression problem:

$$\min_{w} \|y - X^\top w\|_2^2 + \lambda \|w\|_2^2,$$

where $\lambda > 0$ is the regularization strength. Use grid search to find λ such that $\|w^* - w_0^*\|_2$ is small where w^* is the minimizer to the Ridge regression problem and w_0^* is the minimizer to the previous part. Plot $\|w^* - w_0^*\|_2$ versus λ. Use constraints=None for cvxpy when solving unconstrained problems.

3. Consider the following LASSO regression problem:

$$\min_{w} \|y - X^\top w\|_2^2 + \lambda \|w\|_1,$$

where λ is the regularization strength. Observe the sparsity pattern as you vary λ. Plot the number of nonzero elements in w^* versus λ. Consider an element as nonzero only when its absolute value is larger than 10^{-5}.

4. Now find an appropriate δ such that solving the following problem via cvxpy generates the same minimizer as the previous part with a value of λ that gives a minimizer with only *one* nonzero element. Report the δ you find and the norm of difference between the two minimizers.

$$\min_{w} \|y - X^\top w\|_2 : \|w\|_1 \leq \delta.$$

Hint: Evaluate $\|w^*\|_1$ of the minimizer in the previous part.

5. Consider the following ℓ_∞ penalized regression problem:

$$\min_{w} \|y - X^\top w\|_2^2 + \lambda \|w\|_\infty,$$

where λ is the regularization strength. Fix $\lambda = 0.1$ and solve the above problem to get w^*. Suppose we were given a subsequence x we have not seen before – we would predict the difference in log price tomorrow as $\hat{y} = x^\top w^*$. Provide an interpretation of what $x^\top w^*$ is.

Exercise 6.4 *SVD and regression*
Let $X = [x^{(1)}, \ldots, x^{(m)}] \in \mathbb{R}^{n,m}$ be a data matrix with the m data points $x^{(1)}, \ldots, x^{(m)}$ as its columns. Consider the linear regression problem

$$\min_{w \in \mathbb{R}^n} \|y - X^\top w\|_2^2,$$

as well as the SVD of $X^\top$, $X^\top = U\Sigma V^\top$, $U \in \mathbb{R}^{m,m}$, $V \in \mathbb{R}^{n,n}$.

1. Solve the linear regression problem by using the SVD of $X^\top$; that is, find a w^* that minimizes the objective, which you may write in terms of SVD parts. *Hint:* Note $\|U^\top x\|_2 = \|x\|_2$ since $U^\top$ is norm preserving, and perform a variable transformation $\tilde{w} = V^\top w$.

2. Consider the minimizer set $\mathcal{O} = \{w^* : \|y - X^\top w^*\|_2 = \min_w \|y - X^\top w\|_2\}$. What is $\mathcal{O}$, exactly, in terms of SVD parts? Describe also the minimizer set of the problem

$$\min_{w \in \mathcal{O}} \|w\|_2.$$

3. Similarly to question (1), solve the ridge regression problem
$$\min_{w \in \mathbb{R}^n} \|y - X^\top w\|_2^2 + \gamma \|w\|_2^2,$$
where $\gamma > 0$.

4. Let $Z \doteq X^\top V_k \in \mathbb{R}^{m,k}$, where $V_k \in \mathbb{R}^{n,k}$ consists of the first k columns of V, and $k \leq \text{rank}(X)$. Interpret Z; what do the columns represent? What do the rows represent?

5. Show that Z defined above has orthogonal columns, and that when any matrix $Z \in \mathbb{R}^{n,k}$ has orthogonal nonzero columns, the solution to
$$\min_{\tilde{w} \in \mathbb{R}^k} \|y - Z\tilde{w}\|_2^2$$
is obtained by the computationally straightforward formula $\tilde{w}_i^* = \frac{z^{(i)\top} y}{\|z^{(i)}\|^2}$, $i = 1, \ldots, k$, where $z^{(i)}$ is the ith column of Z.

The interpretation is that the coefficient for the ith feature is simply the scalar product between the ith feature and the response, divided by the squared norm of the ith feature.

Hint: You can either use part (a) to show $\tilde{w}^*$ is the solution, or set the gradient of the objective with respect to $\tilde{w}$ to 0.

6. Given a new data point $x \in \mathbb{R}^n$ and your solution $\tilde{w}^*$ from the previous part, how would you construct an output prediction $\hat{y}$ corresponding to x?

Exercise 6.5 *Housing price prediction* In the file USA_Housing.csv in the online resources[13] you are provided data about USA housing sales. This data set contains information about $m = 5{,}000$ house sales, where each house is described by $n = 6$ features: average area income, average area house age, number of rooms, number of bedrooms, area population, and address. In addition, the data set contains also the price at which each house has been sold. In this exercise, we will build a regression model that predicts the house price based on the other features.

[13] This data set is also available at www.kaggle.com/datasets/vedavyasv/usa-housing.

1. Load the data set. Center and normalize the data (do not consider non-numerical features).

2. Split the data set into a training set and a testing set. The size of the training set should be 70% of the entire data set. The input feature vector x is composed of the numerical features contained in the data set (we will not consider the Address feature because it only has text information). Consider as output response y the house price.

3. Train a linear regression model (without regularization term) using the training set obtained in the previous point. Use the MSE criterion as loss function for training.

4. Use the testing set to evaluate the model obtained in the previous point. Compute the prediction error using the root mean square error (RMSE) and the MAE.

5. Repeat points 3 and 4 using Ridge regression. Use 10-fold cross-validation on the training set to find the best value of the regularization strength λ in the interval $(0.01, 100)$.

6. Repeat the previous point using the Lasso regression problem.

7. Repeat the previous point using the Elastic Net regression problem. Use 10-fold cross-validation on the training set to find the best value of the regularization strength λ and the best value of the parameter α.

8. Compare the results obtained in the previous points.

Appendix B

In this Appendix we illustrate how some of the regression models discussed in Section 6.2 can be formally obtained by following a Bayesian approach, and how different models reflect different hypotheses on the noise affecting the observations and the prior knowledge on the parameters.

B.1 *Bayesian learning*

We assume that the DGM function f in equation (6.1) belongs to the class of functions of the form (6.3). This means that it is assumed that the model (6.3) is *rich enough* to be potentially able to capture precisely the structure of the unknown function f. The DGM is therefore of the form

$$y = w^\top \phi(x) + \epsilon, \tag{B.1}$$

where ϵ has zero mean and (unknown) variance σ^2. The vector w of parameters in the DGM is unknown, and the regression problem amounts to finding a suitable estimate $\hat{w}$ of w to be used in the regression model $\hat{y} = \hat{w}^\top \phi(x)$. Note that the assumption that ϵ has zero mean is essentially made without loss of generality, since a nonzero mean additive term in (B.1) could be readily absorbed by an offset term, as done in (6.7). For simplicity, and contrary to the derivations in Section 6.2, we here assume that no offset term b is present. This can be done without loss of generality, by assuming that the constant term is included in the basis functions.

We next assume that there exist an unknown probability distribution $p(x)$ on the inputs which, together with the independent noise term ϵ, induces some joint probability distribution $p(x,y)$ on the input–output pairs of model (B.1). This model has two parameters that need be estimated, namely w and σ^2, which we collect into a parameter vector $\theta \doteq (w, \sigma^2)$. The estimation of the parameters is next made according to a Bayesian approach on the basis of (a) prior information on θ, expressed by means of a prior $p_\theta(\theta)$, and (b) data

$\mathcal{D} = \{(x^{(1)}, y_1), \ldots, (x^{(m)}, y_m)\}$ coming from m i.i.d. realizations of the input–output pairs.

The likelihood of the data is, by definition, the conditional distribution of $\mathcal{D}$ given the parameter θ, that is,

$$p(\mathcal{D}|\theta) = = \prod_{i=1}^{m} p((x^{(i)}, y_i)|\theta) = \prod_{i=1}^{m} p(y_i|\theta, x^{(i)}) p(x^{(i)})$$
$$\propto \prod_{i=1}^{m} p(y_i|\theta, x^{(i)}),$$

where we used the independence assumption and the definition of conditional distribution. According to the Bayes' rule, the posterior distribution on θ is proportional to the likelihood times the prior, therefore

$$p(\theta|\mathcal{D}) \propto p_\theta(\theta) \cdot \prod_{i=1}^{m} p(y_i|\theta, x^{(i)}),$$

and the log-posterior is therefore

$$\log p(\theta|\mathcal{D}) \propto \sum_{i=1}^{m} \log p(y_i|\theta, x^{(i)}) + \log p_\theta(\theta). \tag{B.2}$$

A point estimate for the unknown parameter θ can now be computed according to the maximum a-posteriori principle (MAP) as the peak of the log-posterior. To proceed further, however, we must at this point select a prior, and make additional assumptions on the noise model. Indeed, observe that, for given θ and given x, $y = w^\top \phi(x) + \epsilon$ is simply a random variable with mean $w^\top \phi(x)$ and distribution given by the distribution of ϵ, shifted by the mean. Therefore $p(y_i|\theta, x^{(i)})$ coincides with the shifted distribution of the noise terms ϵ_i. Notice that in our current setting the variance of the noise terms ϵ_i remains the same σ^2 for all $i = 1, \ldots, m$, a situation known with the name of *homoscedasticity* in the literature.

B.1.1 Gaussian noise

A first standard and important case arises when we assume the noise distribution to be Gaussian, with zero mean and unknown variance σ^2, that is, $\epsilon \sim \mathcal{N}(0, \sigma^2)$. In this case, we have that

$$\log p(y_i|\theta, x^{(i)}) = \log\left[\left(\frac{1}{2\pi\sigma^2}\right)^{1/2} \exp\left(-\frac{1}{2\sigma^2}(y_i - w^\top \phi(x^{(i)}))^2\right)\right]$$
$$= -\frac{1}{2}\log(2\pi\sigma^2) - \frac{1}{2\sigma^2}(y_i - w^\top \phi(x^{(i)}))^2.$$

We next plug this expression back into (B.2) and consider several cases, depending on the choice of the prior $p_\theta(\theta)$.

Flat prior Assume first that we have no prior knowledge about θ, hence a flat prior $p_\theta(\theta) = $ constant. Then, this constant term becomes irrelevant for the purpose of the maximization of $\log p(\theta|\mathcal{D})$ in (B.2), and the MAP estimate is found by simply maximizing the log-likelihood or, equivalently, by minimizing its negative, that is, by solving $\min_{\theta=(w,\sigma^2)} \psi(w,\sigma^2)$ where

$$\begin{aligned}\psi(w,\sigma^2) &\doteq -\sum_{i=1}^m \log p(y_i|\theta, x^{(i)}) \\ &= \frac{m}{2}\log(2\pi\sigma^2) + \frac{1}{2\sigma^2}\sum_{i=1}^m (y_i - w^\top \phi(x^{(i)}))^2 \\ &= \frac{m}{2}\log(2\pi\sigma^2) + \frac{1}{2\sigma^2}\|\Phi^\top w - y\|_2^2,\end{aligned}$$

where we defined

$$\Phi \doteq \begin{bmatrix} \phi(x^{(1)}) & \cdots & \phi(x^{(m)}) \end{bmatrix}, \quad y \doteq \begin{bmatrix} y_1 & \cdots & y_m \end{bmatrix}^\top.$$

The point here is that by following a Bayesian approach we naturally end up with the specific MSE loss in (B.3). The minimization of (B.3) with respect to w is a plain least-squares (LS) problem. This problem can be solved by taking the gradient of $\psi(w,\sigma^2)$ with respect to w and setting it to zero:

$$\nabla_w \psi(w,\sigma^2) = \frac{1}{\sigma^2}\left(\Phi\Phi^\top w - \Phi y\right) = 0,$$

which amounts to solving a system of linear equations known as the *normal equations*,

$$\Phi\Phi^\top w = \Phi y.$$

These equations can be proved to *always* admit a solution, which can be written as

$$\hat{w} = (\Phi^\top)^\dagger y,$$

where $(\cdot)^\dagger$ denotes the Moore–Penrose pseudoinverse of its matrix argument. When $\Phi\Phi^\top$ is invertible, the solution is unique and can be written in terms of the standard inverse as $\hat{w} = (\Phi\Phi^\top)^{-1}\Phi y$.

Similarly, the minimization of (B.3) with respect to σ^2 can be done by taking the gradient of $\psi(\hat{w},\sigma^2)$ with respect to σ^2 and setting it to zero:

$$\nabla^2_\sigma \psi(\hat{w},\sigma^2) = \frac{m}{2\sigma^2} - \frac{1}{2\sigma^4}\text{RSS}(\hat{w}) = 0,$$

which yields

$$\hat{\sigma}^2 = \frac{\text{RSS}(\hat{w})}{m} = \text{MSE}(\hat{w}),$$

where we defined the residual sum-of-squares $\text{RSS}(w) \doteq \|\Phi^\top w - y\|_2^2$, and the mean squared error $\text{MSE}(w) \doteq \text{RSS}(w)/m$.

Assuming $\Phi\Phi^\top$ invertible, from $\hat{w} = (\Phi\Phi^\top)^{-1}\Phi y$, substituting $y = \Phi^\top w + \epsilon$, where $\epsilon \doteq [\epsilon_1 \cdots \epsilon_m]^\top$, it can be easily verified that $E\{\hat{w}\} = w$, which implies that $E\{\hat{f}\} = f$, which means that the estimator is unbiased. As expected, since the regression model captures the structure of the unknown function, we obtain zero bias error, and the only contributions to the prediction error are the estimator variance and the irreducible error. The variance error is given by

$$\begin{aligned}\mathrm{var}\{\hat{f}\} &= E\{(\hat{f} - E\{\hat{f}\})^2\} \\ &= E\{\phi^\top(x)(\hat{w} - w)(\hat{w} - w)^\top \phi(x)\} \\ &= \phi^\top(x) Q \phi(x),\end{aligned}$$

where $\hat{w} - w = (\Phi\Phi^\top)^{-1}\Phi\epsilon$, and

$$\begin{aligned}Q &\doteq E\{(\hat{w} - w)(\hat{w} - w)^\top\} = E\{(\Phi\Phi^\top)^{-1}\Phi\epsilon\epsilon^\top\Phi^\top(\Phi\Phi^\top)^{-1}\} \\ &= \sigma^2 (\Phi\Phi^\top)^{-1},\end{aligned}$$

where we used the fact that $E\{\epsilon\epsilon^\top\} = \sigma^2 I$, as a consequence of homoscedasticity. Observe that since x is assumed to be random with distribution $p(x)$, also the feature vector $\phi(x)$ is random, with second-order moment matrix $\Omega \doteq E\{\phi(x)\phi^\top(x)\}$. Considering then that the empirical mean

$$\frac{1}{m}\sum_{i=1}^{m}\phi(x^{(i)})\phi^\top(x^{(i)})$$

converges to its true mean Ω as $m \to \infty$, we have that, under the assumption that Ω is invertible,

$$\begin{aligned}Q &= \sigma^2 (\Phi\Phi^\top)^{-1} = \sigma^2 \left(\sum_{i=1}^{m}\phi(x^{(i)})\phi^\top(x^{(i)})\right)^{-1} \\ &\to \sigma^2 (m\Omega)^{-1} = \frac{\sigma^2}{m}\Omega^{-1}.\end{aligned}$$

This shows that, under the stated assumptions, the covariance matrix Q of $\hat{w}$ goes to zero as the number m of observations increases, whence the estimator variance error $\mathrm{var}\{\hat{f}\}$ also goes to zero as σ^2/m.

Gaussian prior and the Ridge regression We consider now the case in which our prior on $\theta = (w, \sigma^2)$ is that w is concentrated around zero, according to a Gaussian distribution with given covariance matrix $\lambda^2 I_d$, while the prior on σ^2 is flat. In such a case, the prior $p_\theta(\theta)$ is proportional to $(2\pi)^{-d/2}\lambda^{-d}\exp(-w^\top w/2\lambda^2)$ and the minimization of the negative of the log posterior in (B.2) is written as $\min_{\theta=(w,\sigma^2)} \psi_2(w,\sigma^2)$, where

$$\psi_2(w,\sigma^2) \doteq \frac{m}{2}\log(2\pi\sigma^2) + \frac{1}{2\sigma^2}\|\Phi^\top w - y\|_2^2 + \frac{1}{2\lambda^2}\|w\|_2^2.$$

While the estimation of σ^2 proceeds as in the previous case, the minimization with respect to w now amounts to solving an ℓ_2-regularized least-squares problem of the form

$$\min_w \|\Phi^\top w - y\|_2^2 + \gamma\|w\|_2^2, \tag{B.3}$$

where $\gamma \doteq \sigma^2/\lambda^2 \geq 0$ is a tradeoff parameter. The problem in the form (B.3) has the structure of the Ridge regression problem discussed in Section 6.3.1. For any $\gamma > 0$ we now have that the regularized matrix $(\Phi\Phi^\top + \gamma I)$ is always invertible, whence the Ridge estimate of w is unique and given by

$$\hat{w} = (\Phi\Phi^\top + \gamma I)^{-1}\Phi y. \tag{B.4}$$

We see now that by substituting $y = \Phi^\top w + \epsilon$ in the above equation, and taking the expectation, we have $E\{\hat{w}\} \neq w$, hence the Ridge estimator is biased, in general.

Laplace prior and the Lasso Suppose next that the prior on $\theta = (w, \sigma^2)$ is that w is concentrated around zero, according to a Laplace distribution with given scale parameter $\lambda > 0$. In such case, the prior $p_\theta(\theta)$ is proportional to $(2\lambda)^{-d}\exp(-\|w\|_1/\lambda)$ and the minimization of the negative of the log posterior in (B.2) is written as $\min_{\theta=(w,\sigma^2)} \psi_1(w,\sigma^2)$, where

$$\psi_1(w,\sigma^2) \doteq$$
$$\frac{m}{2}\log(2\pi\sigma^2) + \frac{1}{2\sigma^2}\|\Phi^\top w - y\|_2^2 + \frac{1}{\lambda}\|w\|_1.$$

The minimization with respect to w now amounts to solving an ℓ_1-regularized least-squares problem of the form

$$\min_w \|\Phi^\top w - y\|_2^2 + \gamma\|w\|_1, \tag{B.5}$$

where $\gamma \doteq 2\sigma^2/\lambda \geq 0$ is a tradeoff parameter. Problem (B.2) has the structure of the Lasso problem discussed in Section 6.3.2.

B.1.2 *Laplacian noise: robust linear regression*

In Appendix B.1.1 we considered the linear regression model (B.1) under the hypothesis that ϵ was a zero-mean Gaussian noise. This standing assumption produced as a consequence estimation criteria for the parameter w that were based on minimization of the sum of squared errors $\|\Phi^\top w - y\|_2^2$, in the case of flat prior, or on the minimization of the sum of squared errors plus an ℓ_2 or ℓ_1 norm regularization term in the case of Gaussian or Laplacian prior on the parameters, respectively. We next discuss what happens under a different assumption on the noise distribution. Specifically, we consider a Laplace distribution on the error ϵ, that is, $\epsilon \sim \mathcal{L}(0, \beta)$, where $\mathcal{L}(x; 0, \beta) = (2\beta)^{-1} \exp(-|x|/\beta)$, being $\beta > 0$ the scale parameter of the distribution, whose variance is $2\beta^2$. We have indeed that

$$\log p(y_i | \theta, x^{(i)}) = \log \left[(2\beta)^{-1} \exp\left(-\frac{1}{\beta} |y_i - w^\top \phi(x^{(i)})|\right) \right]$$
$$= -\log(2\beta) - \frac{1}{\beta} |y_i - w^\top \phi(x^{(i)})|,$$

whence minimization of the negative of the log posterior in (B.2) amounts to solving

$$\min_{\theta=(w,\beta)} -\sum_{i=1}^m \log p(y_i|\theta, x^{(i)}) - \log p_\theta(\theta)$$
$$= \min_{\theta=(w,\beta)} m \log(2\beta) + \frac{1}{\beta} \sum_{i=1}^m |y_i - w^\top \phi(x^{(i)})| - \log p_\theta(\theta)$$
$$= \min_{\theta=(w,\beta)} m \log(2\beta) + \frac{1}{\beta} \|\Phi^\top w - y\|_1 - \log p_\theta(\theta).$$

Similar to the cases explored in Appendix B.1.1, the above problem can be made explicit by specifying the prior $p_\theta(\theta)$. In the simplest case of a flat prior, this term does not influence the minimization and hence the optimal estimate $\hat{w}$ solves

$$\min_w \|\Phi^\top w - y\|_1, \tag{B.6}$$

which has the structure of the ℓ_1 regression problem discussed in Section 6.3.4, and the optimal estimate of β is given by the mean absolute error (MAE)

$$\hat{\beta} = \text{MAE}(\hat{w}).$$

7
Linear classifiers

CLASSIFIERS are an important family of supervised learning models, in which one tries to predict the class or label of a given input feature vector. In most respects, classification models are similar to regression models, with the only difference that the output to be predicted is of discrete categorical nature in classifiers (e.g., "positive" or "negative," or "sunny"/"cloudy"/"rainy"), while it is of continuous nature in regression models. In this chapter we shall provide the basics of binary classification and of the corresponding performance metrics. In particular, we focus on linear classifiers (i.e., classifiers whose discrimination surfaces are hyperplanes) such as the logistic regression model and the (linear) support vector machine (SVM) model. We then extend the scope of classification to the multi-class case, and discuss issues related to regularization, sparsity, robustness, and the problem of class imbalance.

7.1 Basics of binary classification

Similar to the case of regression models, in binary classification we are given a data set $\mathcal{D}$ of m observations composed of pairs $(x^{(i)}, y_i)$, $i = 1, \ldots, m$, where $x^{(i)} \in \mathbb{R}^n$ are input feature vectors and $y_i \in \{-1, 1\}$ are the output *labels*, usually referred to as the *negative* and the *positive* class. The fact that the output labels are in the set $\{-1, 1\}$ is only one convention; we can as well say that $y_i \in \{0, 1\}$, or $y_i \in \{1, 2\}$, or $y_i \in \{A, B\}$, and so on. The point is that the output can be one of two given classes, examples are given in Table 7.1. Using the data set $\mathcal{D}$, our goal is to find a classification rule $\hat{y} = f(x)$ that allows us to predict the label $\hat{y}$ of a new data point x. A plethora of different classification models exist (to name a few: naive Bayes, logistic regression, SVMs, decision trees and random forests, neural networks,

Feature vectors	Labels
Companies' corporate info	default/no default
Stock price data	price up/down
News data	price up/down
News data	sentiment (positive/negative)
Emails	presence of a keyword
Genetic measures	presence of disease

Table 7.1 Examples of features/labels in binary classification problems.

etc.), each method proposing its specific form for the discrimination function $f(x)$ and specific method for its estimation (learning) from data. In this chapter we focus on two classifiers (namely, the logistic regression classifier and the SVM classifier) for which the discrimination rule is based on the sign of an affine function, that is,

$$\hat{y} = \text{sgn}(w^\top x + b), \qquad (7.1)$$

where $w \in \mathbb{R}^n$ and $b \in \mathbb{R}$ are parameters of the classifier, which need be estimated from data. For fixed w and b, the classifier in (7.1) assigns a predicted ± 1 output label to an input vector x.

7.1.1 Misclassification errors and classifier's performance

In the case of regression it was quite natural to evaluate the predictive performance of the model in the test set on the basis of the prediction error, for example, by considering the mean squared error $(1/N) \sum_{i=1}^{N} (y_i - \hat{y}_i)^2$, where N denotes the number of observations available for validation/testing the model, post-training. This type of error still makes sense in the case of classifiers, and it is called *accuracy*:

$$\text{accuracy} = \frac{\text{number of correct predictions}}{\text{total number of predictions made}}.$$

However, the accuracy weights the positive and the negative errors symmetrically, and it can be a quite misleading indicator of a classifier's performance in certain cases. Indeed, in most classification tasks the misclassification errors in predicting the positives and those in predicting the negatives may have different consequences. For this reason, in binary classification we consider:

- type I errors, also called *false positives* (FP), which correspond to individuals (input vectors) that are classified as positive, while they are negative in reality;

- type II errors, also called *false negatives* (FN), which correspond to individuals that are classified as negative, while they are positive in reality.

For example, in a missile pointing device that classifies a target as positive when it is an enemy and negative otherwise, a FP (type I) error may lead to destroying a civilian or allied target, while a FN error may have less disastrous consequences. On the other hand, in a medical test for HIV that classifies a patient as positive if they may have contracted HIV, a FN (type II) error may result in the patient missing vital therapy opportunities and/or lead to further transmission of infection, while a FP error will typically just lead to further medical analysis.

Example 7.1 (*Accuracy and imbalanced classes*) In many situations the sample population is highly *unbalanced*, that is, the negative class vastly outnumbers the positive class. In such situations, the accuracy can be a misleading indicator of a classifier's predictive performance. For example, there is about a 0.5% global prevalence of HIV infections among the world population (2019 data). That is, the world population is 99.5% in the negative class and only 0.5% in the positive class. Suppose we build a very "naive" HIV test that, for any patient, always returns the result "negative." Such a naive test, on average, will be correct 99.5% of the times, that is, its accuracy is 99.5%, which is very high indeed! This example makes a clear case for the fact that accuracy alone may be a very poor indicator of a classifier's performance. Of course, what counts in this case is the so-called *sensitivity* of the test, that is, the measure of the proportion of positives that are correctly identified as positive, which is zero for our naive HIV test.

The previous discussion shows that, for correctly evaluating the performance of a classifier, we need to assess false positives and false negatives, and we can use the corresponding error rates (evaluated on the *test* set) as performance criteria, see Figure 7.1.

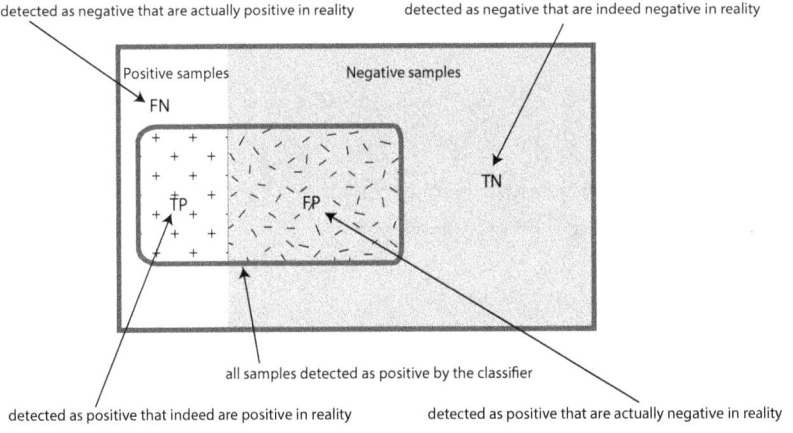

Figure 7.1 The four classification outcomes: true positive (TP), false positive (FP), true negative (TN), and false negative (FN).

7.1.2 Performance metrics

Classifiers' performance metrics are suitably defined in terms of conditional probabilities. We assume there exist an unknown probability distribution over the input vectors x and some true underlying unknown function that assigns the true class y to each given x. There is also our known classifier function f, which assigns a predicted class $\hat{y}$ to each given x. This type of mechanism induces the existence of a probability distribution on the true class y and on the predicted class $\hat{y}$. In such setting, y can be interpreted as the hidden (true and unknown) state of an individual x, with $y = -1$ for a negative individual and $y = 1$ for a positive individual, and $\hat{y}$ denotes the classifier output, with $\hat{y} = -1$ for a negative-classified individual and $\hat{y} = 1$ for a positive-classified individual.

We then define the *precision* (or positive predictive value – PPV) of the classifier as

$$p \doteq \text{Prob}\{y = 1 | \hat{y} = 1\},$$

that is, the probability that the true state is positive, given that the sample is classified as positive. We define the *sensitivity* (true positive rate – TPR, or *recall*) of the classifier as

$$r \doteq \text{Prob}\{\hat{y} = 1 | y = 1\},$$

that is, the probability that the classifier returns positive, given that the sample is positive in reality. We further define the *specificity* (true negative rate – TNR) as

$$s \doteq \text{Prob}\{\hat{y} = -1 | y = -1\},$$

that is, the probability that the classifier returns negative, given that the sample is negative in reality. Since $\text{Prob}\{\hat{y} = 1 | y = -1\} + \text{Prob}\{\hat{y} = -1 | y = -1\} = 1$, we have that

$$\text{FPR} \doteq \text{Prob}\{\hat{y} = 1 | y = -1\} = 1 - \text{Prob}\{\hat{y} = -1 | y = -1\} = 1 - \text{TNR},$$

that is, the false positive rate (FPR, or type I error rate) is the complement of the specificity (TNR). Similarly, since $\text{Prob}\{\hat{y} = 1 | y = 1\} + \text{Prob}\{\hat{y} = -1 | y = 1\} = 1$, we have that

$$\text{FNR} \doteq \text{Prob}\{\hat{y} = -1 | y = 1\} = 1 - \text{Prob}\{\hat{y} = 1 | y = 1\} = 1 - \text{TPR},$$

that is, the false negative rate (or type II error rate) is the complement of the sensitivity (TPR). Precision p and recall r (TPR) are related via Bayes' rule:

7.1 BASICS OF BINARY CLASSIFICATION

$$p = \text{Prob}\{y = 1 | \hat{y} = 1\} = \frac{\text{Prob}\{\hat{y} = 1 | y = 1\}\text{Prob}\{y = 1\}}{\text{Prob}\{\hat{y} = 1\}}$$

$$= \frac{\text{TPR} \cdot \text{Prob}\{y = 1\}}{\text{Prob}\{\hat{y} = 1 | y = -1\}\text{Prob}\{y = -1\} + \text{Prob}\{\hat{y} = 1 | y = 1\}\text{Prob}\{y = 1\}}$$

$$= \frac{\text{TPR} \cdot \text{Prob}\{y = 1\}}{\text{FPR} \cdot (1 - \text{Prob}\{y = 1\}) + \text{TPR} \cdot \text{Prob}\{y = 1\}},$$

where $\text{Prob}\{y = 1\}$ is the so-called baseline risk.

Estimation of the performance metrics Assuming $y, \hat{y} \in \{-1, 1\}$ and given a batch of testing or validation data $\mathcal{D}_v = \{(x^{(1)}, y_1), \ldots, (x^{(N)}, y_N)\}$, we produce the predictions of a classifier f by evaluating $\hat{y}_i = f(x^{(i)})$, $i = 1, \ldots, N$. We can then estimate the accuracy of the classifier as

$$\text{Accuracy} = 1 - \hat{p}_e,$$

where $\hat{p}_e$ is the estimated probability of misclassification:

$$\hat{p}_e = \frac{1}{2N} \sum_{i=1}^{N} |y_i - \hat{y}_i|.$$

Similarly, precision, recall, and specificity can be estimated by simply *counting* the occurrences in the test/validation set of the true positive, false positive, true negative, and false negative events:

$$\text{TP} = \sum_{\{i:y_i=1\}} \max(\hat{y}_i, 0), \qquad \text{FP} = \sum_{\{i:y_i=-1\}} \max(\hat{y}_i, 0),$$

$$\text{TN} = \sum_{\{i:y_i=-1\}} -\min(\hat{y}_i, 0), \quad \text{FN} = \sum_{\{i:y_i=1\}} -\min(\hat{y}_i, 0).$$

These counts are usually organized into a 2×2 matrix known as the *confusion matrix*, see Figure 7.2. Once the counts are obtained, we estimate:

- *precision, p*: the number of TP divided by the number of all classified positive results:

$$p \doteq \text{Prob}\{y = 1 | \hat{y} = 1\} \simeq \frac{\text{TP}}{\text{TP} + \text{FP}};$$

- *recall, r*: the number of TP divided by the number of total actual positives:

$$\text{TPR} = r \doteq \text{Prob}\{\hat{y} = 1 | y = 1\} \simeq \frac{\text{TP}}{\text{TP} + \text{FN}};$$

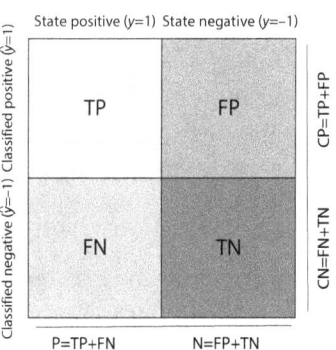

Figure 7.2 Confusion matrix.

- *specificity, s*: the number of TN divided by the number of total actual negatives:

$$\text{TNR} = s \doteq \text{Prob}\{\hat{y} = -1|y = -1\} \simeq \frac{\text{TN}}{\text{FP} + \text{TN}}.$$

Multi-criteria metrics To capture different aspects of a classifier's performance, for instance precision and recall, in a single number, suitable composite indices have been proposed, notably:

- the F_1 score

$$F_1 \doteq \frac{2}{\frac{1}{r} + \frac{1}{p}} = 2\frac{pr}{p+r},$$

which is the harmonic mean of precision and recall; and

- the *balanced accuracy*

$$\text{BA} = \frac{\text{TPR} + \text{TNR}}{2},$$

which is the average of TPR and TNR.

When the classifier depends on some hyper-parameter λ representing, for instance, a decision threshold, the tradeoff between the true positive rate against the false positive rate as a function of λ is captured by the so-called receiver operating characteristic (ROC) curve, which is a parametric plot (λ being the parameter) of the TPR versus the FPR, see Example 7.3.

7.2 The logistic classifier model

In the binary logistic classifier model the output variable $y \in \{-1, 1\}$, given the input $x \in \mathbb{R}^n$, is assumed to be a Bernoulli random variable with probability $\eta(x)$ for the positive outcome and probability $1 - \eta(x)$ for the negative outcome, that is,

$$\text{Prob}\{y = 1|x\} = \eta(x), \quad \text{Prob}\{y = -1|x\} = 1 - \eta(x).$$

Further, in the logistic model one assumes that the log-odds ratio is an affine function of x, that is,

$$\log \frac{\text{Prob}\{y = 1|x\}}{\text{Prob}\{y = -1|x\}} = \log \frac{\eta(x)}{1 - \eta(x)} = w^\top x + b, \quad (7.2)$$

where $w \in \mathbb{R}^n$ and $b \in \mathbb{R}$ are model parameters to be learnt from data. As a consequence of the affine log-odds ratio assumption, we have that

$$\eta(x, w, b) = \frac{\exp(w^\top x + b)}{1 + \exp(w^\top x + b)}$$

$$1 - \eta(x, w, b) = \frac{1}{1 + \exp(w^\top x + b)}.$$

The conditional probability density of $y \in \{-1,1\}$ given x and $\theta = (w,b)$ can be written compactly as

$$p(y|x,\theta) = \eta(x,\theta)^{(1+y)/2}(1-\eta(x,\theta))^{(1-y)/2}$$
$$= \frac{1}{1+\exp\left(-y(w^\top x + b)\right)}.$$

Once we constructed a logistic model (i.e., once we have w and b), for a given input sample x we can compute the corresponding probability of positive outcome $\eta(x,\theta)$ in (7.3). This model is often referred to as the logistic *regression* model. Such terminology, which may be a little confusing since the model is used mainly for classification rather than for regression, is due to the fact that the logistic model is based on estimating the real-valued function η describing the conditional positive outcome probability, given x. The actual binary output y of the model is obtained next by comparing $\eta(x,\theta)$ with a given threshold $\gamma \in (0,1)$, which is typically (but not necessarily) set to $\gamma = 0.5$: if $\eta(x,\theta) \geq \gamma$ the model classifies the input sample as positive, otherwise it classifies the sample as negative. In formulae, the predicted output is

$$\hat{y} = \text{sign}(\eta(x,\theta) - \gamma), \tag{7.3}$$

where we let $\text{sign}(0) \doteq 1$. Elaborating on (7.3), we observe that the predicted output can be expressed equivalently as

$$\hat{y} = \text{sign}(w^\top x + b - \alpha), \tag{7.4}$$

where $\alpha \doteq \log \frac{\gamma}{1-\gamma}$. We see from (7.4) that the logistic classifier is a linear classifier, since the discrimination surface between the positive and the negative predicted samples is given by the hyperplane of equation $w^\top x + b - \alpha = 0$. In particular, all input points x belonging to the half-space $\{x \in \mathbb{R}^n : w^\top x + b - \alpha \geq 0\}$ are classified as positive (i.e., $\hat{y} = 1$), and all input points belonging to the half-space $\{x \in \mathbb{R}^n : w^\top x + b - \alpha < 0\}$ are classified as negative (i.e., $\hat{y} = -1$).

Remark 7.1 (*Hyperplanes and half-spaces*) An hyperplane in $\mathbb{R}^n$ is an $(n-1)$-dimensional affine set defined as the level set of an affine function. Letting the affine function be represented as $f(x) = w^\top x + w_0$, with $w \neq 0$, we describe an hyperplane $\mathcal{H}$ as the locus of points where $f(x) = 0$, that is,

$$\mathcal{H} = \{x \in \mathbb{R}^n : w^\top x + w_0 = 0\}.$$

Geometrically, vector $w \in \mathbb{R}^n$ represents the *normal direction* to the hyperplane, that is, a direction which is orthogonal to the surface of $\mathcal{H}$, see Figure 7.3.

The scalar w_0 is instead related to the signed Euclidean distance $d(0,\mathcal{H})$ to be travelled along direction w in order to reach $\mathcal{H}$ from 0:

$$d(0,\mathcal{H}) = -\frac{w_0}{\|w\|_2}.$$

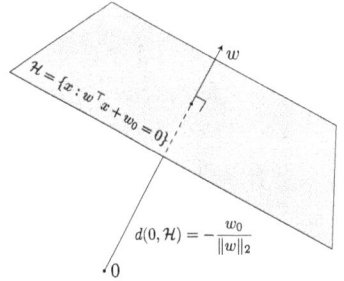

Figure 7.3 Hyperplane geometry.

172   7 LINEAR CLASSIFIERS

To an hyperplane $\mathcal{H}$ we associate two half-spaces $\mathcal{H}_+$ and $\mathcal{H}_{--}$ corresponding, respectively, to the sets of points lying above and below $\mathcal{H}$. More precisely,

$$\mathcal{H}_+ \doteq \{x \in \mathbb{R}^n : w^\top x + w_0 \geq 0\},$$
$$\mathcal{H}_{--} \doteq \{x \in \mathbb{R}^n : w^\top x + w_0 < 0\}.$$

Figure 7.4 depicts an hyperplane in $\mathbb{R}^2$, along with its associated half-spaces.

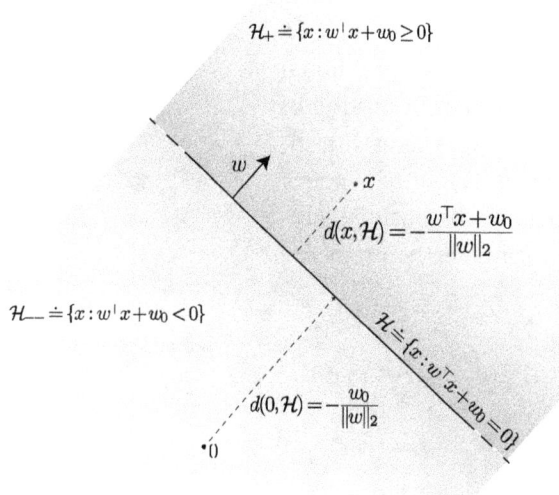

Figure 7.4 Hyperplane and half-spaces.

7.2.1 Learning the logistic classifier

We next discuss how to estimate the parameters w and b of the logistic model on the basis of a set $\mathcal{D}$ of m training observations composed of i.i.d. pairs $(x^{(i)}, y_i)$, $i = 1, \ldots, m$, where $x^{(i)} \in \mathbb{R}^n$ are the input feature vectors and $y_i \in \{-1, 1\}$ are the output labels. Considering that the conditional distribution of the output is given by (7.3), and that the outputs are independent, we have that the probability of $y_1, \ldots, y_m$ given the inputs $x^{(1)}, \ldots, x^{(m)}$ and the model parameters $\theta = (w, b)$, that is, the likelihood, is given by

$$\prod_{i=1}^{m} p(y_i | x^{(i)}, \theta|) = \prod_{i=1}^{m} \frac{1}{1 + \exp\left(-y_i(w^\top x^{(i)} + b)\right)}.$$

The log-likelihood is thus

$$\ell(w,b) = -\sum_{i=1}^{m} \log\left(1 + \exp\left(-y_i(w^\top x^{(i)} + b)\right)\right). \quad (7.5)$$

If a maximum-likelihood approach is taken for learning the model, then the optimal w, b are found by maximizing the log-likelihood. To this purpose, we observe that, neglecting b (without loss of generality, since it can be absorbed into w, by augmenting x with a one), the gradient of $\ell(w)$ is

$$\nabla_w \ell(w) = \sum_{i=1}^{m} \frac{y_i}{1 + \exp\left(y_i(w^\top x^{(i)} + b)\right)} x^{(i)},$$

and the Hessian is

$$\nabla_w^2 \ell(w) = -\sum_{i=1}^{m} \alpha_i x^{(i)} x^{(i)\top},$$

where

$$\alpha_i \doteq \frac{\exp\left(y_i(w^\top x^{(i)} + b)\right)}{\left(1 + \exp\left(y_i(w^\top x^{(i)} + b)\right)\right)^2}, \quad i = 1, \ldots, m.$$

Since $\alpha_i \geq 0$ for any w, it follows that $\nabla_w^2 \ell(w) \preceq 0$, that is, the Hessian of $\ell(w)$ is negative semidefinite, which implies that $\ell(w)$ is a *concave* function. Concavity implies that a global maximum of $\ell(w)$ can be found efficiently via numerical methods, for example, by applying a Newton-type algorithm.

Loss function for the logistic classifier We see from equation (7.5) that the logistic classifier is trained by minimizing a loss function of the type

$$\mathcal{L}(w,b) = \sum_{i=1}^{m} \log(1 + \exp(-z_i)), \quad (7.6)$$

where $z_i \doteq y_i(w^\top x^{(i)} + b)$. Observe that $z_i > 0$ for points that are correctly classified (i.e., for which y_i and $w^\top x^{(i)} + b$ have the same sign), while $z_i < 0$ for points that are misclassified. Further, the absolute value of z_i is proportional to how far the point $x^{(i)}$ is from the discrimination hyperplane $w^\top x + b = 0$. We see from Figure 7.5 that $\log(1 + \exp(-z_i))$ (solid line) is a smooth approximation of the so-called *perceptron* loss (dashed line), defined as $[-z_i]_+ \doteq \max(0, -z_i)$.

In practice, the loss function (7.6) is often augmented with a regularization term, such as

$$\mathcal{L}(w,b) = \sum_{i=1}^{m} \log(1 + \exp(-z_i)) + \lambda \|w\|_p, \quad (7.7)$$

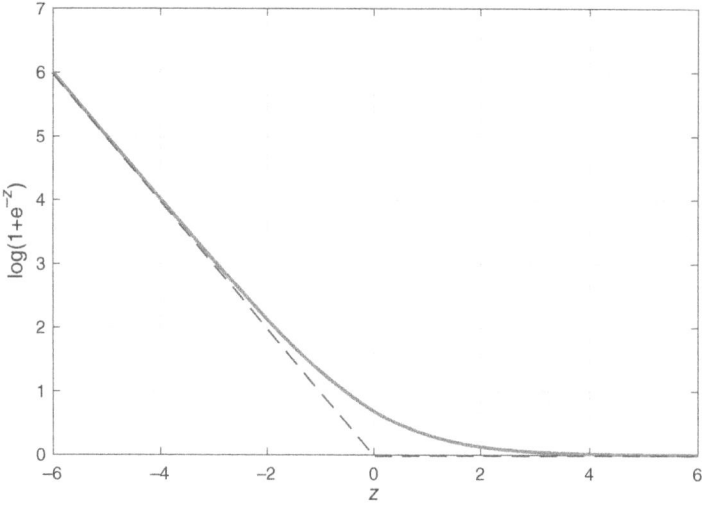

Figure 7.5 Logistic regression loss (solid line) and perceptron loss (dashed line).

where $\lambda \geq 0$ is a regularization parameter that is typically selected via cross validation, and $p = 2$ for the ℓ_2 regularization, or $p = 1$ if one wants to promote sparsity in the w vector.

Example 7.2 (*Predicting volatility direction*) We use a logistic model to predict if a stock's volatility is going up or down in the next period. We select a period duration of 15 days. The volatility is evaluated as the sample standard deviation of the stock log-returns over the considered period. For each period, we extract from the data the following features:

1. the mean return over the current period;
2. the mean return over the previous period;
3. the volatility over the current period;
4. the volatility over the previous period;
5. average price level over the current period;
6. the average price level over the previous period.

To construct the test set, we compute the volatility over the next period and set label $y = +1$ if it increased with respect to the current period, and label $y = -1$ otherwise. We used AAPL (Apple) stock data, from Dec. 5, 2011 to Dec. 1, 2017. Seventy percent of the data, that is, $m = 68$ periods, was used for training a logistic classifier, and the remaining 30%, that is, $m' = 30$ periods, was used for testing and evaluating the out-of-sample performance. A MATLAB® code for this example is found in file logistic_volatility_ex.m in the online resources. We obtained the following confusion matrix:

7.2 THE LOGISTIC CLASSIFIER MODEL

	Condition positive (15)	Condition negative (15)	
Predicted Positive (21)	True Positive TP = 15	False Positive FP = 6	Positive Predictive Value PPV = TP/(TP+FNP) = 0.71
Predicted Negative (9)	False Negative FN = 0	True Negative TN = 9	Negative Predictive Value NPV = TN/(FN+TN) = 1
	Sensitivity TPR = TP/(TP+FN) = 1	Specificity TNR = TN/(FP+TN) = 0.6	

From the matrix we find that the test has an estimated precision $p = 71\%$ and recall $r = 100\%$.

Example 7.3 (*ROC curve*) The ROC curve characterizes a classifier model in function of its decision threshold γ; two different classifier models can be compared by comparing their ROC curves. In particular, the area under the ROC curve (AUC) gives a single index for comparison, whereby the highest AUC value corresponds to the best classifier model. Consider for instance the use a logistic regression model to predict the credit risk of a loan applicant. For each applicant we are provided a set of demographic and socio-economic information, such as sex, age, job, housing, purpose of the loan, and so on. On the basis of such information we can predict if the applicant is a bad or good credit risk for the bank.

We used the data set contained in the file german_credit_risk.csv.[1] We subdivide the data set in a training set, containing 75% of the samples of the data set, and a test set, containing the remaining 25% of the data set. We train the logistic regression model using the training set and then we use the test set to evaluate the performance of the learned model. A Python code for this example is found in the file german_credit_risk.ipynb in the online resources. Figure 7.6 shows the obtained ROC curve, which gives an AUC value of 0.744.

[1] This is a cleaned version of the data set available at www.kaggle.com/datasets/uciml/german-credit.

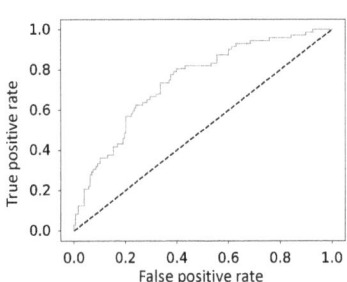

Figure 7.6 ROC curve of a logistic regression classifier for credit risk prediction.

7.2.2 Multi-class logistic classifier

In the previous sections we discussed the logistic classifier in the context of binary classification, that is, when the output may belong to one of two possible classes. There are, however, problems in which there are $K \geq 2$ possible output classes and corresponding output labels. For example, if we want to predict the rating of a movie based on a five-star rating systems the classifier should have five output labels, or if a financial analyst's recommender system provides a positive, neutral or negative predicted sentiment the classifier should have three output labels. A general approach to multi-class classification is to transform the problem into a sequence of binary classification problems, as discussed in Remark 7.2. Logistic classifiers, however, are capable of handling multi-class models directly. The idea is to model the log-odds of each class $k = 1, \ldots, K$ against the last class as affine functions of x, similar to what was done in (7.2) for the two-class case. Namely, for the output label $y \in \{1, \ldots, K\}$ we postulate that

$$\log \frac{\text{Prob}\{y = 1|x\}}{\text{Prob}\{y = K|x\}} = b_1 + w^{(1)\top} x,$$

$$\log \frac{\text{Prob}\{y = 2|x\}}{\text{Prob}\{y = K|x\}} = b_2 + w^{(2)\top} x,$$

$$\vdots$$

$$\log \frac{\text{Prob}\{y = K-1|x\}}{\text{Prob}\{y = K|x\}} = b_{K-1} + w^{(K-1)\top} x.$$

A simple calculation shows that

$$\text{Prob}\{y = k|x\} = \frac{\exp(b_k + w^{(k)\top} x)}{1 + \sum_{i=1}^{K-1} \exp(b_i + w^{(i)\top} x)}, \quad k = 1, \ldots, K-1,$$

$$\text{Prob}\{y = K|x\} = \frac{1}{1 + \sum_{i=1}^{K-1} \exp(b_i + w^{(i)\top} x)},$$

and it can be verified that all probabilities are nonnegative and sum to one. We may represent the output via a vector $y \in \mathbb{R}^K$ containing the one-hot encoding of y, that is, $y_k = 1$ if $y = k$ and it is zero otherwise. Similarly, we let $p \in \mathbb{R}^K$ contain the class probabilities predicted by the model, that is, $p_k = \text{Prob}\{y = k|x\}$, $k = 1, \ldots, K$. Since there is nothing special about the Kth class, we can actually express the probabilities $\text{Prob}\{y = k|x\}$ as

$$p_k(x) \doteq \text{Prob}\{y = k|x\} = \frac{\exp(z_k)}{\sum_{i=1}^{K} \exp(z_i)}, \quad k = 1, \ldots, K,$$

where

$$z_k \doteq b_k + w^{(k)\top} x, \quad k = 1, \ldots, K.$$

The vector of predicted probabilities $p(x) = (p_1(x), \ldots, p_K(x))$ can be written compactly in terms of the so-called soft-max function as

$$p(x) = S(W^\top x + b), \tag{7.8}$$

where $S \colon \mathbb{R}^K \to \mathbb{R}^K_+$ is defined elementwise as $S_k(z) = \frac{\exp(z_k)}{\sum_{i=1}^{K} \exp(z_i)}$, and the model parameters are given by the vector of bias terms $b = (b_1, \ldots, b_K)$ and matrix $W = [w^{(1)} \cdots w^{(K)}]$. Now, the probability mass of y can be expressed in terms of the multinomial density

$$\text{Prob}\{y|x\} = \prod_{k=1}^{K} p_k(x)^{y_k},$$

and the log-likelihood of m observed samples is

$$\ell(W,b) = \log \prod_{i=1}^{m} \text{Prob}\{y^{(i)}|x^{(i)}\} = \log \prod_{i=1}^{m} \prod_{k=1}^{K} p_k(x^{(i)})^{y_k^{(i)}}$$
$$= \sum_{i=1}^{m} \sum_{k=1}^{K} y_k^{(i)} \log p_k(x^{(i)}). \tag{7.9}$$

It can be noticed that the negative of the log-likelihood corresponds to the sum over samples of the cross entropy between the predicted probability distribution $p(x^{(i)})$ and its observed one-hot counterpart $y^{(i)}$. Substituting (7.8) into (7.9) one obtains an explicit expression for the log-likelihood as a function of the model parameters W, b. This expression turns out to be concave in the parameters, and hence amenable to efficient numerical solution.[2] Similar to the binary case, for a given input x the multi-class logistic classifier returns a vector $p(x) \in \mathbb{R}^K$ of estimated probabilities for the classes. The actual assignment of x to a specific class happens by selecting the entry in $p(x)$ having the highest value.

[2] Details of the derivation may be found in Section 8.3.7 of K. P. Murphy, *Machine Learning: A probabilistic perspective*, Cambridge, MA: MIT Press, 2012.

Remark 7.2 (*OvA and OvO classifiers*) There exist general approaches that can be used for doing multi-class classification using as a basis a binary classifier. One approach, named one-vs-all (OvA) or one-vs-rest (OvR), consists in splitting the multi-class problem into K binary classification subproblems. In each subproblem k, we treat class k as the positive class and all other classes as negative, and we train a binary classifier f_k on the subproblem. At the end of the training, we have K binary classifiers f_k, $k = 1, \ldots, K$, each of which provides as output a class-membership probability or score $f_k(x)$ for input x. The classification decision is hence made by assigning x to the class which has the highest score, that is, $y = \arg\max_{k=1,\ldots,K} f_k(x)$.

A similar approach is the so-called one-vs-one (OvO) approach, in which the multi-class problem is split into $K(K-1)/2$ binary classification problems, one for every pair of classes, where one class is considered against every other class. This approach requires more classifiers than the OvA approach, but can be convenient in cases where it is advantageous to use smaller datasets for training the model, since each model is trained only with a subset of the data corresponding to one specific pair of classes. Similar to the OvA case, each binary classifier predicts a class-membership probability or score. For each class we then compute the sum of all received scores, and finally assign to the input the class which has the maximum sum of scores. Popular machine learning libraries such as scikit-learn implement both the OvR and the OvO strategies as dedicated classes for their binary classification models.

7.2.3 Class imbalance

Class imbalance is an issue that arises in binary classification problems when the cardinality of the classes is substantially different. Indeed, in many cases of practical interest the negative class is prevailing, being the positive class only a small fraction of the total population. This is the case for instance in tests for a particular desease (the vast majority of the population is *not* affected by the disease), in fraud detection for credit card transactions (the majority of transactions are legitimate), and in general in all endeavors where the goal is the detection of rare events.

Suppose there are m_- negative samples (i.e., samples for which $y_i = -1$), and m_+ positive samples (i.e., samples for which $y_i = 1$), where $m_- + m_+ = m$, the total number of training samples. Let $\mathcal{P}$ denote the indices of the positive samples, and $\mathcal{N}$ the indices of the negative samples. We see from (7.6) that the loss function to be minimized can be split as

$$\mathcal{L}(w,b) = \sum_{i \in \mathcal{N}} \log(1 + \exp(-z_i)) + \sum_{i \in \mathcal{P}} \log(1 + \exp(-z_i)),$$

where $z_i \doteq y_i(w^\top x^{(i)} + b)$. In the above loss function the first term essentially quantifies the misclassifications in the negative class (i.e., the false positives), and the second term quantifies the misclassifications in the positive class (i.e., the false negatives). Intuitively, if the cardinality of the positive class is very small compared to that of the negative class, then the corresponding term in the loss function will have little relative weight, which implies that the classifier resulting from minimization of the loss function may allow a high false negative rate (FNR) and hence, since FNR $= 1 -$ TPR, a low true positive rate (sensitivity). If class imbalance is not dealt with correctly, therefore, the resulting classifier may provide bad (low) sensitivity.

A simple remedy is to modify the loss function so to suitably counter-balance the small cardinality of the positive class. This can be done, for instance, by letting the loss function be of the form

$$\mathcal{L}(w,b) \doteq \frac{m_+}{m} \sum_{i \in \mathcal{N}} \log(1 + \exp(-z_i)) + \frac{m_-}{m} \sum_{i \in \mathcal{P}} \log(1 + \exp(-z_i)). \tag{7.10}$$

Notice that other choices are possible for the weights of the negative and positive classes; in particular, weights could be treated as hyperparameters of the model and set by means of suitable cross-validation experiments.

Other approaches for dealing with class imbalance are known as "external" methods, which work at the level of the preprocessing of the data. For instance, balanced data sets can be obtained by resampling the original data so to under-represent the negative class (under-sampling), or so to over-represent the positive class by sampling with replacement the positive samples (over-sampling). Both techniques have advantages and drawbacks; most prominently, under-sampling implies a loss of information due to incomplete exploitation of the available data and it is thus generally not suggested.[3]

[3] There is an extensive literature on learning with imbalanced classes; a discorsive presentation with pointers to relevant literature is available for instance in B. Krawczyk, "Learning from Imbalanced Data: Open Challenges and Future Directions," *Progress in Artificial Intelligence*, vol. 5, 2016. A reference book is also *Imbalanced Learning: Foundations, Algorithms, and Applications*, by H. He and Y. Ma, Hoboken, NJ: John Wiley & Sons, 2012.

7.3 Linear support vector machines

Linear support vector machines (SVMs) are a class of binary classifiers that are based on the idea of geometric separation of points via hyperplanes. In order to understand the idea behind SVMs, we shall start with a simpler ancestor model called the *perceptron*.

7.3.1 Perceptron classifier

Classifiers that compute a linear combination of the input features and return the sign, were called *perceptrons* in the engineering literature in the late 1950s.[4] Perceptrons set the foundations for the neural network models of the 1980s and 1990s. The perceptron learning algorithm tries to find a discriminating hyperplane by minimizing the distance of misclassified points to the decision boundary $\mathcal{H}$. Consider the situation sketched in Figure 7.7, where data points are depicted as dots for the negative class and as crosses for the positive class. The

[4] See F. Rosenblatt, "The Perceptron: A Probalistic Model for Information Storage and Organization in the Brain," *Psychological Review*, vol. 65(6), pp. 386–408, 1958.

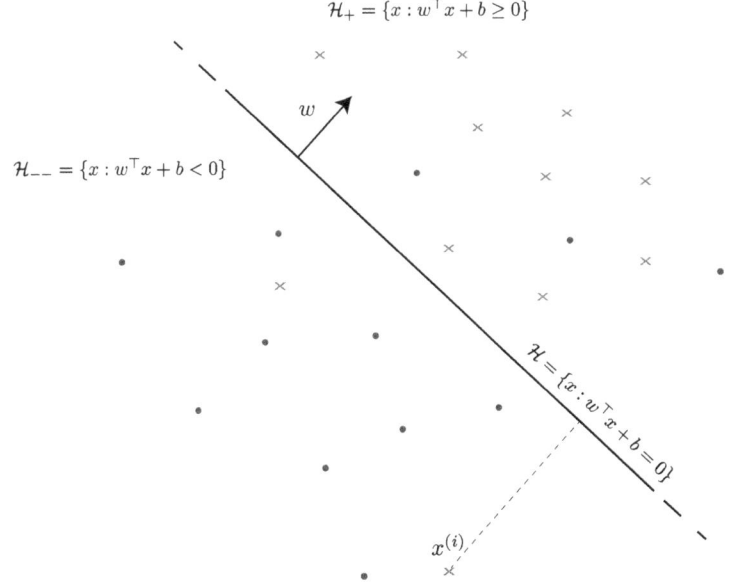

Figure 7.7 Hyperplane and labeled data points (dots: negative, crosses: positive).

goal of the perceptron classifier is to classify feature vectors based on an hyperplane $\mathcal{H}$ as a discrimination surface: we say say a point x is in class +1 if it belongs to $\mathcal{H}_+$ or it is in class -1 if it belongs to $\mathcal{H}_{--}$. The training phase of the perceptron amounts to finding a discrimination hyperplane by minimizing a suitable loss function. Given a tentative discrimination hyperplane $\mathcal{H} = \{x \colon w^\top x + b = 0\}$, there will be some of the training data points lying in the positive half-space $\mathcal{H}_+$ and some lying in the negative half-space $\mathcal{H}_{--}$. Among the points that fall in $\mathcal{H}_+$ some are crosses, and hence are correctly classified as positive and some (three in this example) are dots, and hence are misclassified as positive while being negative in reality. Similarly, some of the points that fall in $\mathcal{H}_{--}$ are dots, and hence are correctly classified as negative and some (two in this example) are crosses, and hence are misclassified as negative while being positive in reality. In the perceptron approach, the points that are correctly classified give zero error contribution to the loss function, while the points that are misclassified give an error which is proportional to the distance from the point to the discrimination surface. In formulas, this corresponds to minimizing (over b and nonzero w) a loss function of the type

$$\mathcal{L}(w,b) = \sum_{i=1}^{m}[-z_i]_+ = \sum_{i=1}^{m}[-y_i(w^\top x^{(i)} + b)]_+, \qquad (7.11)$$

where $[\cdot]_+$ denotes the positive part of its argument. A perceptron classifier is found by minimizing the loss function in (7.11), which can be achieved, for instance, by means of sub-gradient type algorithms. Observe that in the case when the data is *linearly separable*[5] the perceptron approach is guaranteed to find a separating hyperplane for the positive and negative classes. There are, however, some issues with perceptron learning, one of these being that the minimizer of (7.11), and hence the discriminating hyperplane, may not be unique. This fact is particularly evident when the data is linearly separable, in which case there are typically infinitely many discriminating hyperplanes for the data, see Figure 7.8.

In the presence of multiple discriminating hyperplanes, it seems natural to add further criteria so that a unique hyperplane is singled out. One of such criteria is for instance to seek for an hyperplane which maximizes the *margin of separation* between the data classes. In the separable case, the margin associated with a given hyperplane $\mathcal{H}$ is defined as the maximum distance to which we can move $\mathcal{H}$ up and down parallel to itself before "touching" a data point. Clearly, the larger the margin the better the two point classes are separated.

[5] A binary labeled data set $(x^{(i)}, y_i)$, $i = 1, \ldots, m$, is said to be linearly separable if there exists an hyperplane $\mathcal{H} = \{x \colon w^\top x + b = 0\}$ such that all the positive-labeled points lie in the half-space $\{x \colon w^\top x + b \geq 0\}$ and all the negative-labeled points lie in the half-space $\{x \colon w^\top x + b \leq 0\}$. Checking linear separability amounts to solving a linear programming feasibility problem in the variables $w \in \mathbb{R}^n$ and $b \in \mathbb{R}$ with constraints $y_i(w^\top x^{(i)} + b) \geq 0$, $i = 1, \ldots, m$. The data set is said to be *strictly* linearly separable if there exist w and b such that $y_i(w^\top x^{(i)} + b) > 0$, $i = 1, \ldots, m$.

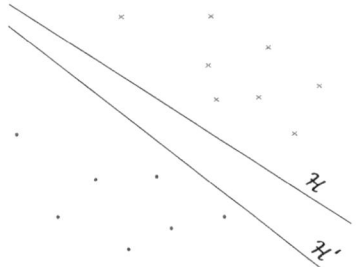

Figure 7.8 Non-uniqueness of the discriminating hyperplane.

7.3.2 Maximum margin classifier

In Figure 7.9, for example, the separating hyperplane $\mathcal{H}$ has a lower separation margin with respect to the separating hyperplane $\mathcal{H}'$.

This idea of separation margin leads to a classifier model named *maximum margin classifier*, which is discussed next.

Suppose that the training data set $(x^{(i)}, y_i)$, $i = 1, \ldots, m$, is strictly linearly separable. Our goal is to find a separating hyperplane that provides the maximum margin of separation between the classes, see Figure 7.10.

Let the hyperplane be represented as $\mathcal{H} = \{x \colon w^\top x + b = 0\}$. Points in the positive class should lie in $\mathcal{H}_+$, and their distance from the boundary is $d_i = (w^\top x^{(i)} + b)/\|w\|_2$. Points in the negative class should lie in $\mathcal{H}_-$ and their distance from the boundary is $d_i = -(w^\top x^{(i)} + b)/\|w\|_2$. The conditions below thus guarantee that all points lie in their required half-space, at distance no smaller than $M \geq 0$ from the discriminating surface $\mathcal{H}$:

$$d_i = \frac{y_i(w^\top x^{(i)} + b)}{\|w\|_2} \geq M, \quad i = 1, \ldots, m.$$

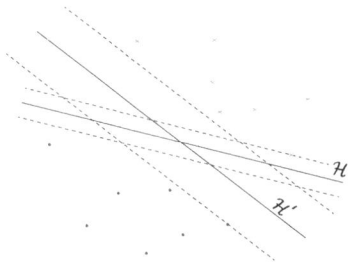

Figure 7.9 Two separating hyperplanes and their margins.

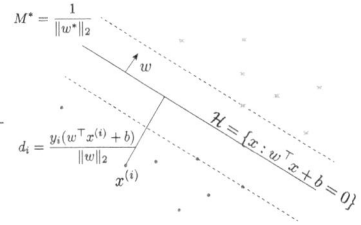

Figure 7.10 Geometric sketch for the maximum margin classifier.

The maximum margin problem can hence be posed as

$$\max_{w \in \mathbb{R}^n, b, M \in \mathbb{R}} M,$$
$$\text{s.t.:}\ y_i(w^\top x^{(i)} + b) \geq \|w\|_2 M,\ i = 1, \ldots, m.$$

We observe that the above problem is positively homogeneous in the (w, b) variables, in the sense that if (w, b) are optimal for the problem, then also $(\alpha w, \alpha b)$ are optimal, for any $\alpha > 0$. We can therefore impose an arbitrary normalization on w, without loss of generality and with the advantage of removing the homogeneity from the problem. In particular, it is convenient to set $\|w\|_2 = M^{-1}$, so that $\|w\|_2 M = 1$ and $M = 1/\|w\|_2$. The problem can thus be restated equivalently as

$$\max_{w \in \mathbb{R}^n, b \in \mathbb{R}} 1/\|w\|_2,$$
$$\text{s.t.:}\ y_i(w^\top x^{(i)} + b) \geq 1,\ i = 1, \ldots, m.$$

Further, since maximizing $\|w\|_2$ is the same as minimizing $\|w\|_2$ or $\|w\|_2^2$, we rewrite the problem again in the form of a convex quadratic program:

$$\min_{w \in \mathbb{R}^n, b \in \mathbb{R}} \|w\|_2^2, \tag{7.12}$$
$$\text{s.t.:}\ y_i(w^\top x^{(i)} + b) \geq 1,\ i = 1, \ldots, m.$$

7.3.3 The linear SVM

In the general situation in which the data is *not* linearly separable, problem (7.12) is infeasible and hence returns no solution. The linear SVM overcomes this issue by allowing violations of the separation constraints, while penalizing the overall amount of violation. More precisely, the constraints in (7.12) are relaxed to

$$y_i(w^\top x^{(i)} + b) \geq 1 - \xi_i,$$

where $\xi_i \geq 0$, $i = 1, \ldots, m$, denote the amount of violation of the constraints. This allows for some points to be on the "wrong" side of the discrimination surface or inside the margin slab, see Figure 7.11.

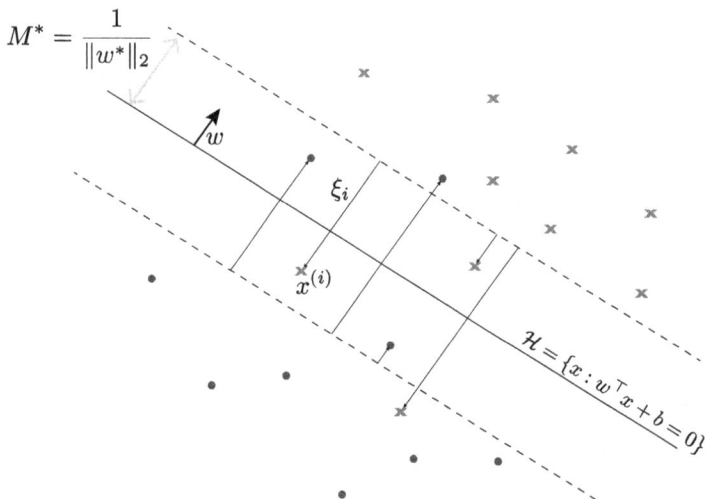

Figure 7.11 Margin separation with violations allowed.

Problem (7.12) becomes

$$\min_{w \in \mathbb{R}^n, \xi \in \mathbb{R}^m, b \in \mathbb{R}} \|w\|_2^2 + c \sum_{i=1}^m \xi_i, \qquad (7.13)$$

$$\text{s.t.: } y_i(w^\top x^{(i)} + b) \geq 1 - \xi_i, \ i = 1, \ldots, m,$$

$$\xi \geq 0,$$

where $c > 0$ is a parameter that regulates the amount of penalty due to constraint violations. Problem (7.13) is again a convex quadratic program, and the relaxed constraints formulation ensures that this problem is *always* feasible.

7.3 LINEAR SUPPORT VECTOR MACHINES

Linear SVM in unconstrained form Observe that the constraints in (7.13) require that, for $i = 1, \ldots, m$,

$$\xi_i \geq \max(1 - y_i(w^\top x^{(i)} + b), 0) = [1 - y_i(w^\top x^{(i)} + b)]_+,$$

and that the sum of the ξ_is is to be minimized in the objective, which implies that at the optimum it will always be that $\xi_i = [1 - y_i(w^\top x^{(i)} + b)]_+$. We can therefore eliminate the constraints and substitute the expressions of ξ_i in the objective, obtaining the following equivalent representation of the linear SVM problem in unconstrained form:

$$\min_{w \in \mathbb{R}^n, b \in \mathbb{R}} \|w\|_2^2 + c \sum_{i=1}^{m} [1 - y_i(w^\top x^{(i)} + b)]_+. \quad (7.14)$$

In this format the problem takes the typical form of a regularized loss minimization problem, where the loss here is

$$\mathcal{L}(w, b) = \sum_{i=1}^{m} h(z_i),$$

with $z_i = y_i(w^\top x^{(i)} + b)$ and h is the *hinge loss* function

$$h(z) \doteq [1 - z]_+ = \max(0, 1 - z). \quad (7.15)$$

Cost based on the hinge loss serves as an approximation to the number of errors made on the training set, see Figure 7.12.

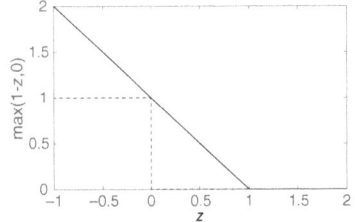

Figure 7.12 Hinge loss function (solid).

Remark 7.3 (*Comparison of perceptron, logistic and hinge loss*) We have encountered so far in the discussion of classifiers three loss functions:

- the logistic loss $\mathcal{L}_{\text{logit}} = \sum_{i=1}^{m} \log(1 + \exp(-z_i))$;
- the perceptron loss $\mathcal{L}_{\text{percpt}} = \sum_{i=1}^{m} [-z_i]_+$;
- the hinge loss $\mathcal{L}_{\text{hinge}} = \sum_{i=1}^{m} [1 - z_i]_+,$

where $z_i = y_i(w^\top x^{(i)} + b)$ is positive for correctly classified training points and it is negative otherwise; in both cases, the absolute value of z_i is proportional to the distance of point $x^{(i)}$ from the discrimination hyperplane $\mathcal{H} = \{x : w^\top x + b = 0\}$. The shape of the basic component of these three loss functions is shown in Figure 7.13. We see in particular that

$$\log(1 + \exp(-z)) \gtrsim \max(0, -z)$$

hence the logistic loss is an upper bound to the perceptron loss. The two losses behave differently for small $|z|$ but are similar when $|z|$ is large, a regime where both minimize the sum of miscalssifications, or "equation errors" from the decision boundary $\mathcal{H}$.

184   7 LINEAR CLASSIFIERS

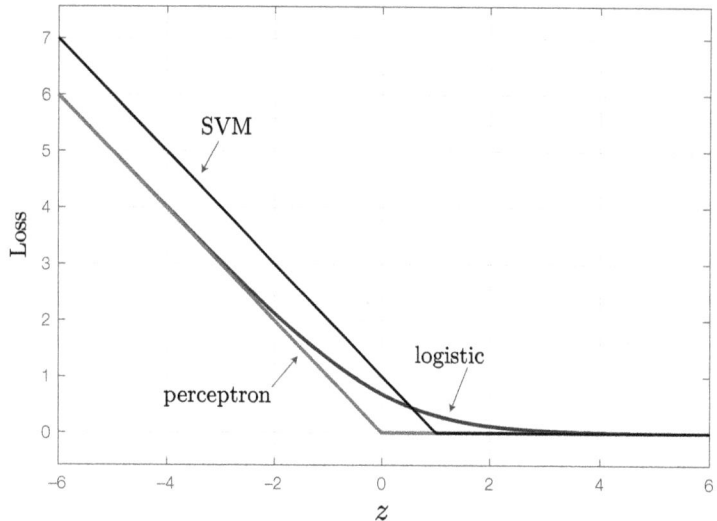

Figure 7.13 Classifiers' loss functions: logistic, perceptron, and SVM.

Remark 7.4 (*Class imbalance in SVMs*) The class imbalance issue discussed in Section 7.2.3 for the logistic classifier applies also to the SVM classifier. Besides external methods based on resampling the data, a direct method based on rebalancing the loss function can be used, thus modifying (7.14) to

$$\min_{w\in\mathbb{R}^n, b\in\mathbb{R}} \|w\|_2^2 + c\frac{m_-}{m}\sum_{i\in\mathcal{P}} h(z_i) + c\frac{m_+}{m}\sum_{i\in\mathcal{N}} h(z_i). \qquad (7.16)$$

Example 7.4 (*Credit risk prediction*) We consider the same data set considered in Example 7.3. We split the data set into a training set and a test set using the same method described in Example 7.3. We then train a linear SVM classifier with $c = 10^{-4}$. Then, we test the performance of the classifier on the test set. Figure 7.14 shows the obtained ROC curve, which corresponds to an AUC value of 0.742. We can observe that the performance of this classifier are similar to the one of the logistic regression model presented in Example 7.3. A Python code for this example can be found in the file german_credit_risk_lsvm.ipynb in the online resources.

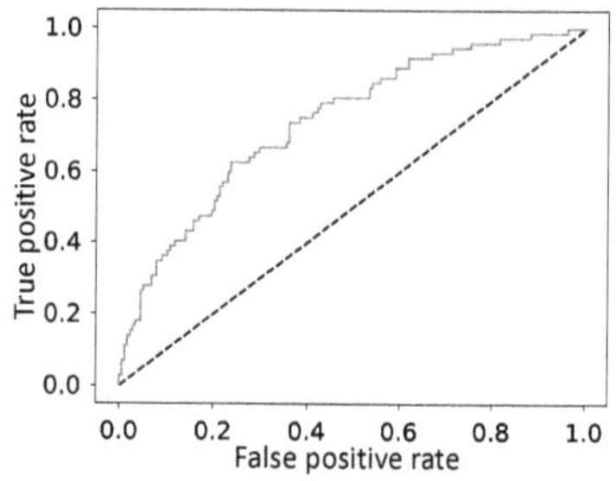

Figure 7.14 ROC curve of a linear SVM classifier for credit risk prediction.

7.4 Regularization and robustness

The three classifiers discussed so far (logistic, perceptron, linear SVM) share a common structure for their training procedure, which amounts to solving a regularized loss minimization problem of the type

$$\min_{w,b} \mathcal{L}(X^\top w + b\mathbf{1}, y) + \lambda \psi(w), \qquad (7.17)$$

where $\mathcal{L}$ is a convex loss function that encodes the error between the observed value and the predicted value, (w, b) are the model parameters, ψ is a regularization function in the regression parameters, and $\lambda \geq 0$ is a penalty parameter. This structure is indeed identical to the one considered in (6.10) for regression problems, only with different types of loss function. Changing loss functions allows us to cover perceptron, SVMs, and logistic regression problems. Typical penalties include the ℓ_1 norm to enforce sparsity, the ℓ_2 norm (often, squared) to control statistical noise and improve prediction error, or sum-block norms to enforce whole blocks of w to be zero.

In this section we discuss the topic of data robustness and its consequences on the training of a classifier. In many applications the input data points for training $x^{(i)}$, $i = 1, \ldots, m$, can be the result of observations or measurements which are possibly prone to errors. The point of view that we take is that the observed quantities, which we now denote with $\hat{x}^{(i)}$, $i = 1, \ldots, m$, represent nominal values of the underlying observations $x^{(i)}$, whose value is only known to be in some neighborhood $\mathcal{X}_i$ of the nominal value $\hat{x}^{(i)}$, see examples in Figure 7.15.

For given uncertainty sets $\mathcal{X}_i$, $i = 1, \ldots, m$, for the data points, we let
$$\mathcal{X} \doteq \{[x^{(1)} \cdots x^{(m)}] : x^{(i)} \in \mathcal{X}_i, i = 1, \ldots, m\}$$
denote the resulting uncertainty on the data matrix X, and we define the nominal data matrix as $\hat{X} \doteq [\hat{x}^{(1)} \cdots \hat{x}^{(m)}]$. A robust classifier is obtained by solving a robust version of the training problem, which amounts to minimizing the worst-case regularized loss:

$$\min_{w,b} \max_{X \in \mathcal{X}} \mathcal{L}(X^\top w + b\mathbf{1}, y) + \lambda \psi(w). \quad (7.18)$$

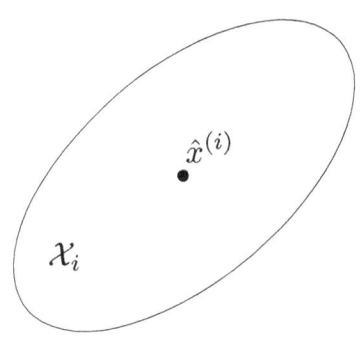

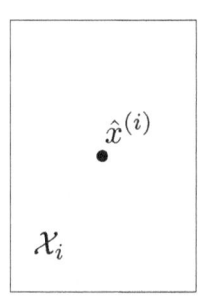

Figure 7.15 Uncertainty sets: ellipsoidal (top) and interval or box (bottom).

Different robust models are obtained depending on the choice of the loss function and on the description of the uncertainty set. We next discuss the linear SVM loss and two types of uncertainty descriptions: interval and ellipsoidal.

7.4.1 Robust SVM under interval uncertainty

In the interval uncertainty model we assume that each entry of the data matrix X is bound into an interval, independent of other entries
$$X_{ij} \in [\hat{X}_{ij} - R_{ij}, \hat{X}_{ij} + R_{ij}],$$
with $\hat{X}_{ij}, R_{ij} > 0$ given, $1 \leq i \leq n$, $1 \leq j \leq m$. This corresponds to the robust model (7.18) with $\mathcal{X} = [\hat{X} - R, \hat{X} + R]$ an interval matrix (here $R = [R_{ij}]$). The following key fact holds: for given $\hat{x} \in \mathbb{R}^n$, $\rho \in \mathbb{R}^n_+$,
$$\max_{x \,:\, |x - \hat{x}| \leq r} w^\top x = w^\top \hat{x} + r^\top |w|,$$

and
$$\min_{x:\ |x-\hat{x}|\leq r} w^\top x = w^\top \hat{x} - r^\top |w|,$$

where $|z|$ denotes the vector of magnitudes of elements in vector z.

Consider then the hinge function $h(z) = \max(0, 1 - z)$, and let $x^{(i)} = \hat{x}^{(i)} + \delta^{(i)}$, where $|\delta^{(i)}| \leq r^{(i)}$, being $r^{(i)}$ the ith column of R. We have that

$$\min_{w,b} \lambda \psi(w) + \max_{|\delta^{(i)}|\leq r^{(i)}} \sum_{i=1}^m h(y_i(w^\top x^{(i)} + b))$$

$$= \min_{w,b} \lambda \psi(w) + \sum_{i=1}^m \max_{|\delta^{(i)}|\leq r^{(i)}} h(y_i(w^\top x^{(i)} + b))$$

$$= \min_{w,b} \lambda \psi(w) + \sum_{i=1}^m h(\min_{|\delta^{(i)}|\leq r^{(i)}} y_i(w^\top x^{(i)} + b))$$

$$= \min_{w,b} \lambda \psi(w) + \sum_{i=1}^m h(y_i(w^\top \hat{x}^{(i)} + b) - r^{(i)^\top}|w|)$$

$$= \min_{w,b} \lambda \psi(w) + \sum_{i=1}^m \max(0, 1 - y_i(w^\top \hat{x}^{(i)} + b) + r^{(i)^\top}|w|),$$

where the second passage in the equations above follows from the fact that h is convex and non-increasing, thus

$$\max_{\delta \in \Delta} h(z(\delta)) = h(\min_{\delta \in \Delta} z(\delta)).$$

The robust linear SVM training problem thus amounts to minimizing the modified regularized loss

$$\min_{w,b} \lambda \|w\|_2^2 + \sum_{i=1}^m h(z_i - r^{(i)^\top}|w|), \quad (7.19)$$

where $z_i = y_i(w^\top \hat{x}^{(i)} + b)$. Further, by observing that

$$h(z - \varrho) \leq h(z) + \varrho$$

for any $\varrho \geq 0$, we may approximate (7.19) with its upper bound

$$\min_{w,b} \lambda \|w\|_2^2 + \bar{r}^\top |w| + \sum_{i=1}^m h(z_i), \quad (7.20)$$

where $\bar{r} = \sum_{i=1}^m r^{(i)}$. The new term $\bar{r}^\top |w|$ that appears in the objective is a weighted ℓ_1 norm term which can therefore be interpreted as an additional regularization term that promotes sparsity in the w parameter.

7.4.2 Robust SVM under ellipsoidal uncertainty

An alternative uncertainty model involves a spherical (or, more generally, ellipsoidal) uncertainty, where for each data point we assume that
$$x^{(i)} = \hat{x}^{(i)} + r_i D u^{(i)},$$
with $D = \mathrm{diag}\,(\sigma_1,\ldots,\sigma_n)$ is a positive-definite diagonal scaling matrix, $\|u^{(i)}\|_2 \leq 1$, and $r_i > 0$. The intuition behind this model is that up to a point-dependent scaling factor r_i, variances are the same across the data points. Following the same approach illustrated in the previous section, and recalling that $\min_{u:\|u\|_2\leq 1} y^\top u = -\|y\|_2$, we obtain that

$$\min_{w,b} \lambda\|w\|_2^2 + \max_{|u^{(i)}|\leq 1} \sum_{i=1}^m h(y_i(w^\top x^{(i)} + b))$$
$$= \min_{w,b} \lambda\|w\|_2^2 + \sum_{i=1}^m \max_{|u^{(i)}|\leq 1} h(y_i(w^\top x^{(i)} + b))$$
$$= \min_{w,b} \lambda\|w\|_2^2 + \sum_{i=1}^m h(\min_{|u^{(i)}|\leq 1} y_i(w^\top x^{(i)} + b))$$
$$= \min_{w,b} \lambda\|w\|_2^2 + \sum_{i=1}^m h(y_i(w^\top \hat{x}^{(i)} + b) + \min_{|u^{(i)}|\leq 1} y_i r_i w^\top D u^{(i)})$$
$$= \min_{w,b} \lambda\|w\|_2^2 + \sum_{i=1}^m h(y_i(w^\top \hat{x}^{(i)} + b) - r_i\|Dw\|_2)$$
$$= \min_{w,b} \lambda\|w\|_2^2 + \sum_{i=1}^m \max(0, 1 - y_i(w^\top \hat{x}^{(i)} + b) + r_i\|Dw\|_2),$$

which represents the modified cost function to be minimized for training a robust linear SVM with ellipsoidal uncertainty. Using again the upper bound $h(z - \varrho) \leq h(z) + \varrho$ we obtain an alternative robust SVM criterion of the form

$$\min_{w,b} \lambda\|w\|_2^2 + \bar{r}\|Dw\|_2 + \sum_{i=1}^m h(z_i), \qquad (7.21)$$

where $\bar{r} = \sum_{i=1}^m r_i$.

7.5 Exercises

Exercise 7.1 *SVM for text classification*

In this exercise, we consider an application of SVM in text classification for volatility prediction. In the file NewsData.mat in the online resources, you are given a 1,470 × 971 matrix where each row corresponds to a published article and each column to the frequency of a word that appears in the article (i.e. this archive contains 1,470 articles and our dictionary contains 971 keywords). Each article is about a certain company. You are also given a 1,470 × 1 vector of labels. An article's label is +1 if the article caused an immediate and significant change (positive or negative) to the company's stock. Otherwise, the label is −1. The data has been divided into a training set which will be used to train you SVM and a validation set which will be used to test the SVM's prediction accuracy.

1. Consider the linear SVM problem in unconstrained form presented in (7.14). To address the text classification problem, implement a linear SVM in the computational environment of your choice (e.g., Python or MATLAB®), tune the penalty parameter c from 0 to 200, and fit w and b for each value of parameter c on the training data set. Plot the training accuracy versus c curve and the validation accuracy versus c curve. Briefly comment on the result.

2. Perform feature selection to reduce the problem scale. In this case, we want to find the keywords that are most important for classification. Let $c = 20$, perform SVM to fit w on the training set. Sort elements of w by their absolute value in descending order. Then perform SVM on part of features. Sweep the number of (the most significant) features to be used in the model from 10 to 200, with increments of 10. Evaluate the performances on the validation set. Comment on the result.

3. Try a different approach to do feature selection. Substitute the ℓ_2 norm penalty with an ℓ_1 norm penalty as

$$\min_{w,b} \|w\|_1 + c \sum_{i=1}^{1470} [1 - y_i(w^\top x^{(i)} + b)]_+.$$

Sweep the parameter c, plot the number of nonzero elements in w versus c curve. Note that due to limited numerical precision, zero elements may not be exactly 0; use a threshold $|w_i| > 10^{-6}$ to detect nonzero elements. Comment on the prediction quality and compare with the results from the previous point.

Exercise 7.2 *Robust hyperplanes for separating boxes and spheres*
Let us consider the case of a maximum margin classifier where the training data set is strictly linearly separable.

- Spherical uncertainty model: assume that the data points are actually unknown but bounded in hyperspheres: for $i = 1, \ldots, m$,

$$x^{(i)} \in \mathcal{X}_i = \{\hat{x}^{(i)} + r u^{(i)} : \|u^{(i)}\|_2 \leq 1\},$$

where $r > 0$, and $\hat{x}^{(i)}$ represent the nominal values of the underlying observations $x^{(i)}$ (see Section 7.4 for further details). Let us suppose that the nominal values $\hat{x}^{(i)}$ are known, instead $u^{(i)}$ is unknown. Find the separating hyperplane that separates the spheres (and not just the points) with the maximum possible radius r.

- Box uncertainty model: Let us now consider a box uncertainty model: for $i = 1, \ldots, m$,

$$x^{(i)} \in \mathcal{X}_i = \{\hat{x}^{(i)} + u^{(i)} : \|u^{(i)}\|_\infty \leq r\},$$

where $r > 0$, and $\hat{x}^{(i)}$ represent the known nominal values of the underlying observations $x^{(i)}$. Find the separating hyperplane that separates the boxes maximizing r.

Exercise 7.3 *Logistic regression for credit card fraud detection*
In this exercise we consider a data set[6] containing transactions made by credit cards in two days of September 2013 by European cardholders. In this data set, we have 492 frauds out of 284,807 transactions. The data set is thus highly unbalanced, since the positive class (frauds) accounts for only 0.172% of all transactions. The data set contains 28 numerical features that are the principal components obtained via PCA on the original features. Due to confidentiality issues, the original features and more background information about the data are not provided. In addition to these transformed features, the time (computed as the number of seconds elapsed between each transaction and the first transaction in the dataset) and the amount of each transaction are also provided.

1. Load the data set contained in the file `creditcard.csv` in the online resources. Discard the information regarding the time of the transactions, since it is uninformative for our purpose.

2. Split the data set into a training set and a testing set. The size of the training set should be 75% of the entire data set.

3. Train a logistic regression model using the training set and then evaluate the performance of the trained model on the test set. Since

[6] This data set is also available at www.kaggle.com/datasets/mlg-ulb/creditcardfraud.

the data set is highly unbalanced with the positive class vastly outnumbered by the negative class, using the accuracy to evaluate the performance of the classifier can be misleading (as discussed in Example 7.1). A more informative evaluation of the performance can be obtained by computing the confusion matrix and the sensitivity of the trained model.

4. Since the training set contains a much higher number of negative samples with respect to positive ones, the training of the model is skewed to predict more accurately the negative samples and be less precise when handling the positive samples. This may result in a higher number of false negatives and a low value of the sensitivity score. However, a false negative (i.e., ignoring an actual fraud) is more dangerous than a false positive (i.e., a normal transaction that is erroneously detected as a fraud). In order to address this problem, we can introduce class weights as shown in (7.10).[7] Compute the confusion matrix and the sensitivity of the model in this case. Compare the results with the ones obtained in the previous point.

[7] In the LogisticRegression function of the sklearn package, you can use the option class_weight='balanced' to introduce weights inversely proportional to class frequencies in the input data.

5. Another method to decrease the number of false negatives, and thus increase the sensitivity, is to decrease the decision threshold γ. By default the decision threshold is usually set to 0.5, try to decrease γ and evaluate the performance in terms of sensitivity of the classifier in this case. Compare the results with the ones obtained in the previous two points.

Exercise 7.4 *Multi-class logistic regression*
In this exercise, we consider a problem of predictive maintenance, where we have to predict if a machine will fail and the type of failure. We consider a data set with 10,000 data points, where each one is described by six features (namely: product type, air temperature, process temperature, rotational speed, torque, and tool wear). The target variable may assume six possible values: no failure or five different types of failures. Hence, this is a multi-class classification problem with six classes.

1. Load the data set contained in the file predictive_maintenance.csv in the online resources.[8] Split the data set into a training set and a testing set. The size of the training set should be 80% of the entire data set.

[8] This is a cleaned version of the data set available at https://archive.ics.uci.edu/dataset/601/ai4i+2020+predictive+maintenance+dataset

2. In Section 7.2.2, we have introduced the multi-class logistic classifier. Train this model[9] using the training set and then evaluate the performance of the trained model on the test set. As in the

[9] In Python you can use the LogisticRegression function with the option multi_class='multinomial'.

previous exercise, also in this case the data set is highly unbalanced since the class zero (i.e., no failure) significantly outnumbers the other classes. For this reason, the performance evaluation should take in consideration not only the accuracy of the classifier, but also the confusion matrix and the sensitivity of the trained model. Try to introduce class weights in order to improve the performance of the classifier.

3. In Remark 7.2, we have seen a general approach, called OvR, for doing multi-class classification using as a basis a binary classifier. Apply this approach to the problem considered in this exercise using a binary logistic regression model.[10] Compare the results with the ones obtained in the previous point.

[10] In Python you can use the LogisticRegression function with the option multi_class='ovr'.

Exercise 7.5 *SVM for credit card fraud detection*

In this exercise we consider the same data set used in Exercise 7.3. We are now interested in training a linear support vector machine to solve this problem.

1. Split the data set into a training set and a testing set. The size of the training set should be 75% of the entire data set. Train a support vector machine using the training set and then evaluate the performance of the trained model on the test set. Compute the confusion matrix and the sensitivity of the trained model to evaluate the performance of the model. As discussed in Exercise 7.3, since the classes are highly unbalanced the confusion matrix and the sensitivity score provide a more informative evaluation than just the accuracy score. Compare the results with the ones obtained in point 3 of Exercise 7.3.

2. Use class weights to address the problem of class imbalance,[11] as shown in Remark 7.4. Compute the confusion matrix and the sensitivity of the model in this case. Compare the results with the ones obtained at the previous point and in point 4 of Exercise 7.3.

[11] In the SVC function of the sklearn package, you can use the option class_weight='balanced' to introduce weights inversely proportional to class frequencies in the input data.

8
Nonlinear classifiers and kernel methods

NONLINEAR CLASSIFIERS EXTEND the linear classification approaches introduced in the previous chapter so to make them able to capture nonlinear patterns in data. Kernel methods achieve this goal via a nonlinear mapping of the data in a higher-dimensional space. We first show why mapping the data in such higher-dimensional space can make the nonlinear separation problem easier. Next, we introduce the so-called kernel trick, which allows us to operate in the original lower-dimensional domain without explicitly computing the mapping of the data points in the higher-dimensional space, thus keeping the computational complexity under control. Then we discuss the solution of the "kernelized" SVM problem via a dual approach.

In the second part of this chapter we discuss *decision trees*, which offer a different approach to nonlinear classification based on partitioning the feature space according to successive dichotomous decisions on the problem features. Decision trees are among the most popular, effective, and interpretable classification algorithms, although they suffer from problems related to data sensitivity and over-fitting. These issues are addressed via *ensembles techniques*, which are discussed at the end of this chapter.

8.1 Motivation

In Section 3.9.2 and Example 6.3 we discussed a linear autoregressive process of order p, named AR(p), for time series:

$$y_t = \sum_{k=1}^{n} w_k y_{t-k} + b + \epsilon_t, \quad t = 1, \ldots,$$

where ϵ_t is a zero-mean white noise process. Let us focus for simplicity on a linear autoregressive process of order 2:

$$y_t = b + w_1 y_{t-1} + w_2 y_{t-2} + \epsilon_t, \quad t = 3, \ldots.$$

Here, y_t is a linear function of y_{t-1}, y_{t-2}, which can be written as $y_t = w^\top x^{(t)} + b + \epsilon_t$, where $w = [w_1, w_2]^\top$ and $x^{(t)}$ are the "feature vectors":

$$x^{(t)} \doteq [y_{t-1}, y_{t-2}]^\top, \quad t = 3, \ldots.$$

Given m observations of the feature vector $x^{(t)}$, we define $X \in \mathbb{R}^{2,m}$ as the matrix where the $(t-2)$th column is equal to $x^{(t)}$, with $3 \leq t \leq m+2$. We can thus estimate w and b by solving, for example, the following least-squares regression problem:

$$\min_{w,b} \|y - X^\top w - b\mathbf{1}\|_2^2,$$

where $y = [y_3, \ldots, y_m]^\top$. After having estimated w and b, we define the prediction rule as follows:

$$\hat{y}_{t+1} = b + w_1 y_t + w_2 y_{t-1} = w^\top x^{(t+1)} + b.$$

Let us now consider a nonlinear autoregressive model, where y_t is a *quadratic* function of y_{t-1}, y_{t-2}:

$$y_t = w_1 y_{t-1} + w_2 y_{t-2} + w_3 y_{t-1}^2 + w_4 y_{t-1} y_{t-2} + w_5 y_{t-2}^2 + b + \epsilon_t,$$

where ϵ_t is a zero-mean white noise process. We can rewrite this model in a linear form as $y_t = w^\top \phi(x^{(t)}) + b + \epsilon_t$, where $w = [w_1, w_2, w_3, w_4, w_5]^\top$ and $\phi(x^{(t)})$ are the augmented feature vectors

$$\phi(x^{(t)}) \doteq [y_{t-1}, y_{t-2}, y_{t-1}^2, y_{t-1} y_{t-2}, y_{t-2}^2].$$

We can then define the prediction rule as

$$\hat{y}_{t+1} = w^\top \phi(x^{(t+1)}) + b.$$

With respect to the linear case, we can observe that the dimension of the feature space has increased in order to make the model linear.

The same behavior can be observed also for classification. Let us consider the following nonlinear decision boundary $\mathcal{H}$ for $x \in \mathbb{R}^2$:

$$\mathcal{H} = \{x \in \mathbb{R}^2 : w_1 + w_2 x_1 + w_3 x_2 + w_4 x_1^2 + w_5 x_1 x_2 + w_6 x_2^2 = 0\}.$$

This can be written in a linear form as $\mathcal{H} = \{x : w^\top \phi(x) = 0\}$, with $\phi(x) \doteq [1, x_1, x_2, x_1^2, x_1 x_2, x_2^2]^\top$ and $w = [w_1, w_2, w_3, w_4, w_5, w_6]^\top$. Also in this case, the dimension of the feature space has increased in order to make the separation surface linear in the augmented space.

Example 8.1 Figure 8.1 shows two examples of datasets with classes that are not linearly separable in the original space (see Figure 8.1(a)), but that can become linearly separable by applying a suitable nonlinear feature mapping ϕ (see Figure 8.1(b)). In this example, the feature mapping has been selected ad-hoc and, in particular, we used the mapping

$$\phi(x) = (x_1^2, \sqrt{2}x_1x_2, x_2^2)$$

for the data in the first row in Figure 8.1, and

$$\phi(x) = \left(x_1, x_2, \sqrt{x_1^2 + x_2^2}\right)$$

for the data in the second row.

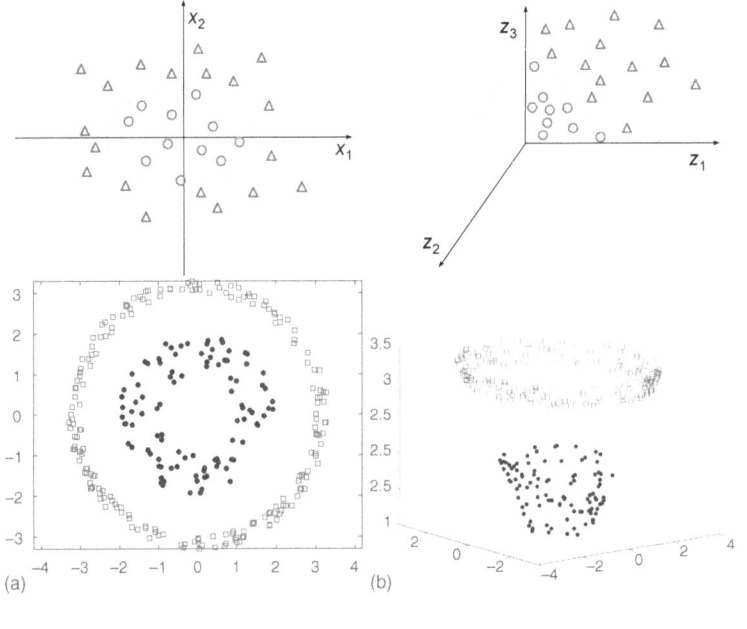

Figure 8.1 Datasets that are not linearly separable (a), but that can become linearly separable by applying a suitable nonlinear mapping (b).

In principle, we can always perform a nonlinear mapping and augment the dimension of the feature space so to make the data approximately linearly separable, as illustrated in Figure 8.2. As already discussed in Section 6.2, this amounts to introducing a feature map function $\phi : \mathbb{R}^n \to \mathbb{R}^d$ and considering the mapped data points

$$\Phi \doteq \left[\phi(x^{(1)}) \cdots \phi(x^{(m)})\right] \in \mathbb{R}^{d,m}$$

in place of the original data points $X = [x^{(1)} \cdots x^{(m)}] \in \mathbb{R}^{n,m}$, where typically we have $d \gg n$. The problem, however, is how to determine a suitable mapping ϕ. In the remainder of this chapter we discuss an indirect approach for determining such a transformation, which also provides an efficient computational method for the training of the ensuing classifier.

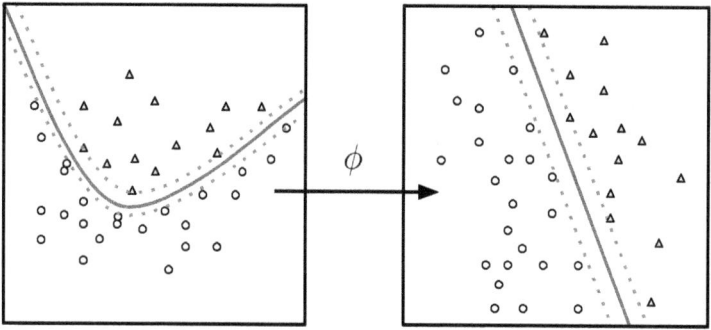

Figure 8.2 The mapping $\phi(x)$ maps the data points in a space where they are linearly separable.

8.2 Kernel trick

Suppose next we are given a feature mapping $\phi\colon \mathbb{R}^n \to \mathbb{R}^d$, although we shall soon see that it is not necessary to know this mapping explicitly, since it can be derived indirectly from the kernel. We use the simplified notation $\phi^{(i)}$ to denote the transformed feature vector $\phi(x^{(i)})$.

In Chapters 6 and 7, we have seen that many linear classification and regression problems can be written as a regularized learning problem[1] of the form

$$\min_{w} \mathcal{L}(\Phi^\top w, y) + \lambda \|w\|_2^2, \qquad (8.1)$$

where $\Phi = [\phi^{(1)}, \ldots, \phi^{(m)}] \in \mathbb{R}^{d,m}$, $y \in \mathbb{R}^m$ contains output response values or labels, $w \in \mathbb{R}^d$ contains the parameters of the classifier or regression model, $\lambda \geq 0$ is a regularization parameter, and $\mathcal{L}$ is a loss function that depends on the problem considered. Some of the most common choices for $\mathcal{L}$ that we have seen in Chapters 6 and 7 are recalled here:

[1] For simplicity, in this chapter we consider the ℓ_2 norm penalty only and omit the offset term b.

- Squared loss for least-squares based regularized regression,

$$\mathcal{L}(z, y) = \|z - y\|_2^2 = \sum_{i=1}^{m} (x_i - y_i)^2.$$

- Hinge loss for SVMs:

$$\mathcal{L}(z, y) = \sum_{i=1}^{m} \max(0, 1 - y_i z_i)$$

- Logistic loss for logistic regression:

$$\mathcal{L}(z, y) = -\sum_{i=1}^{m} \log(1 + e^{-y_i z_i}).$$

In all cases, given a new data point $x \in \mathbb{R}^n$, the prediction or classification rule depends only on $w^\top \phi(x)$.

We now focus on solving problem (8.1) by introducing the following key result.

Theorem 8.1 *For the generic problem* (8.1) *the optimal w lies in the span of the data points* $(\phi^{(1)}, \ldots, \phi^{(m)})$, *that is,*

$$w = \Phi v$$

for some vector $v \in \mathbb{R}^m$.

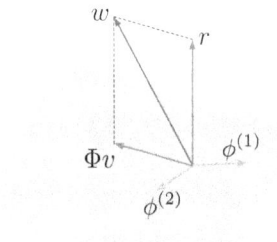

Figure 8.3 Vector $w \in \mathbb{R}^3$ can be represented as the sum of two orthogonal vectors: $w = \Phi v + r$, where $\Phi = [\phi^{(1)}, \phi^{(2)}]$, $v \in \mathbb{R}^2$, and $\Phi^\top r = 0$.

The above theorem can be easily proved by considering that, for any matrix $\Phi \in \mathbb{R}^{d,m}$, every $w \in \mathbb{R}^d$ can be written as the sum of two orthogonal vectors, one in the range of Φ and the other orthogonal to it:

$$w = \Phi v + r,$$

where $v \in \mathbb{R}^m$, and $\Phi^\top r = 0$ (i.e., r is in the nullspace $\mathcal{N}(\Phi^\top)$). Figure 8.3 depicts an example of the above decomposition. Thanks to this result, known as the *fundamental theorem of linear algebra*, we can rewrite problem (8.1) as

$$\min_{v \in \mathbb{R}^m, r \in \mathcal{N}(\Phi^\top)} \mathcal{L}(\Phi^\top(\Phi v + r), y) + \lambda \|\Phi v + r\|_2^2.$$

Since r and Φv are orthogonal, we have that $\|\Phi v + r\|_2^2 = \|\Phi v\|_2^2 + \|r\|_2^2$. We thus obtain the following problem:

$$\min_{v \in \mathbb{R}^m, r \in \mathcal{N}(\Phi^\top)} \mathcal{L}(\Phi^\top \Phi v, y) + \lambda \|\Phi v\|_2^2 + \|r\|_2^2,$$

from which we observe that the optimal r is equal to 0, whence the optimal w will be of the form $w = \Phi v$, which is the claim of Theorem 8.1.

Given the above result, we observe that the mapped feature matrix Φ appears in (8.1) only through the $m \times m$ positive semidefinite matrix $K \doteq \Phi^\top \Phi$, that is, we have

$$\min_v \mathcal{L}(Kv, y) + \lambda v^\top K v. \tag{8.2}$$

Matrix $K = \Phi^\top \Phi$ is called the *kernel* matrix and it contains the scalar products between all data point pairs:

$$K_{ij} = \phi^{(i)\top} \phi^{(j)}, \quad 1 \leq i, j \leq m. \tag{8.3}$$

Reformulating the learning problem as in (8.2) can result in a significant computational advantage. Indeed, once K is formed (this takes

$O(m^2 d)$ operations), then the training problem has only m variables. When $d \gg m$, this leads to a dramatic reduction in problem size.

We can use the result of Theorem 8.1 to reformulate also the prediction/classification rule, since

$$w^\top \phi(x) = v^\top \Phi^\top \phi(x) = v^\top k,$$

where $k \doteq \Phi^\top \phi(x) = [\phi(x)^\top \phi^{(1)}, \ldots, \phi(x)^\top \phi^{(m)}]^\top$, which shows that also the prediction can be made by evaluating the inner products $\phi(x)^\top \phi^{(i)}$, $i = 1, \ldots, m$.

As discussed in the previous section, the dimension d of the feature space is usually much higher than the dimension of the original data space n. Therefore, reformulating the learning problem as in (8.2) provides a significant advantage, since the dimension of the learning problem does not depend on d but only on m. Hence, the main computational effort involved in solving the learning problem and making a prediction depends on our ability to quickly evaluate the scalar products needed to compute the kernel matrix. To this end, we introduce a *kernel function* defined as

$$k(x,z) \doteq \phi(x)^\top \phi(z). \tag{8.4}$$

This function can be used to compute the kernel matrix:

$$K_{ij} = k(x^{(i)}, x^{(j)}), \tag{8.5}$$

and the classification rule:

$$\begin{aligned} v^\top \Phi^\top \phi(x) &= [\phi(x)^\top \phi(x^{(1)}), \ldots, \phi(x)^\top \phi(x^{(m)})]v \\ &= [k(x, x^{(1)}), \ldots, k(x, x^{(m)})]v. \end{aligned} \tag{8.6}$$

The kernel function also provides information about the metric in the feature space, since we can compute the distance between two points in the feature space using the kernel function:

$$\|\phi(x) - \phi(z)\|_2^2 = k(x,x) - 2k(x,z) + k(z,z). \tag{8.7}$$

If we consider the classification problem with quadratic decision boundary discussed above, given two vectors $x, z \in \mathbb{R}^2$ the kernel function is defined as

$$k(x,z) = \phi(x)^\top \phi(z) = (1 + x^\top z)^2.$$

More generally, when $\phi(x)$ is the vector formed with all the products between the components of $x \in \mathbb{R}^n$, up to degree q, then for any two vectors $x, z \in \mathbb{R}^n$,

$$k(x,z) = \phi(x)^\top \phi(z) = (1 + x^\top z)^q.$$

We observe that, using such polynomial kernel, the computational effort grows linearly in n. This represents a dramatic reduction in speed over the "brute force" approach[2] where we first form $\phi(x)$, $\phi(z)$ and then evaluate $\phi(x)^\top \phi(z)$, resulting in a computational effort that grows as $\binom{n+q}{q}$. This is the so-called *kernel trick*: kernel methods use only pairwise similarity comparisons between the data observations in the original domain, instead of explicitly applying the mapping $\phi(x)$ and representing the data in the higher-dimensional feature space. Such pairwise similarities can be computed using the kernel function as shown in (8.5), (8.6), and (8.7) without even requiring to know the feature map explicitly. Indeed, it is only necessary to know the kernel function. This results in a dramatic reduction of the computational effort.

It is important to underline that we cannot choose the kernel function arbitrarily, since it has to satisfy property (8.4) for some ϕ. There is a large variety of kernels that can be used and some of them are adapted to specific structure of data (e.g., text, images, etc). One of the most common is the Gaussian kernel function (also called the radial basis kernel):

$$k(x,z) = \exp\left(-\frac{\|x-z\|_2^2}{2\sigma^2}\right),$$

where $\sigma > 0$ is a scale parameter. This kernel allows us to ignore points that are too far apart and it corresponds to a nonlinear mapping ϕ to an infinite-dimensional feature space.

8.3 Nonlinear SVM

In Section 7.3 we discussed a class of binary classifiers called linear support vector machines (SVMs), which are based on the idea of geometric separation of points via hyperplanes. However, as we discussed in the previous sections, it is often the case that the data is far from being linearly separable, see, for example, Figure 8.4.

In order to extend SVMs to the case where the two classes are structurally not linearly separable, we can follow the same approach presented in the previous sections and introduce a nonlinear mapping ϕ such that point classes that were not linearly separable in the original input space may become approximately linearly separable in the higher-dimensional feature space defined by ϕ. Using this approach, the decision boundary in the original input space is now nonlinear, and determined by the equation

$$w^\top \phi(x) + b = 0.$$

[2] We recall that the number of monomials in n variables of degree less than or equal to q is $\binom{n+q}{q}$. This number rapidly becomes large as n and q grow. For example, for $n = 10$, $q = 5$ we have $d = \binom{n+q}{q} = 3{,}003$, while for $n = 100$, $q = 5$ we have $d = \binom{n+q}{q} = 96{,}560{,}646$.

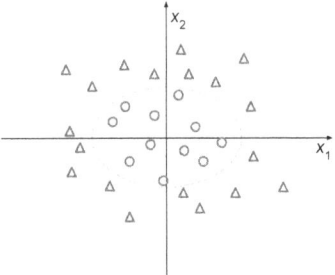

Figure 8.4 An example of two classes that are not linearly separable: no hyperplane of the form $w^\top x + b = 0$ can separate them.

8 NONLINEAR CLASSIFIERS AND KERNEL METHODS

The SVM problem in the mapped space retains the very same structure as in the linear case. It is just needed to replace x with $\phi(x)$, and hence pose the learning problem on the new *mapped* feature data $(\phi^{(i)}, y_i), i = 1, \ldots, m$, where

$$\phi^{(i)} \doteq \phi(x^{(i)}), \quad i = 1, \ldots, m.$$

Therefore, we can rewrite the SVM problem (7.13) as follows:

$$\min_{w,b,\xi} \quad \|w\|_2^2 + c \sum_{i=1}^m \xi_i, \tag{8.8}$$

$$\text{s.t.: } y_i(w^\top \phi^{(i)} + b) \geq 1 - \xi_i \quad i = 1, \ldots, m,$$

$$\xi_i \geq 0 \quad i = 1, \ldots, m.$$

It can be proved that the above problem admits a Lagrangian dual problem achieving the same objective value (i.e., strong duality holds). The dual problem is formulated as

$$\max_{\alpha, \mu} \quad g(\alpha, \mu), \tag{8.9}$$

$$\text{s.t.: } \alpha \geq 0,$$

$$\mu \geq 0,$$

where $\alpha, \mu \in \mathbb{R}^m$, and

$$g(\alpha, \mu) = \min_{w,b,\xi} L(w, b, \xi, \alpha, \mu)$$

$$\doteq \min_{w,b,\xi} \|w\|_2^2 + c \sum_{i=1}^m \xi_i$$

$$- \sum_{i=1}^m \alpha_i \left(y_i(b + w^\top \phi^{(i)}) - (1 - \xi_i) \right) - \sum_{i=1}^m \mu_i \xi_i.$$

The Lagrangian L is convex in b, w, ξ. The minimum of L is thus obtained for

$$\frac{\partial L}{\partial b} = -\sum_{i=1}^m \alpha_i y_i = 0,$$

$$\nabla_w L = w - \sum_{i=1}^m \alpha_i y_i \phi^{(i)} = 0,$$

$$\nabla_\xi L = c - \alpha - \mu = 0.$$

Substituting this into the objective function of (8.9) we eliminate μ and obtain

$$g(\alpha) = -\sum_{i=1}^m \sum_{j=1}^m \alpha_i \alpha_j y_i y_j \phi^{(i)\top} \phi^{(j)} + \sum_{i=1}^m \alpha_i$$

$$= -(\alpha \odot y)^\top K (\alpha \odot y) + \mathbf{1}^\top \alpha,$$

where $\alpha \odot y$ is the vector whose ith entry is $\alpha_i y_i$, and $K \in \mathbb{R}^{m,m}$ is the kernel matrix as defined in (8.3). We thus find the solution of problem (8.8) by first solving the dual problem

$$\max_{\alpha} \ -(\alpha \odot y)^\top K(\alpha \odot y) + \mathbf{1}^\top \alpha, \tag{8.10}$$

$$\text{s.t.:} \quad y^\top \alpha = 0,$$
$$\alpha \geq 0,$$
$$\alpha \leq c.$$

Once we have the optimal α from (8.10), we use equation $\nabla_w L = 0$ to obtain the optimal primal vector w as

$$w = \sum_{i=1}^{m} \alpha_i y_i \phi^{(i)}. \tag{8.11}$$

The features $\phi^{(i)}$ for which $\alpha_i > 0$ are called support vectors, since the solution w can be expressed as a linear combination of these vectors only. We also observe that among the support vectors there are some lying on the margin slab, that is, some for which $\xi_i = 0$, or otherwise we could increase the margin. Then, we can use any of these support vectors to solve problem (8.8) for the bias term b.[3] For numerical stability, typically it is better to compute an average of all the ensuing b values.

As already discussed in the previous section, there is a strong computational advantage in solving problem (8.10) instead of (8.8). Indeed, if we suppose to use a very high-dimensional feature map, that is, $d \gg m$, problem (8.8) has a much higher number of decision variables with respect to problem (8.10): the first one has $d + m + 1$ variables, whereas the latter one only has m decision variables. Further, we recall that solving (8.10) does not even require knowing the feature map ϕ explicitly, since ϕ appears in (8.10) only via the kernel matrix.

Once the model is learned, the classification rule is based on the sign of $b + w^\top \phi(x)$. Since the optimal w is given by (8.11), we can rewrite this rule as

$$b + \sum_{i=1}^{m} \alpha_i y_i k(x^{(i)}, x),$$

where k is the kernel function as defined in (8.7). Hence, all that is needed for classification is knowledge of the kernel function: instead of explicitly deciding which feature map ϕ to use, we can just decide what kernel function to use, and then the feature map is indirectly implied by this choice. In practice, standard kernel types (such as Gaussian or polynomial) are available in most numerical platforms, and the actual choice is made via trials and validation.

[3] This is because $\alpha_i > 0$ implies, by the so-called complementary slackness property, that the corresponding constraint is saturated at the optimum, which means that $y_i(w^\top \phi^{(i)} + b) = 1 - \xi_i$. Futher, since $\xi_i = 0$ holds for the points on the boundary of the margin slab (these can be determined, using again complementary slackness, in correspondence with the indices where $\mu_i > 0$), we can determine b algebraically as $b = (1 - y_i w^\top \phi^{(i)})/y_i$.

Example 8.2 (*Nonlinear SVM with Gaussian kernel*) We consider a dataset with two classes as depicted in Figure 8.5. We observe that the default

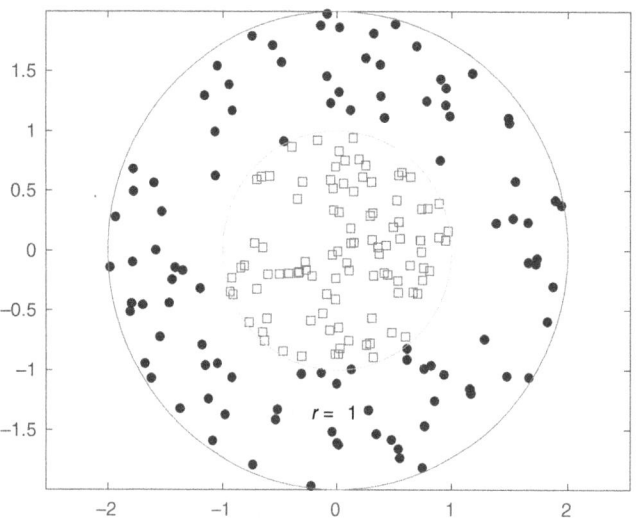

Figure 8.5 A dataset with two classes that are not linearly separable.

linear classifier is obviously unsuitable for this problem, since the model is circularly symmetric. We can thus train an SVM with Gaussian kernel and obtain the result shown in Figure 8.6; these results can be reproduced by running the MATLAB® file ex_svm.m in the online resources. We observe that such nonlinear SVM can correctly classify the data points of this dataset, with no error on the training set.

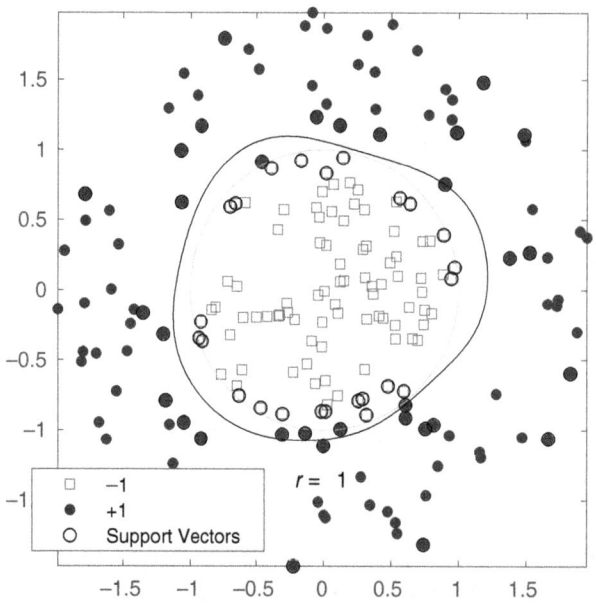

Figure 8.6 A nonlinear SVM with Gaussian kernel.

8.4 Decision trees and random forests

8.4.1 Decision trees

A treatment of nonlinear classifiers would not be complete without discussion of one of the most popular classes of classification methods, that is, *decision trees* (DTs). The popularity of such methods comes from the fact that a decision tree leads to the classification output by guiding the user through a sequence of flow-chart-like decision steps where, starting at a root node, at each step one particular feature is examined and the tree is branched depending on the possible feature values, until one reaches the decision leaves containing the classification output. The logical structure of a decision tree can be described by means of a nested if-then-else architecture, and also displayed graphically for easy interpretability.

In this section we illustrate the classical *Iterative Dichotomizer 3* (ID3) algorithm[4] for building decision trees on problems with discrete features. The ID3 approach works iteratively, starting from the whole data set and selecting an initial feature according to a specific information criterion that is discussed next. The data set is then partitioned based on the possible discrete values of the selected feature, and the algorithm is recursively applied to each subset (split) of the data, by selecting each time a new feature that has not been selected before. The recursion on a subset terminates when a subset is found in which all elements belong to the same output class, or when there are no more features to be considered, or yet when the subset is empty. Throughout this section we shall use the data set shown in Table 8.1 as a working example to illustrate the concepts involved in the construction of a decision tree. The data in Table 8.1 is a subset of $m = 20$ samples taken from the loan data set described in Section 2.1, with features reduced to the following $n = 4$ discrete attributes (the originally continuous "Applicant Income" feature has been here discretized into three classes):

[4] See J. R. Quinlan, "Induction of Decision Trees," *Machine Learning*, vol. 1(1), pp. 81–106, 1986.

$$\begin{aligned}
\text{Education} &\in \{\text{Not Graduate, Graduate}\} \\
\text{Applicant Income} &\in \{\text{Low, Medium, High}\} \\
\text{Credit History} &\in \{\text{Bad, Good}\} \\
\text{Property Area} &\in \{\text{Rural, Semiurban, Urban}\},
\end{aligned}$$

and with the output class being the "Loan Status," which can be approved (Y), or not approved (N).

We now come to the main question in ID3: how to select which feature to test at each given iteration? The rationale behind this choice in

Applicant ID	Education	Applicant Income	Credit History	Property Area	Loan Status
1	Graduate	High	Good	Urban	Y
2	Graduate	Medium	Good	Rural	N
3	Graduate	Medium	Good	Urban	Y
4	Not Graduate	Medium	Good	Urban	Y
5	Graduate	High	Good	Urban	Y
6	Graduate	High	Good	Urban	Y
7	Not Graduate	Medium	Good	Urban	Y
8	Graduate	Low	Bad	Semiurban	N
9	Graduate	High	Good	Urban	Y
10	Graduate	High	Good	Semiurban	N
11	Graduate	Medium	Good	Urban	Y
12	Graduate	Medium	Good	Urban	Y
13	Graduate	Medium	Good	Urban	Y
14	Graduate	Low	Good	Rural	N
15	Graduate	Low	Good	Urban	Y
16	Graduate	High	Good	Urban	Y
17	Graduate	Medium	Bad	Urban	N
18	Not Graduate	Medium	Good	Rural	N
19	Not Graduate	High	Bad	Urban	N
20	Graduate	Medium	Bad	Semiurban	Y

Table 8.1 Extract from loan data set with reduced features, see `loan_data_set_reduced.xlsx` in the online resources.

ID3 is to assess how well a given attribute alone separates the training examples according to their target classification, and it does so by using the statistical concept of *entropy* of the empirical distribution of the output class in a subset of the samples. Supposing the output label has K classes, and observing the empirical frequencies $p_1, \ldots, p_K$[5] of these K classes in a given subset S of the data, with $p_i \geq 0$ and $\sum_i p_i = 1$, the entropy is defined as

$$H \doteq -\sum_{i=1}^{K} p_i \log_2 p_i, \qquad (8.12)$$

which is a quantity in the range $[0, \log_2 K]$ that provides a measure of the *impurity* of the samples. Low entropy means that members of S tends to belong to the same class. For instance, in our loan example we have $K = 2$ output classes, hence the entropy takes values in the range $[0, 1]$; if all elements in a subset S have the Y label, then the entropy is zero. Contrary, high entropy means that members of S tends to be well mixed (impure); for example, if the elements in S are 50% Y and 50% N, then the entropy is one, see Figure 8.7.

[5] Here, p_i denotes the empirical probability with which the ith label appears among the samples in the considered subset. If there are N samples in the subset, and N_i of them have i as the output label, then $p_i \doteq N_i/N$, $i = 1, \ldots, K$.

Given a subset S of the data, we may compute its entropy $H(S)$ as well as the entropy of S conditional on the knowledge of a feature f, which we shall denote by $H(S|f)$, where

$$H(S|f) = \sum_{v \in \text{val}(f)} \text{Prob}\{f = v\} H(S|f = v), \quad (8.13)$$

where $\text{val}(f)$ denotes the set of discrete values that feature f may take, $\text{Prob}\{f = v\} = |S_v|/|S|$, where S_v is the subset of S where $f = v$, and $H(S|f = v)$ denotes the entropy of the subset of S for which feature f is equal to v, that is,

$$H(S|f = v) = -\sum_{i=1}^{K} \text{Prob}\{y_i|f = v\} \log_2 \text{Prob}\{y_i|f = v\}.$$

Now, the *difference* between the original entropy $H(S)$ of S and the conditional entropy $H(S|f)$ defines the *information gain* (IG)

$$IG(S, f) \doteq H(S) - H(S|f), \quad (8.14)$$

which indeed measures the expected reduction in entropy caused by knowing the value of attribute f. The idea is then to evaluate the information gain for all the available features, and select the feature that yields the largest IG value.

To exemplify the above approach, we consider the data in Table 8.1, and start by evaluating the entropy of the entire data set S. Counting the output labels in each of the two classes we obtain that

$$p_Y = \frac{13}{20} = 0.65, \quad p_N = \frac{7}{20} = 0.35,$$

hence

$$H(S) = -p_Y \log_2 p_Y - p_N \log_2 p_N = 0.9341.$$

Next, we consider all features one by one, and for each feature we compute the information gain. Starting with $f = $ Education, which has two possible outcomes {Not Graduate, Graduate}, we have that:

- if $f = $ Education and $v = $ Not Graduate, there are $|S_v| = 4$ elements in the subset of the data for which $f = v$ and, among these, 2 have output Y and 2 have output N, therefore

$$H(S|\text{Education} = \text{Not Graduate}) = -\frac{2}{4} \log_2 \frac{2}{4} - \frac{2}{4} \log_2 \frac{2}{4} = 1;$$

- if $f = $ Education and $v = $ Graduate, there are $|S_v| = 16$ elements in the subset of the data for which $f = v$ and, among these, 11 have output Y and 5 have output N, therefore

$$H(S|\text{Education} = \text{Graduate}) = -\frac{11}{16} \log_2 \frac{11}{16} - \frac{5}{16} \log_2 \frac{5}{16} = 0.8960.$$

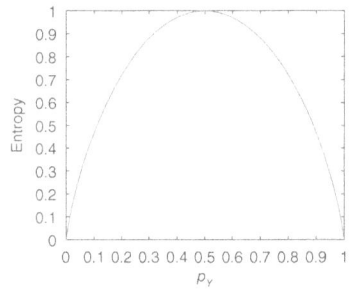

Figure 8.7 Entropy as a function of probability p_Y of the positive class (the probability of the negative class being $p_N = 1 - p_Y$).

Using (8.13) we then compute the conditional entropy as

$$H(S|\text{Education}) = \frac{|\text{Not Graduate}|}{20} H(S|\text{Education} = \text{Not Graduate})$$
$$+ \frac{|\text{Graduate}|}{20} H(S|\text{Education} = \text{Graduate})$$
$$= \frac{4}{20} 1 + \frac{16}{20} 0.8960 = 0.9168,$$

and then the IG as

$$\text{IG}(S, \text{Education}) = H(S) - H(S|\text{Education}) = 0.9341 - 0.9168 = 0.0173.$$

We next proceed to compute the IG for the other features. Considering $f = $ Applicant Income, which has three possible outcomes {Low, Medium, High}, we have that:

- if $f = $ Income and $v = $ Low, there are $|S_v| = 3$ elements in the subset of the data for which $f = v$ and, among these, one has output Y and two have output N, therefore

$$H(S|\text{Income} = \text{Low}) = -\frac{1}{3} \log_2 \frac{1}{3} - \frac{2}{3} \log_2 \frac{2}{3} = 0.9183;$$

- if $f = $ Income and $v = $ Medium, there are $|S_v| = 10$ elements in the subset of the data for which $f = v$ and, among these, 7 have output Y and 3 have output N, therefore

$$H(S|\text{Income} = \text{Medium}) = -\frac{7}{10} \log_2 \frac{7}{10} - \frac{3}{10} \log_2 \frac{3}{10} = 0.8813;$$

- if $f = $ Income and $v = $ High, there are $|S_v| = 7$ elements in the subset of the data for which $f = v$ and, among these, 5 have output Y and 2 have output N, therefore

$$H(S|\text{Income} = \text{Medium}) = -\frac{5}{7} \log_2 \frac{5}{7} - \frac{2}{7} \log_2 \frac{2}{7} = 0.8631.$$

Using (8.13) we then compute the conditional entropy as

$$H(S|\text{Income}) = \frac{|\text{Low}|}{20} H(S|\text{Income} = \text{Low})$$
$$+ \frac{|\text{Medium}|}{20} H(S|\text{Income} = \text{Medium})$$
$$+ \frac{|\text{High}|}{20} H(S|\text{Income} = \text{High})$$
$$= \frac{3}{20} 0.9183 + \frac{10}{20} 0.8813 + \frac{7}{20} 0.8631 = 0.8804,$$

and then the IG as

$$\text{IG}(S, \text{Income}) = H(S) - H(S|\text{Income}) = 0.9341 - 0.8804 = 0.0537.$$

Proceeding in an analogous way we obtain

$$\text{IG}(S, \text{Credit}) = H(S) - H(S|\text{Credit}) = 0.9341 - 0.8113 = 0.1228,$$

and

$$\text{IG}(S, \text{Property}) = H(S) - H(S|\text{Property}) = 0.9341 - 0.5519 = 0.3822.$$

We thus conclude that the feature with the largest IG is the Property Area, and therefore the ID3 algorithm chooses this feature as the root node of the tree for performing the first data split; the corresponding initial part of the decision tree is shown in Figure 8.8.

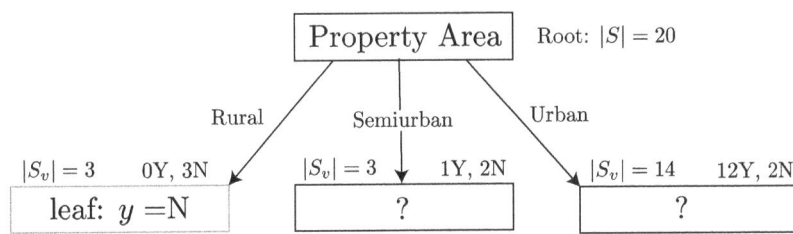

Figure 8.8 Root node and first level of the decision tree on the loan data.

After the first split using the Property Area feature we find three level-1 nodes corresponding to data subsets where the considered feature has a specific attribute. The first node on the left is actually a leaf node, since all elements in its subset have the same label N, hence the recursion stops for that node. For each of the other two nodes in level 1, we repeat the previous procedure and compute the IG relative to the not-yet-explored features and select the feature yielding the maximum IG. For the subset at level 1 with Property Area=Semiurban, we have IG= 0 for the Education feature, IG= 0.9183 for the Applicant Income feature, and IG= 0.2516 for the Credit History feature. We then choose the Applicant Income feature, which splits the node into three subsets, one set corresponding to Applicant Income=Low, consisting of one element with N label, one set corresponding to Applicant Income=Medium, consisting of one element with Y label, and one set corresponding to Applicant Income=High, consisting of one element with N label. All these nodes are therefore leaf nodes. For the subset at level 1 with Property Area=Urban, we have instead IG= 0.0496 for the Education feature, IG= 0.0173 for the Applicant Income feature, and IG= 0.5917 for the Credit History feature. So, we choose Credit History for splitting this node, which result into two leaf nodes, one set corresponding to Credit History=Bad, consisting of two N labels, and one set corresponding to Credit History=Good, consisting of 12 Y labels. The full decision tree is shown in Figure 8.9.

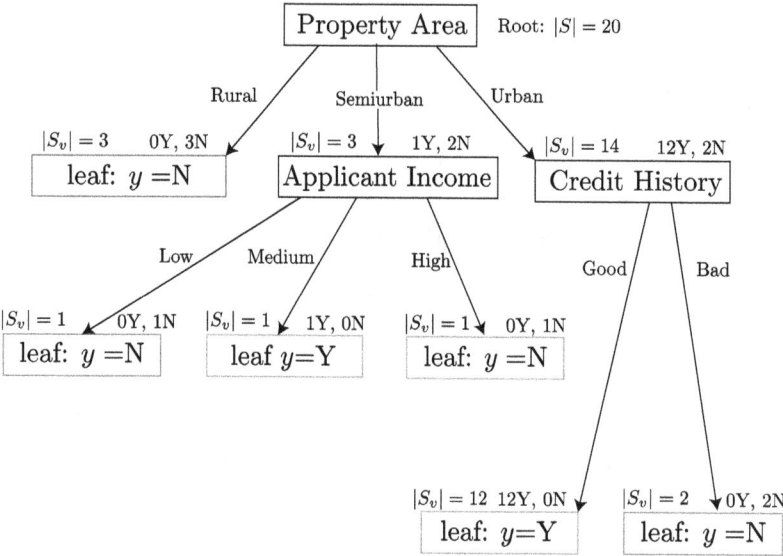

Figure 8.9 Full decision tree on the loan data.

Our working example clearly has too few data points to be of any practical significance. However, this simple example is sufficient for highlighting some key features of the ID3 approach. First, ID3 is a *greedy* algorithm, which at each stage maximizes the local IG, and thus tends to create "short," or simple, trees that reach the decision leaves as soon as possible. In our example, it does so by even neglecting completely one of the features (Education). The construction in the example also highlight a second aspect, that is, that the described approach tends to *overfit* the data, by creating a structure that is highly dependent on the observed data set. Indeed, it could be easily verified that by perturbing the data set slightly, or adding a few further observations, the structure of the decision tree is subject to change drastically. Overfitting in plain decision trees is also due to the fact that the hypothesis space is *complete*, meaning that *any* finite discrete-valued function can be described by a decision tree. ID3 is therefore not affected by bias. Instead, the risk of overfitting is ever present. Also, ID3 maintains only a single current hypothesis (tree), as it searches through the space of decision trees. It does not have the ability of determining how many alternative DTs are consistent with the training data. Further, ID3 in its basic form performs no backtracking: once it selects an attribute to test at some level, it proceeds without ever reconsidering this choice, which yields local (but possibly not global) optimality of the constructed tree. As already observed, ID3 prefers shorter trees over longer ones, that is, trees that place high information gain close to the root are preferred over those that do not. ID3's inductive bias (i.e., the a-priori structure that should permit generalization) is hence towards simple trees (Occam's razor).

Reduced-error pruning As we discussed, plain DTs present the risk of overfitting, that is, of resulting in a model with small (or zero) error on the training set, which then may perform poorly on unseen test instances. An effective approach for reducing overfitting is the so-called *reduced-error pruning*, whereby the DT is first constructed using a training set of examples. Then, each node of the resulting tree is checked as a candidate for pruning (i.e., for removing the entire subtree rooted at that node) and assigning to that node the majority vote of the training examples associated with that node. Candidates for pruning are evaluated on the basis of a validation set of examples: a node may be pruned if the resulting reduced tree has a better or the same accuracy than the tree before reduction, on the validation data set. Nodes are pruned iteratively, always choosing the node whose removal most increases the model accuracy over the validation set.

Extensions The basic ID3 approach can be modified and extended in several ways. First, other criteria besides the Information Gain can be used for selecting which feature to test. Common alternatives used in the literature include for instance the Gain Ratio, the Gini Index (used in the CART algorithm), and the Kolmogorov–Smirnov index. Also, different pruning approaches can be employed, such as minimum error pruning or cost-complexity pruning.[6]

A direct descendant of ID3 is the C4.5 algorithm by Quinlan,[7] which extends ID3 by, among other things, handling both continuous and discrete attributes and being able to deal with training data with missing attribute values. A simple idea for using a decision tree in the presence of continuous features is to discretize the range of real-valued values of each continuous feature by splitting it into a given number of intervals and assigning a discrete label to each of these intervals. In the C4.5 algorithm the number of splits and the split thresholds are determined internally by the algorithm so to provide optimized performance.

8.4.2 Ensemble methods and random forests

We observed that DT classifiers may overfit and thus exhibit high variance on validation data. Also, the structure of the tree may be

[6] See, for example, L. Rokach and O. Maimon, "Decision Trees," 2005.
[7] See J. R. Quinlan, *C4.5: Programs for Machine Learning*, Burlington, MA: Morgan Kaufmann, 1993.

sensitive to the specific data set. Ideally, if we had several (say, N) independent training data sets, we could improve the variance of the prediction by constructing N models, one per training data set, and then *averaging* the individual models' predictions $\hat{y}_i$ to obtain a fused prediction $\bar{y}$. This multiple-model, or ensemble, approach is not limited to decision trees, but can actually be applied to any regression or classification model. In the case of regression models, the fused prediction is properly the sample average of the individual model predictions, that is, $\bar{y} = \frac{1}{N}\sum_i \hat{y}_i$, while in the case of classifiers and categorical predictions the fused prediction is given by the majority label of the individual models' predicted labels. In the case of regression and independent predictions having each the same variance, we know that the variance of the averaged prediction $\bar{y}$ is reduced by a factor of $1/N$ with respect to the variance of the individual models' predictions, this suggesting more generally that fusing multiple models is indeed effective for reducing the prediction variance.

A key practical problem, however, arises from the fact that we typically have only *one* training set of data to work with, and not multiple ones. This difficulty can be circumvented by using bootstrap sampling in a model aggregation approach known as *bootstrap aggregation*, or *bagging*:[8]

[8] L. Breiman, "Bagging Predictors," Machine Learning, vol. 24, pp. 123–140, 1996.

1. Bootstrap sampling: given a training data set X containing n training examples, create data set X_i, $i = 1, \ldots, N$, by drawing n examples uniformly at random with replacement from X.

2. Train N distinct models, one on each X_i.

3. Classify a new instance by majority vote or average of the individual models' outputs.

The bagging approach to ensemble model construction is schematized in Figure 8.10.

Random forests Random forests (RFs) are ensemble methods designed specifically for decision tree classifiers. Random forests introduce two sources of randomness into the model construction: bagging and random input vectors. Via bagging, each tree model is grown using a bootstrap sample of the original training data. Additional randomness is introduced since in the construction of each tree the best split at nodes is chosen using a random sample of $r < n$ attributes, instead of all n attributes. Once trained, each tree in the forest provides an output label in correspondence to a given input vector of features, and the overall output of the RF model is obtained by taking the majority vote of the output labels. Interestingly, in RFs the user also has

8.4 DECISION TREES AND RANDOM FORESTS

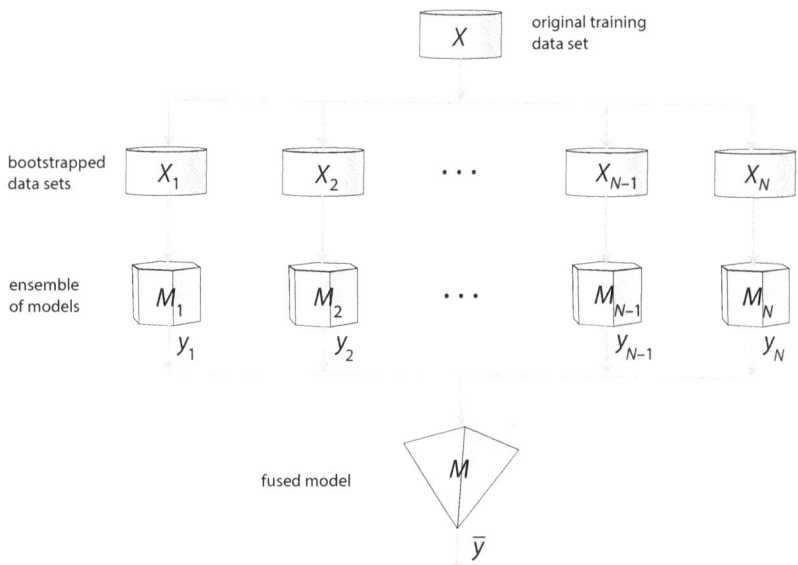

Figure 8.10 Schematic representation of a fused model construction via bootstrap aggregation.

access to the whole distribution of the output labels from the N component trees. Thus, not only we have a point prediction (the majority label), but also an assessment of its reliability coming from the distribution of the labels. To exemplify, in a binary classification problem of positive and negative samples, it is different if we know that the output is positive because 51% of the trees voted for positive, or if it positive because 90% of the trees voted for positive.

Example 8.3 Random forest algorithms are available in most numerical computing environments, including MATLAB® and Scikit-learn under Python. For example, we run a RF classifier on the full loan data set,[9] randomly splitting the data into a training set containing 70% of the data (i.e., $m = 336$ training samples), and a test set containing the remaining 30% of the data. Using the RandomForestClassifier of sklearn, we grew a forest of $N = 100$ trees and we obtained on the test set the confusion matrix shown in Figure 8.11, which shows that the designed RF classifier has precision $96/(96 + 16) = 0.8571$, sensitivity (recall) $96/(96 + 9) = 0.9143$, and specificity $23/(23 + 16) = 0.5897$, see code in loan_data_random_forest.py in the online resources.

[9] We used the data in the file loan_data_set_num.csv in the online resources, which contains the original loan data features converted into numeric format. All data records containing NaN were dropped, resulting in a data set of 480 samples and $n = 11$ features.

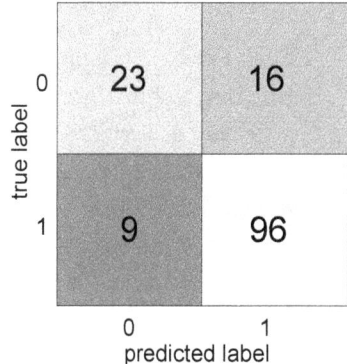

Figure 8.11 Confusion matrix for the RF prediction of the loan request outcome.

8.5 Exercises

Exercise 8.1 *Kernel SVM for credit card fraud detection*
In this exercise, we examine the credit card fraud detection problem already introduced in Exercises 7.3 and 7.5 and implement a solution based on kernel SVM.

1. Load the data set contained in the file creditcard.csv in the online resources. Discard the information regarding the time of the transactions, since it is uninformative for our purpose. Split the data set into a training set and a testing set. The size of the training set should be 75% of the entire data set.

2. Train an SVM with RBF (Gaussian) kernel using the training set and use the test set to evaluate the performance computing the confusion matrix and the sensitivity score.

3. Since the data set is highly unbalanced, use class weights to address the problem of class imbalance,[10] as shown in Remark 7.4. Compute the confusion matrix and the sensitivity of the model in this case. Compare the results with the ones obtained at the previous point.

4. Compare the results obtained in the previous points with the ones obtained in Exercise 7.5.

[10] In the SVC function of the sklearn package, you can use the option class_weight='balanced' to introduce weights inversely proportional to class frequencies in the input data.

Exercise 8.2 *Predicting customer churn*
Customer churn occurs when a customer stops using a company's products or services. The goal of this exercise is to predict customer churn for a bank using kernel SVM. We will use the data set contained in the file customer_churn.csv in the online resources.[11] This data set contains information from about 10,000 customers, where each customer is described using 17 features. Among these customers, 2,037 have churned (i.e., the "Exited" feature is equal to 1) and 7,963 have not churned. Therefore, the data set is slightly unbalanced with only 20% of samples belonging to the positive class.

[11] This data set is a pre-processed version of the dataset available at https://www.kaggle.com/datasets/adammaus/predicting-churn-for-bank-customers.

1. Load the data set contained in the file customer_churn.csv. The size of the training set should be 75% of the entire data set.

2. Train an SVM with polynomial kernel using the training set and use the test to evaluate the performance computing the confusion matrix and the accuracy and recall of the model. Use a polynomial kernel of degree 3 and set the regularization parameter c equal to 100.[12] Since the data set is unbalanced, use class weights[13] to address this problem.

[12] In the SVC function of the sklearn package, you have to set C=100.

[13] In the SVC function of the sklearn package, you can use the option class_weight='balanced' to introduce weights inversely proportional to class frequencies in the input data.

3. Try to to use a polynomial kernel with a higher degree. Compare the results with the ones obtained in the previous point (compare also the training accuracy).

4. Train an SVM with RBF (Gaussian) kernel (use the same settings defined in point 2) and compare the results with the ones obtained at the previous points.

5. Compare the results obtained at the previous points with the ones obtained using a linear SVM.

Exercise 8.3 *SVM with Gaussian kernel*
In this exercise, we will focus on a nonlinear SVM model with Gaussian kernel $k(x,z) = \exp\left(-\frac{\|x-z\|_2^2}{2\sigma^2}\right)$, where the parameter σ defines the width of the Gaussian kernel. We show that as long as there are no identical points in the training set, we can always find a value of the parameter σ such that the SVM achieves zero training error.
In Section 8.3, we have seen that the classification rule can be rewritten using the kernel function. Let us now suppose that the data samples in the training set are separated by at least a distance ϵ (i.e., $\|x^{(i)} - x^{(j)}\|_2 \geq \epsilon$ for any $i \neq j$). Find the values of the parameters α and b and of the Gaussian kernel width σ such that all the points of the training set are correctly classified.
Hint: Start by setting $\alpha_i = 0$ for all $i = 1, \ldots, m$ and $b = 0$, then notice that the prediction on the sample $x^{(i)}$ will be correct if $|z_i - y_i| < 1$, where $z_i = b + \sum_{j=1}^{m} \alpha_j y_j k(x^{(j)}, x^{(i)})$ is the model prediction for the data sample $x^{(i)}$. Finally, find a value of σ that satisfies such inequality for all i.

Exercise 8.4 *Decision tree for car evaluation*
In this exercise, we build a decision tree to evaluate if the price of a car is acceptable or not. We use the data set car_evaluation.csv in the online resources,[14] which contains 1,728 entries. Each entry describes a car using six features (namely: buying price, price of maintenance, number of doors, capacity in terms of person to carry, size of luggage boot, estimated safety of the car) and for each entry it is indicated if the car is considered unacceptable, acceptable, good, or very good. All the features are categorical features.

[14] This data set is also available at https://archive.ics.uci.edu/dataset/19/car+evaluation.

1. Load the data set contained in the file car_evaluation.csv. Since all the categorical features are ordinal, transform them into ordinal integers.[15] Split the data set into a training set and a testing set. The size of the training set should be 75% of the entire data set.

[15] In Python, you can use the OrdinalEncoder function from the package category_encoders The class DecisionTreeClassifier of the sklearn package does not support categorical features that have not been converted to ordinal integers.

2. Train a decision tree using the training set and use the test set to evaluate the performance (compute the classification accuracy and the confusion matrix of the trained model). Use as classification criterion the Information Gain[16] as discussed in Section 8.4.1.

3. Plot the decision tree obtained at the previous point.[17]

4. Compare the training and testing accuracies of your model. If there is a significant gap between these two scores it means that the model is overfitting. An easy way to reduce this issue is to limit the dimension of the tree, this can be done by setting a minimum number of samples required to be at a leaf node. Train a decision tree imposing that there are at least five training samples at each leaf node.[18] Plot the decision tree and compare it with the one obtained at the previous point. Compute the training and testing performance of this model and compare them with the ones of the model obtained at point 2.

5. At the end of Section 8.4.1, we have seen that other criteria besides the Information Gain can be used for building the decision tree. Train a decision tree using as classification criterion the Gini Index[19] and compare its performance with the ones obtained at points 2 and 4.

Exercise 8.5 *Random forest for car evaluation*

In this exercise we consider the same data set as the previous exercise. We are now interested in training a random forest classifier to solve the problem.

1. Load the data set and split it into a training set and test set (the size of the training set should be 75% of the entire data set). As in the previous exercise, transform the categorical features into ordinal integers.

2. Train a random forest classifier composed of 50 decision trees with the same settings defined at point 4 of the previous exercise. Use the test set to evaluate the performance of the trained model (compute the train and test accuracy and the confusion matrix of the trained model).

3. Compare the results obtained at the previous point with the ones obtained in the previous exercise.

---

[16] In Python, if you use the class DecisionTreeClassifier of the sklearn package you should set the option criterion='entropy'.

[17] In Python, you might use the graphviz package in order to obtain a nicer visualization.

[18] In Python, if you use the class DecisionTreeClassifier of the sklearn package this can be done by setting the option min_samples_leaf=5.

[19] In Python, if you use the class DecisionTreeClassifier of the sklearn package you should set the option criterion='gini'

9
Neural networks and deep learning

NEURAL NETWORKS AND DEEP LEARNING have revolutionized the field of artificial intelligence and machine learning. At their core, neural networks are computational models inspired by the human brain's neural structure. Deep learning, a subset of neural networks, involves training large, multi-layered networks to learn patterns and make predictions from vast amounts of data. These networks consist of interconnected nodes, or "artificial neurons," that process and transmit information. By iteratively adjusting the strength of connections between neurons, deep learning algorithms can discover intricate and abstract relationships within data, enabling remarkable tasks such as image recognition, natural language processing, and speech synthesis. Deep learning has brought unprecedented breakthroughs in diverse domains, such as, for instance, autonomous vehicles, healthcare, speech recognition, and robotics.

Neural networks have of course also found numerous applications in finance, transforming the way financial institutions analyze and predict market trends, make investment decisions, and manage risk. One prominent use case is in stock market prediction, where neural networks are employed to analyze historical market data, identify patterns, and forecast future return characteristics. These models leverage the power of deep learning to capture intricate relationships between various financial indicators and market behavior, enabling more accurate predictions. Neural networks are used in credit scoring and fraud detection, where they can process transaction data and detect subtle patterns indicative of potential risks or fraudulent activities, and they also find applications in algorithmic trading, portfolio optimization, and risk management.[1]

In this chapter we introduce the main concepts of neural networks (NNs). First, we show that NNs can be interpreted as nonlin-

[1] See, for example, O. Berat Sezer, M. Ugur Gudelek, and A. Murat Ozbayoglu, "Financial Time Series Forecasting with Deep Learning: A Systematic Literature review: 2005–2019," *Applied Soft Computing*, vol. 90, 2020, and N. Nazareth and Y. Venkata Ramana Reddy, "Financial Applications of Machine Learning: A Literature Review," *Expert Systems with Applications*, vol. 219, 2023, for broad surveys into the current uses of neural networks in finance and their impact on this industry.

ear feature extractors. Next, we present the main building blocks of a neural network and discuss the most common training techniques. We then focus on two well-known type of neural networks, namely convolutional neural networks and recurrent neural networks. At the end of the chapter, we briefly introduce some of the most recent deep learning approaches that have led to significant advances in the field.

9.1 Neural networks as feature extractors

In the previous chapter we introduced nonlinear classifiers, such as nonlinear SVMs, which are able to handle data that is not linearly separable. We have shown that we can define a nonlinear mapping ϕ that projects x into a higher-dimensional space, called the feature space, where the data can be approximately separated using a linear classifier. Thanks to the kernel trick we have seen that we can avoid to explicitly define the nonlinear mapping ϕ, but we still have to define a kernel function k.

In this chapter, we still focus on the problem of nonlinear classification, but from a different point of view. In this case, we are interested in explicitly defining the nonlinear mapping ϕ. However, instead of manually defining this function, we want to learn it from data. Indeed, manually defining the function ϕ leveraging some domain-specific knowledge requires a large human effort, where a distinct ϕ has to be hand-crafted for each specific application task. In the last decade, this approach has been overcome by deep learning models, where ϕ is learned from the data. In deep learning, ϕ represents a neural network, which can be defined as a broad class of non-linear functions parameterized by a set of parameters θ whose values are learned from the training data. We can thus define a deep learning model as a parameterized mapping $y = \phi(x; \theta)$, where $x \in \mathbb{R}^n$ is the input feature vector, θ is the vector of parameters of the NN model, and $y \in \mathbb{R}^p$ is the output, which can be a scalar or a vector depending on the specific application; we discuss this in more detail in Section 9.2.2.

Thanks to the fact that we use a very rich family of functions $\phi(x; \theta)$, we obtain a very generic feature mapping with a large capacity to fit the training data. This is the advantage of deep learning with respect to previous approaches. In the next sections we introduce different families of nonlinear functions $\phi(x; \theta)$, that is, different types of neural networks, each suitable for different tasks.

9.2 Feedforward neural networks

Feedforward networks, also called multilayer perceptrons, are the most well-known deep learning models and they were the first and simplest type of neural network devised. Despite their simplicity, they are still used in many applications and they can be a good starting point for studying how neural networks work.

The goal of a feedforward network, and of neural networks at large, is to approximate an unknown function f that describes the relationship between the input data and the corresponding desired output. As shown also at the beginning of Chapter 6, if we consider a classification or regression problem, given an input sample x, we can assume that there exists a function f that maps x to its output y:

$$y = f(x) + \epsilon,$$

where ϵ represents a noise term. In a supervised setting, the function f is unknown, but we have access to a set $\mathcal{D}$ of labeled training samples $(x^{(i)}, y^{(i)})$, where $i = 1, \ldots, m$, $y^{(i)} = f(x^{(i)}) + \epsilon^{(i)}$, and $\epsilon^{(i)}$ are i.i.d. random noise terms. A feedforward network defines a mapping $\hat{y} = \phi(x; \theta)$, where the value of the parameters θ is learned through a training procedure. Similar to problem (7.17) for linear classification and problem (6.12) for regression, we can define a cost function that should be minimized during the training procedure:

$$\min_{\theta} \mathcal{J}(\mathcal{D}, \theta) = \min_{\theta} \frac{1}{m} \sum_{i=1}^{m} \mathcal{L}(\phi(x^{(i)}, y^{(i)}) + \lambda \psi(\theta), \quad (9.1)$$

where $\mathcal{L}$ is a loss function that encodes the error between the observed value and the predicted value, ψ is a regularization function of the network parameters θ, and $\lambda \geq 0$ is a penalty parameter. Problem (9.1) is usually nonconvex and may exibit many local minima. Therefore, it is not possible to solve it exactly in a global sense, but we can apply gradient-based methods to find a "good" local optimum; these techniques are discussed in detail in Section 9.3.

These models are called networks because they are usually defined by composing together many simpler functions. We can define a feedforward network via a composition of functions

$$\phi(x; \theta) = \phi^{(d)}(\phi^{(d-1)}(\ldots (\phi^{(1)}(x; \theta^{(1)}); \ldots); \theta^{(d-1)}); \theta^{(d)}),$$

where $\phi^{(i)}$ is called the ith layer of the network and $\theta^{(i)}$ are its corresponding parameters. The last layer of the network is called the output layer, while the other layers are called hidden layers. The number of layers d represents the *depth* of the network. The term feedforward refers to the fact that information flows in only one direction from the input layer to the output layer without any feedback connection where outputs of some layer are fed back towards upstream layers.

Each hidden layer of the network performs a linear operation followed by a elementwise nonlinearity:

$$h^{(i)} = \phi^{(i)}(h^{(i-1)}) = \sigma(W^{(i)}h^{(i-1)} + b^{(i)}) \qquad i = 1, \ldots, d-1,$$

where $h^{(i)}$ is the output of the ith layer, σ is a nonlinear elementwise function called activation fuction, $W^{(i)} \in \mathbb{R}^{F^{(i)}, F^{(i-1)}}$ and $b^{(i)} \in \mathbb{R}^{F^{(i)}}$ are the weight matrix and bias of the ith layer, respectively. This type of layer is called *fully connected* because each element of the output $h^{(i)}$ depends on all the elements of the input $h^{(i-1)}$. We refer to the vectors $h^{(i)}$, with $i = 1, \ldots, d-1$, as the hidden feature vectors. The dimension of $h^{(i)}$ is $F^{(i)}$. This represents the number of features of the ith layer and it is called the *width* of the layer. The weight matrix $W^{(i)}$ and bias $b^{(i)}$ are the parameters of the ith layer, that is, $\theta^{(i)} = (W^{(i)}, b^{(i)})$. The operations performed by a feedforward neural network can be schematized as shown in Figure 9.1.

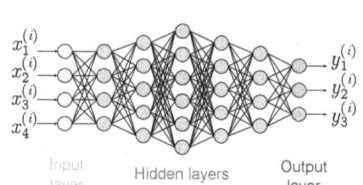

Figure 9.1 Feedforward neural network: each edge in the graph corresponds to an element in one of the weight matrices $W^{(i)}$.

Example 9.1 (*Feedforward neural network*) Let us consider a small feedforward neural network composed by an input layer, a hidden layer, and an output layer. This neural network can be represented as shown in Figure 9.2. Let us suppose that the input of the network has dimension 3, the width of the hidden layer is 3, and the output is a scalar. The parameters of this network are thus $\theta^{(1)} = (W^{(1)}, b^{(1)})$ and $\theta^{(2)} = (W^{(2)}, b^{(2)})$, where $W^{(1)} \in \mathbb{R}^{3,3}$, $W^{(2)} \in \mathbb{R}^3$, $b^{(1)} \in \mathbb{R}^3$, $b^{(2)} \in \mathbb{R}$. Given an input $x^{(i)} \in \mathbb{R}^3$, the network computes the output of the network $y^{(i)} \in \mathbb{R}$ performing the following operation:

$$h_1^{(1)} = \sigma(W_{11}x_1^{(i)} + W_{12}x_2^{(i)} + W_{13}x_3^{(i)} + b_1^{(1)}),$$
$$h_2^{(1)} = \sigma(W_{21}x_1^{(i)} + W_{22}x_2^{(i)} + W_{23}x_3^{(i)} + b_2^{(1)}),$$
$$h_3^{(1)} = \sigma(W_{31}x_1^{(i)} + W_{32}x_2^{(i)} + W_{33}x_3^{(i)} + b_3^{(1)}),$$
$$y^{(i)} = \sigma(W_{11}h_1^{(1)} + W_{12}h_2^{(1)} + W_{13}h_3^{(1)} + b^{(2)}),$$

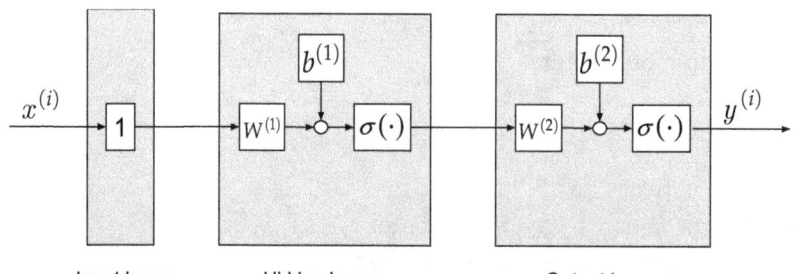

Figure 9.2 Feedforward neural network.

where sigma is an activation function. The above equations can be written more compactly as:

$$h^{(1)} = \sigma(W^{(1)}x^{(i)} + b^{(1)}),$$
$$y^{(i)} = \sigma(W^{(2)}h^{(1)} + b^{(2)}).$$

In the remainder of this section we will focus on the main building blocks of a feedforward neural network, discussing the most common choices.

9.2.1 Activation function

Neural networks can use different types of activation functions. Early NNs typically used smooth activation functions, such as the sigmoid or the hyperbolic tangent, while currently the most common choice is the rectified linear unit (ReLU)

$$\sigma(z) = \max(0, z).$$

The main drawback of this activation function is that training with gradient-based methods is not effective when we have examples for which the output of the activation is zero. This drawback can be overcome by introducing a generalization of the ReLU, called a leaky ReLU:

$$\sigma(z) = \begin{cases} z & \text{if } z > 0, \\ \alpha z & \text{otherwise,} \end{cases}$$

where α is set to a small value, such as 0.01. Using this definition of activation function we allow a small, positive gradient when $z < 0$. Recently, other generalizations have been proposed. Among these, one common choice is the exponential linear unit (ELU):

$$\sigma(z) = \begin{cases} z & \text{if } z > 0, \\ \alpha(\exp(z) - 1) & \text{otherwise.} \end{cases}$$

For negative inputs the ELU smoothly converges to $-\alpha$, where α is a hyper-parameter to be tuned. It has been shown that using ELU as activation function instead of ReLU can speed up the training process and obtain better results. Figure 9.3 shows a graphic representation of the differences between these three activation function.

9.2.2 Output layer

As discussed previously, the last layer of the network is called output layer. This layer is different from the other layers and its definition

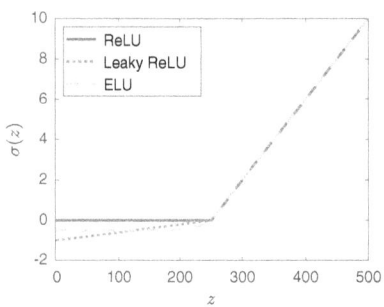

Figure 9.3 Comparison between ReLU, leaky ReLU ($\alpha = 0.1$) and ELU ($\alpha = 0.5$).

depends on the application task. If we are considering a regression problem, the output layer is usually a linear layer without activation function:
$$\hat{y} = W^{(d)} h^{(d-1)} + b^{(d)},$$
where $W^{(d)} \in \mathbb{R}^{p, F^{(d-1)}}$, $\hat{y}$ is the output of the network, and $h^{(d-1)}$ are the hidden features of the $(d-1)$th layer. Instead, in classification problems with K classes the output layer is usually a linear layer followed by a softmax function:
$$\hat{y}_i = \text{softmax}(z)_i = \frac{\exp(z_i)}{\sum_j \exp(z_j)},$$
where z_i is the ith element of $z = W^{(d)} h^{(d-1)} + b^{(d)}$ with $1 \leq i \leq K$, and $W^{(d)} \in \mathbb{R}^{K, F^{(d-1)}}$ is the weight matrix of the linear layer. The linear layer predicts the unnormalized log probabilities, then the softmax function exponentiates and normalizes z in order to obtain the probabilities of each class over all the possible K classes. If we consider a binary classification problem, we can use a linear layer followed by a sigmoid function:
$$\hat{y} = \frac{1}{1 + \exp(-z)},$$
where $z = w^{(d)} h^{(d-1)} + b^{(d)}$ and $w^{(d)} \in \mathbb{R}^{1, F^{(d-1)}}$ is the weight vector of the linear layer. The sigmoid function is used to convert the scalar output of the linear layer into a valid probability value.

9.2.3 Architecture design

We use the term architecture to refer to the overall structure of the network. This includes the number of layers of the network, the size of the output of each layer, the activation function chosen for each layer, and how the layers are connected to each other.

Defining the architecture of a neural network for a given task is a key design choice. In the next section we show that a deep feedforward neural network can in principle fit any training set, provided the number of nodes is sufficiently high. However, networks with a very high number of parameters are usually hard to optimize and the choice of the architecture plays a decisive role. The best architecture can be found experimentally by monitoring the performance of the network on validation data.

9.2.4 Universal approximation theorem

There exist in the literature various results[2] showing that feedforward neural networks can approximate arbitrarily well wide classes

[2] See, for example, K. Hornik, "Approximation Capabilities of Multilayer Feedforward Networks," *Neural Networks*, vol. 4(2), pp. 251–257, 1991.

of functions. This means that feedforward neural networks can be seen as universal approximators. In the following, we present an example of such results.[3]

[3] For more detail on this result, we refer the reader to Z. Lu, "The Expressive Power of Neural Networks: A View from the Width," *Advances in Neural Information Processing Systems*, vol. 30, 2017.

Theorem 9.1 (Universal approximation) *Let us define the width ω of a neural network as the maximal number of features in a layer. For any Lebesgue-integrable function $\Phi : \mathbb{R}^n \to \mathbb{R}$ and any $\epsilon > 0$, there exists a fully connected network ϕ with ReLU activations and width $\omega \leq n + 4$, such that the network ϕ satisfies*

$$\int_{\mathbb{R}^n} |\Phi(x) - \phi(x)| dx < \epsilon.$$

In simple words, the universal approximation theorem guarantees that, whatever function we are trying to learn, a deep feedforward network is able to represent this function. However, it is important to point out that, even if the network is able to represent a function, it does not mean that it will really be able to learn it. In practice, the optimization process can fail due to the fact that the optimization algorithm is not able to find the correct parameters. In addition, it is important to point out that the above result on universal approximation guarantees that there exists a network that achieves any desired accuracy, but it does not provide information on the depth (number of layers) necessary to achieve such accuracy.

9.3 Training a neural network

We next focus on the training procedure of a neural network. Let us consider a training set $\mathcal{D}$ composed of m training examples $(x^{(1)}, y^{(1)})$, ..., $(x^{(m)}, y^{(m)})$. Given a neural network ϕ and its parameters θ, we can measure the mismatch, or training error, between a training example $(x^{(i)}, y^{(i)})$, where $i = 1, \ldots, m$, and the network output $\hat{y} = \phi(x^{(i)}, \theta)$. As shown in (9.1), we can then define an overall measure $\mathcal{J}(\theta, \mathcal{D})$, called the cost function, of mismatch between the observed outputs $y^{(i)}$ and the predicted ones $\phi(x^{(i)}, \theta)$, $i = 1, \ldots, m$.

Different cost functions can be used to train a neural network, in Section 9.3.2 we discuss some of the most common choices. Training a neural network consists in finding values of the network parameters θ that make the training cost small. This is obtained by minimizing the criterion $\mathcal{J}(\theta, \mathcal{D})$ as function of θ. In the next sections we describe in some detail the practical procedures that can be employed to perform this task, including the optimization algorithm itself, the possible choices for the loss function, and the strategies for initializing the parameters.

9.3.1 Optimization algorithm

Stochastic gradient descent (SGD) is the most common optimization algorithm for training deep learning models. The main idea of gradient descent methods is to find the minimum of a function by taking repeated steps in the opposite direction of the gradient of the function at the current point, since this is the direction of steepest descent. In other words, let $\mathcal{J}(\theta, \mathcal{D})$ be the cost function computed on the training set $\mathcal{D}$, then a step of the gradient descent method can be summarized as follows:

$$g = \nabla_\theta \mathcal{J}(\theta, \mathcal{D}),$$
$$\theta \leftarrow \theta - \eta g,$$

where the step size $\eta > 0$ is called the *learning rate*.

The main drawback of the gradient descent method is that it is very expensive to compute the gradient of the cost function over the whole dataset. This problem is particularly relevant when we have a very large training set, as in the most common deep learning problems. The stochastic gradient descent method overcomes this limitation by sampling a subset $\mathcal{B}$, called the mini-batch, of the training set at every step and using this subset to estimate the gradient:

$$g \approx \frac{1}{|\mathcal{B}|} \nabla_\theta \sum_{(x,y) \in \mathcal{B}} \mathcal{J}(\theta, (x, y)),$$

where $|\mathcal{B}|$ represents the number of samples composing the mini-batch $\mathcal{B}$ and $\mathcal{J}(\theta, (x, y))$ is the value of the cost function for the data sample (x, y).

Let us now examine in more detail one iteration of the stochastic gradient descent method for a feedforward neural network. If we consider the lth layer of the network, its corresponding parameters are the weight matrix $W^{(l)}$ and the bias $b^{(l)}$. Then, the stochastic gradient descent algorithm updates these parameters as follows:

$$
\begin{aligned}
W_{ij}^{(\ell)} &\leftarrow W_{ij}^{(\ell)} - \eta \frac{1}{|\mathcal{B}|} \sum_{(x,y) \in \mathcal{B}} \frac{\partial \mathcal{J}(\theta, (x, y))}{\partial W_{ij}^{(\ell)}}, \\
b_i^{(\ell)} &\leftarrow b_i^{(\ell)} - \eta \frac{1}{|\mathcal{B}|} \sum_{(x,y) \in \mathcal{B}} \frac{\partial \mathcal{J}(\theta, (x, y))}{\partial b_i^{(\ell)}},
\end{aligned}
\qquad (9.2)
$$

with $i = 1, \ldots, F^{(\ell)}$, and $j = 1, \ldots, F^{(\ell-1)}$. In order to compute the derivatives in (9.2), we can use an efficient recursive method called the *backpropagation algorithm*.

The backpropagation algorithm Backpropagation is essentially based on the chain rule for computing derivatives. Given a training example

(x, y), we will first run a "forward pass" to compute all the hidden features throughout the network, including the output value $\hat{y}$ of the network. Then, for each layer ℓ we define an error sensitivity term $\delta^{(\ell)}$ as follows:

$$\delta_i^{(\ell)} \doteq \frac{\partial J(\theta, (x, y))}{\partial z_i^{(\ell)}}, \qquad i = 1, \ldots, F^{(\ell)},$$

where $z_i^{(\ell)}$ is the ith element of $z^{(\ell)} = W^{(\ell)} h^{(\ell-1)} + b^{(\ell)}$. This term measures how much the component z_i is responsible for any errors in the output.

Let us first focus on the last layer $\hat{y} = \sigma(W^{(d)} h^{(d-1)} + b^{(d)})$. We can directly measure the value of the cost function $J(\theta, (x, y))$ for the network's output $\hat{y}$ and use that to define the error sensitivity term $\delta_i^{(d)}$ of the ith component of the output:

$$\delta_i^{(d)} \doteq \frac{\partial J(\theta, (x, y))}{\partial z_i^{(d)}} = \frac{\partial J}{\partial \hat{y}_i} \frac{\partial \hat{y}_i}{\partial z_i^{(d)}} \qquad i = 1, \ldots, p.$$

By the chain rule, we can now define the derivative of $J(\theta, (x, y))$ with respect to $W_{ij}^{(d)}$ and $b_i^{(d)}$:[4]

$$\frac{\partial J(\theta, (x, y))}{\partial W_{ij}^{(n)}} = \frac{\partial J(\theta, (x, y))}{\partial z_i^{(n)}} \frac{\partial z_i^{(n)}}{\partial W_{ij}^{(n)}} = \delta_i^{(n)} h_j^{(n-1)},$$

$$\frac{\partial J(\theta, (x, y))}{\partial b_i^{(n)}} = \frac{\partial J(\theta, (x, y))}{\partial z_i^{(n)}} \frac{\partial z_i^{(n)}}{\partial b_i^{(n)}} = \delta_i^{(n)},$$

[4] For simplicity, we will consider a cost function $J(\theta, (x, y))$ without regularization term (i.e., $\lambda = 0$). An analogous derivation can be done if $\lambda > 0$.

with $i = 1, \ldots, p$, and $j = 1, \ldots, F^{(d-1)}$. Let us consider now the hidden layers. For the ℓth layer of the network, we compute the error sensitivity $\delta_i^{(\ell)}$, with $i = 1, \ldots, F^{(\ell)}$, in terms of the error sensitivity in the next layer $\delta^{(\ell+1)}$:

$$\delta_i^{(\ell)} = \frac{\partial J(\theta, (x, y))}{\partial z_i^{(\ell)}} = \sum_{j=1}^{F^{(\ell+1)}} \frac{\partial J(\theta, (x, y))}{\partial z_j^{(\ell+1)}} \frac{\partial z_j^{(\ell+1)}}{\partial z_i^{(\ell)}} = \sum_{j=1}^{F^{(\ell+1)}} \delta_j^{(\ell+1)} \frac{\partial z_j^{(\ell+1)}}{\partial z_i^{(\ell)}}.$$

Since $z^{(\ell+1)} = W^{(\ell+1)} h^{(\ell)} + b^{(\ell+1)} = W^{(\ell+1)} \sigma(z^{(\ell)}) + b^{(\ell+1)}$, it holds that

$$\frac{\partial z_j^{(\ell+1)}}{\partial z_i^{(\ell)}} = W_{ji}^{(\ell+1)} \sigma'\left(z_i^{(\ell)}\right).$$

Therefore, we have that

$$\delta_i^{(\ell)} = \sum_{j=1}^{F^{(\ell+1)}} \delta_j^{(\ell+1)} \frac{\partial z_j^{(\ell+1)}}{\partial z_i^{(\ell)}} = \sum_{j=1}^{F^{(\ell+1)}} \delta_j^{(\ell+1)} W_{ji}^{(\ell+1)} \sigma'\left(z_i^{(\ell)}\right).$$

Starting from $\delta_i^{(d)}$, we can then propagate in a recursive way the sensitivities to all layers, going backwards with $\ell = d, d-1, \ldots, 1$. Once we have the sensitivity $\delta^{(\ell)}$, we can compute the derivatives of the objective with respect to the weights and biases, as follows:

$$\frac{\partial \mathcal{J}(\theta,(x,y))}{\partial W_{ij}^{(\ell)}} = \frac{\partial \mathcal{J}(\theta,(x,y))}{\partial z_i^{(\ell)}} \frac{\partial z_i^{(\ell)}}{\partial W_{ij}^{(\ell)}} = \delta_i^{(\ell)} h_j^{(\ell-1)},$$

$$\frac{\partial \mathcal{J}(\theta,(x,y))}{\partial b_i^{(\ell)}} = \frac{\partial \mathcal{J}(\theta,(x,y))}{\partial z_i^{(\ell)}} \frac{\partial z_i^{(\ell)}}{\partial b_i^{(\ell)}} = \delta_i^{(\ell)},$$

with $i = 1, \ldots, F^{(\ell)}$, and $j = 1, \ldots, F^{(\ell-1)}$. A schematic representation of this algorithm is shown in Figure 9.4.

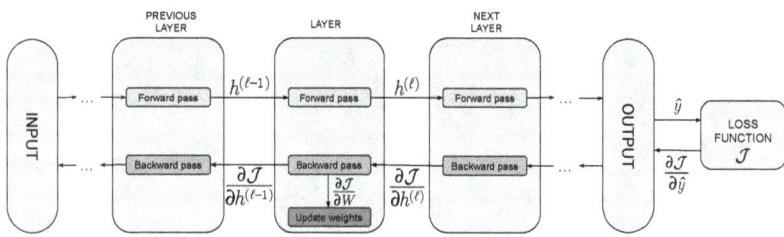

Figure 9.4 A schematic representation of the backpropagation algorihtm.

It is important to point out that $\mathcal{J}(\theta, \mathcal{B})$ is a nonconvex function due to the nonlinearities that are present in the neural network. Therefore, gradient descent methods applied to such problems are susceptible to return local rather than global minima of the objective. However, these methods proved to work fairly well in practice. Several studies provided empirical evidence in order to try to explain the success of gradient descent methods for NNs, even though such problems are nonconvex. For sufficiently large NNs, it has been shown that most local minima have a low cost function value and they perform nearly as well as the global minimum.[5] This means that it may not be so important to find the actual global minimum, if one can find a local minimum with a good (low) cost value. In addition, for many high-dimensional nonconvex problems, local optima are rare and most of the points with zero gradient are saddle points.[6] Empirical studies have shown that gradient descent methods seem to be able to escape such saddle points in many cases.[7]

Several variants of the SGD algorithm have been proposed. One of the most important is the momentum algorithm. This method has been proposed in order to address the problem of noisy gradients. The momentum method computes an exponentially decaying moving average of the past gradients. The update rule is then defined as follows:

[5] See A. Choromanska, M. Henaff, M. Mathieu, G. Ben Arous, and Y. Le Cunn, "The Loss Surfaces of Multilayer Networks," *International Conference on Artificial Intelligence and Statistics (AISTATS)*, 2015.

[6] See Y. N. Dauphin, R. Pascanu, C. Gulcehre, K. Cho, S. Ganguli, and Y. Bengio, "Identifying and Attacking the Saddle Point Problem in High Dimensional Non-convex Optimization," *Advances in Neural Information Processing Systems (NeurIPS)*, vol. 27, 2014.

[7] See I. J. Goodfellow, O. Vinyals, and A. M. Saxe, "Qualitatevely Characterizing Neural Network Optimization Problems," *International Conference on Learning Representations (ICLR)*, 2015.

$$v \leftarrow \beta v - \eta \frac{1}{|\mathcal{B}|} \nabla_\theta \sum_{(x,y)\in\mathcal{B}} J(\theta, (x,y)),$$

$$\theta \leftarrow \theta + v.$$

The term β defines how much the previous gradients affect the direction at the current step.

Another class of SGD variants focuses on the learning rate. This is one of the so-called hyper-parameters of the network, that is, parameters that have to be set manually before training. This hyper-parameter defines the step size of the parameter updates at each iteration of the gradient descent method. Setting a suitable value for the learning rate is a crucial task because it strongly affects the training performance: a large learning rate can make the training unstable, whilst a small learning rate can result in a slow learning process that can be prone to local minima. Moreover, the extremely high number of parameters could result in a different sensitivity on each direction of the parameter space. For instance, a learning rate could be too high for some parameters and too slow for others. In addition, due to their high number, it is impossible to manually set the learning rate for each of the parameters. For this reason, several methods propose to automatically adapt the learning rate for each parameter. Among such methods, AdaGrad is one of the most common. This algorithm automatically adjusts the learning rate of each model parameter based on the sum of all its gradients estimated over the course of training. The rationale behind this method is that parameters with large gradients are compensated with a smaller learning rate, whereas parameters with small gradients will have a larger learning rate. The main drawback of AdaGrad is that the learning rates are monotonically decreasing. This may excessively slow down the training, stopping the learning process too soon before reaching a minimum. Another algorithm with adaptive learning rates called RMSProp tackles this problem by introducing a weighted moving average instead of the gradient accumulation performed by AdaGrad. This allows us to forget the values from the extreme past and focus only on the most recent ones.

9.3.2 *Loss function*

As discussed at the beginning of Section 9.2, the cost function $\mathcal{J}(\theta, \mathcal{B})$ is composed of two terms: a loss function $\mathcal{L}$ that evaluates the error between the observed value and the predicted value, and a regularization function ψ of the network parameters. The choice of the loss function $\mathcal{L}(\theta, \mathcal{B})$ depends on the type of problem we are considering.

If we have a classification problem with K classes and a neural network f with a softmax output layer, the most common cost function is the cross-entropy (CE):

$$\mathrm{CE}(\theta, \mathcal{B}) \doteq -\frac{1}{|\mathcal{B}|} \sum_{t=1}^{|\mathcal{B}|} \sum_{i=1}^{K} \mathbb{I}(i = y^{(t)}) \log(\mathrm{Prob}(y = i | x^{(t)}, \theta)),$$

where $(x^{(t)}, y^{(t)})$ is the tth sample of the mini-batch, y is the categorical output of the network, $\mathrm{Prob}(y = i | x^{(t)}, \theta) = f(x^{(t)}, \theta)_i$ is the ith output of the final softmax layer, and $\mathbb{I}(\cdot)$ is an indicator function which is 1 if its argument is true and 0 otherwise.

Instead, in a regression problem a typical choice for the cost function is the mean squared error (MSE), which, for scalar output, takes the form

$$\mathrm{MSE}(\theta, \mathcal{B}) \doteq \frac{1}{m} \sum_{t=1}^{m} \frac{1}{2} (f(x^{(t)}, \theta) - y^{(t)})^2.$$

An alternative is the mean absolute error (MAE):

$$\mathrm{MAE}(\theta, \mathcal{B}) \doteq \frac{1}{m} \sum_{t=1}^{m} |f(x^{(t)}, \theta) - y^{(t)}|.$$

9.3.3 Regularization

Loss functions measure the mismatch between the training samples and the corresponding network outputs. However, it is important to recall that the true goal of using a neural network is to predict with small error *unseen* outputs. Indeed, since deep neural networks have a very high representational capacity, they are prone to the risk of overfitting, resulting in a very low training error but possibly poor generalization capabilities to unseen data samples. As discussed already in other contexts in this book, an effective approach to prevent overfitting is to add a regularization term ψ to the loss function, as done in (9.1). The purpose of this additional term is to limit the capacity of the neural network by imposing a penalty on the model parameters. The most common choice for the regularization term ψ is the so-called weight decay, which penalizes the magnitude of the weights:

$$\mathcal{J}(\theta, \mathcal{B}) = \mathcal{L}(\theta, \mathcal{B}) + \lambda \sum_{\ell=1}^{d-1} \sum_{i=1}^{} \sum_{j=1}^{} \left| W_{ij}^{(\ell)} \right|^2.$$

We point out that usually this regularization is applied only to the weights of the network, without considering the biases. Applying weight decay to the biases usually does not result in a significant change in performance.

Another regularization method very common in deep learning is *dropout*. As we have seen in the previous chapter, an effective way to reduce overfitting is to use an ensemble of models whereby the prediction is obtained as the average of the predictions of several individual models. However, if the models are large, as is often the case in deep learning, using an ensemble can be very challenging because it requires training multiple models and storing them. Dropout tackles this problem by introducing a method that approximates the training of a large number of neural networks: during training at each iteration of gradient descent some of the layer outputs are randomly ignored or *dropped out*, as shown in Figure 9.5. This means that at each iteration the architecture of the network is slightly different. This makes the training process noisy and reduces the cases in which network layers co-adapt to fix mistakes committed by prior layers, making the model more robust to overfitting.

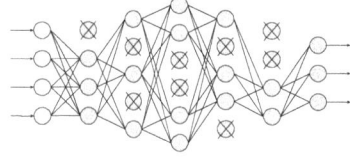

Figure 9.5 Dropout: at each iteration of the training process some of the layer outputs are randomly dropped out.

9.3.4 Initialization

In deep learning models the initialization of the parameters has a relevant role, since the algorithms used for training are usually greatly affected by the choice of initialization. The initial point can influence the convergence rate of the training algorithm and the quality of the point of convergence (e.g., its generalization properties or whether it corresponds to a low or high cost value). In some cases a bad initialization can make the training unstable and cause converge failure.

Several initialization strategies can be adopted. The common property of all these strategies is that the parameters should be randomly initialized. This serves the purpose of symmetry breaking. Indeed, if two hidden layers connected to the same inputs have the same initial parameters, then a deterministic optimization algorithm will end up learning the same parameters for both these layers.

The most common strategy for the initialization of parameters in a deep learning model is the so-called Glorot, or Xavier, initialization. Using this heuristic, the weights of the ith fully connected layer with $F^{(i-1)}$ inputs and $F^{(i)}$ outputs are initialized by sampling each weight from a normal distribution[8] with zero mean and variance $\sigma = \frac{2}{F^{(i-1)}+F^{(i)}}$. This definition is a compromise between the requirement that the variance of the layer outputs is the same across all layers and the requirement that the variance of the gradients is the same across all layers.

[8] An alternative definition of the Glorot initialization considers a uniform distribution between $-\sqrt{\frac{6}{F^{(i-1)}+F^{(i)}}}$ and $\sqrt{\frac{6}{F^{(i-1)}+F^{(i)}}}$. For further details we refer the reader to X. Glorot and Y. Bengio, "Understanding the Difficulty of Training Deep Feedforward Neural Networks," *International Conference on Artificial Intelligence and Statistics (AISTATS)*, 2010.

9.3.5 Batch normalization

As explained in Section 9.3.1, during training the model is updated layer-by-layer backward from the output to the input using an estimate of error that assumes that the weights in the layers prior to the current layer are fixed. However, in practice, at each iteration of the gradient descent algorithm the weights of all the layers are updated simultaneously. This complicates the training, since the distribution of each layer's input changes during training, as the parameters of the previous layers change. In order to address this problem, a technique called batch normalization has been introduced. The main purpose of batch normalization is to help coordinate the update of multiple layers. This is done by scaling the output of the layer. More specifically, it uses the mean and the variance computed on the current mini-batch to rescale the activation vector of the layer in order to have zero mean and standard deviation equal to 1. This helps to stabilize, and thus speed-up, the training process.

9.4 Convolutional neural networks

Convolutional neural networks (CNNs) are a special type of neural networks where there are some layers that use convolution instead of standard matrix-vector multiplication. This type of network has been extremely successful in many applications, especially in computer vision and natural language processing.

The key ingredient of CNNs is the convolution layer. This layer takes input data from the previous layer and averages it locally via a filter or kernel, which defines the weight pattern. More formally, given two functions g, k defined on the set $\mathbb{Z}$ of integers, we can define the convolution operation as follows:

$$(g * k)(t) = \sum_{i=-\infty}^{\infty} g(t-i)k(i), \qquad t \in \mathbb{Z}.$$

In the context of neural networks, h is a multidimensional array representing the input of the layer. Instead, k is the kernel, which is a multidimensional array of learnable parameters. We can thus interpret h and k as functions that are zero everywhere but a finite set of values. Using this interpretation, we can implement the definition of convolution presented above as a sum over a finite set of values. For example, if we consider a one-dimensional input g and a one-dimensional kernel k, the operation performed by a convolutional layer is defined as follows:

$$(g * k)_i = \sum_j g_{i-j} k_j,$$

where j ranges over all valid subscripts for g_{i-j} and k_j. Analogously, if we consider a two-dimensional input G, such as a grayscale image, and a two-dimensional kernel K, the convolution operation is defined as follows:

$$(G * K)_{i,j} = \sum_s \sum_t G_{i-s,j-t} K_{s,t}, \quad (9.3)$$

where s and t range over all valid subscripts for $G_{i-s,j-t}$ and $K_{s,t}$.

The size of the kernel is usually smaller than the input data. The convolution operation can be thus interpreted as sliding a kernel over the input data and computing a dot product between the part of the input the kernel is currently on and the kernel itself. Given the same input, we can repeat the convolution operation with many different kernels and for each kernel we obtain a different output.

Example 9.2 (*Two-dimensional convolution of an RGB image*) Let us consider a convolution operation that takes as input an RGB image of dimension 32×32 pixels and apply a 5×5 kernel. In this case the input has three channels, that is, the three RGB bands. In (9.3), we have defined the two-dimensional convolution operation for an image with a single channel. This operation can be easily extended to the case of an input with u channels as follows:

$$(G * K)_{i,j} = \sum_u \sum_s \sum_t G_{i-s,j-t,u} K_{s,t,u}.$$

The kernel K is thus a three-dimensional tensor with the same number of channels of the input. In this example, the dimension of the kernel is thus $5 \times 5 \times 3$. Figure 9.6 depicts this convolution operation. As shown in the figure, the output of this convolution operation is a single channel of dimension 28×28, where the dimension reduction is due to boundary limitations. We can then repeat this operation with four different kernels, as shown in Figure 9.7. This gives as output four channels of dimension 28×28.

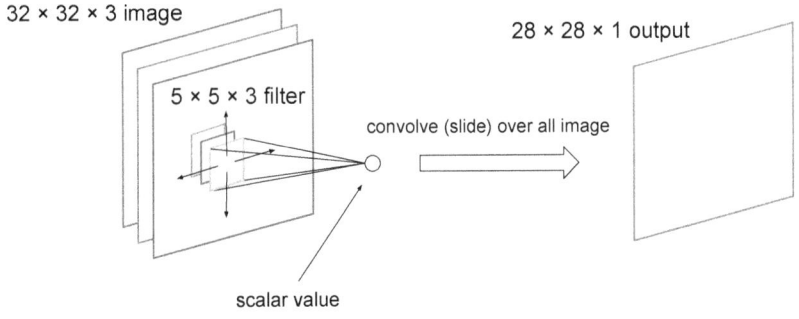

Figure 9.6 Convolution between a RGB image and one 5×5 filter.

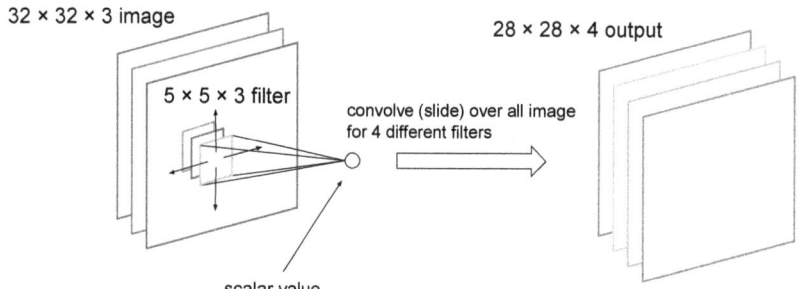

Figure 9.7 Convolution between a RGB image and four 5× 5 filter.

Let us consider an input $h^{(\ell-1)} \in \mathbb{R}^{F^{(\ell-1)},N}$ composed by N data samples with $F^{(\ell-1)}$ features for each sample, we can thus define the operation performed by the ℓth convolutional layer of the network, with $\ell = 1, \ldots, d$, as follows:[9]

$$h_u^{(\ell)} = \sigma \left(\sum_v^{F^{(\ell-1)}} h_v^{(\ell-1)} * k_{u,v}^{(\ell)} + b_u^{(\ell)} \right), \qquad u = 1, \ldots, F^{(\ell)},$$

[9] We consider the case where the layer performs a one-dimensional convolution, an analogous definition can be obtained with the two-dimensional convolution.

where $h_u^{(\ell)}$ is the uth feature channel of the output of the ℓth layer, $k_{u,v}^{(\ell)} \in \mathbb{R}^L$ is the kernel applied to the vth input feature channel in order to obtain the uth output channel, L is the length of the kernel, $b_u^{(\ell)} \in \mathbb{R}^N$ is the bias for the uth output feature channel, and σ is a nonlinear elementwise function. The parameters of this layer are the kernel $k^{(\ell)} \in \mathbb{R}^{K,F^{(\ell)},F^{(\ell-1)}}$ and the bias $b^{(\ell)} \in \mathbb{R}^{N,F^{(\ell)}}$: $\theta^{(\ell)} = (k^{(\ell)}, b^{(\ell)})$.

The convolutional layer can be interpreted as a special type of fully connected layer described in Section 9.2. Indeed, the convolution between an input vector $g \in \mathbb{R}^n$ and a kernel $k \in \mathbb{R}^L$ can be rewritten as a matrix-vector product:

$$g * k = Wg = \begin{bmatrix} k_1 & \ldots & k_L & 0 & 0 & \ldots & 0 \\ 0 & k_1 & \ldots & k_L & 0 & \ldots & 0 \\ \vdots & \ddots & \ddots & \ddots & \ddots & \ddots & \vdots \\ 0 & \ldots & 0 & 0 & k_1 & \ldots & k_L \end{bmatrix} \begin{bmatrix} g_1 \\ g_2 \\ \vdots \\ g_n \end{bmatrix}, \qquad (9.4)$$

where W is a Toeplitz matrix, that is, a matrix where each diagonal from left to right is constant. This representation of the convolutional layer can be useful in order to understand the advantages of this type of layer with respect to a fully connected one. The first observation that can be made from (9.4) is that the weight matrix corresponding to a convolutional layer is sparse. This means that the convolutional layer has far fewer parameters than a fully connected layer, where the weight matrix is dense. This is one of the strong advantages of convolutional layers: using fewer parameters reduces the risk of overfitting and is more efficient in terms of memory requirements and computational complexity. In particular, the convolutional layer exploits spatial locality by enforcing that each element of the output is obtained by a weighted aggregation of a small, local portion of the input. The dimension of this portion is given by the filter size. When many convolutional layers are stacked, we obtain a model that is able to create hierarchical representations of the input data: the first layers learn low-level local features, then the next layers assemble them, obtaining more complex features that represent larger portion of the input data. Figure 9.8 shows an example of such behavior in the context of image classification.

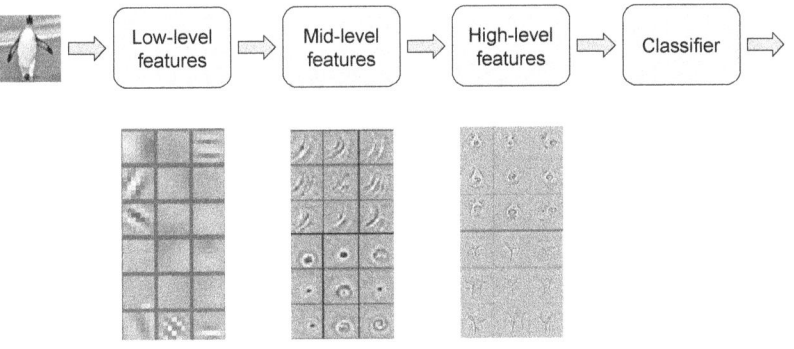

Figure 9.8 Feature visualization of a convolutional neural network trained on Imagenet. Source: The images of the features are from M. D. Zeiler and R. Fergus, "Visualizing and Understanding Convolutional Networks," *European Conference on Computer Vision (ECCV)*, 2014.

Equation (9.4) also shows that the weight matrix corresponding to a convolutional layer has a Toeplitz structure. This is due to the so-called parameter sharing property of convolutional layers. Parameter sharing consists in using the same parameter more than once when computing the output of a layer. Indeed, in a fully connected layer each element of the weight matrix is multiplied by one element of the input and then never used again. Instead, in a convolutional layer each element of the kernel is used at every position of the input. This means that the model has to learn only one set of parameters, rather than a different set for every input position. In addition, parameter sharing makes the convolutional layer equivariant to translation, that is, shifting the input will result in an equal shift of the output of the convolutional layer. In other words, if we consider the ith convolutional layer $h^{(i)} = f^{(i)}(h^{(i-1)})$, translation equivariance means that, if we shift the input $h^{(i-1)}$ by an offset c and then apply the convolutional layer, the result will be equal to shifting by c the output $h^{(i)}$. In many tasks, such as image classification, translation equivariance is a fundamental property that allows the detection of local patterns.

9.4.1 Pooling

In CNNs, it is common to insert pooling layers in order to make the representations smaller and more manageable. A pooling layer is usually inserted after a convolutional layer and it reduces the dimension of the output of the convolutional layer. It operates on each feature channel independently by replacing the value at a certain location with a summary statistic of the nearby values. The most common pooling operation is max pooling, which selects the maximum value within a given neighborhood. Figure 9.9 shows an example of max pooling applied to a two-dimensional input. Another popular pooling operation is average pooling, which computes the average of a given neighborhood.

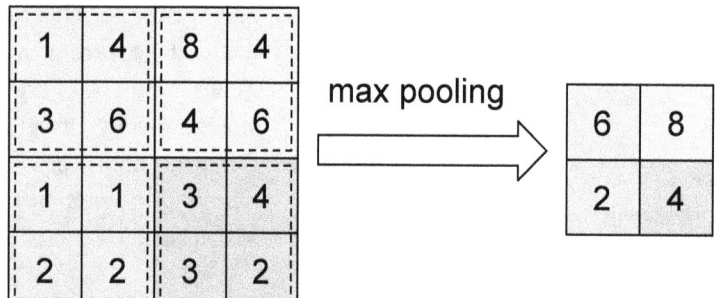

Figure 9.9 Max pooling operation.

9.5 Recurrent neural networks

All types of neural network presented in the previous sections describe a *static* relation between input x and output y. If the input $x(t)$ represents some causative quantity at time t and $y(t)$ a corresponding output, these networks cannot capture possible dynamic relations, in which $y(t)$ actually depends not only on $x(t)$ but also on the past values $x(t-1), x(t-2), \ldots$. Instead, recurrent neural networks (RNN) are designed to capture the recursive *dynamic* nature of certain phenomena (e.g., text, time series, etc.). In these networks, the output at time t depends on the input at time t and on the hidden state, which is a representation of previous inputs, at time $t-1$. This can be schematized as described in Figure 9.10. The feedback loop shown in this representation can be unrolled and we can obtain an architecture that does not involve recurrence as shown in Figure 9.11, where are represented different steps in time of the same RNN. From Figure 9.11, we can observe that the same parameters are used for all time steps.

If we consider a single-layer RNN, at every time step we can describe the operation performed by the RNN as follows:

$$h_t = \sigma_1(W_h x_t + R h_{t-1} + b_h),$$
$$y_t = \sigma_2(W_y h_t + b_y),$$
(9.5)

where h_t is the hidden state at time t, x_t is the input at time t, y_t is the output at time t, σ_1 and σ_2 are nonlinear elementwise functions, and W_h, W_y, R, b_h, b_y are the parameters that are shared temporally. As for feedforward neural networks, also RNNs can be composed by many layers stacked together. More specifically, from Eq. (9.5) we can identify three different types of transformation:

- from the input to the hidden state;
- from the previous hidden state to the next hidden state;
- from the hidden state to the output.

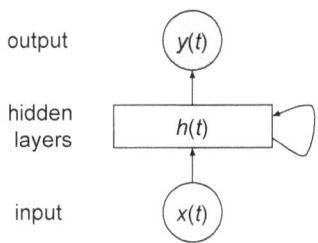

Figure 9.10 Recurrent neural network: at each time step t, the network takes as input $x(t)$ and the hidden state at the previous instant $h(t-1)$.

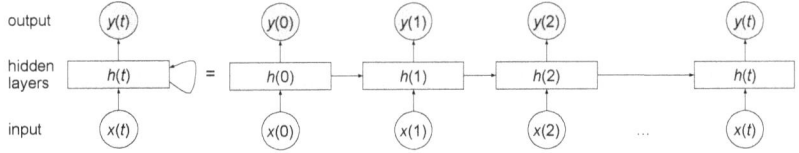

Figure 9.11 An unrolled recurrent neural network.

Each of these three transformations can be composed by multiple layers. It has been shown that increasing the number of layers for each of the three transformations can significantly improve the performance.[10]

Recurrent neural networks require a special training procedure, called backpropagation through time (BPTT). This training procedure is performed on sequences. The BPTT algorithm first unfolds the RNN through all the time steps of the sequence as shown in Figure 9.11. Next, the loss function on the sequence is computed as the sum of the loss at every time step of the sequence. Backpropagation is then used to compute the gradient of the cost function with respect to the network parameters. Notice that the computation of the

[10] See, for example, R. Pascanu, C. Gulcehre, K. Cho, and Y. Bengio, "How to Construct Deep Recurrent Neural Networks," *International Conference on Learning Representations (ICLR)*, 2014.

derivatives is a bit more problematic than in standard feedforward networks, because some of the derivatives depend recursively on the hidden states of all time steps:

$$\frac{\partial h_t}{\partial h_k} = \frac{\partial h_t}{\partial h_{t-1}} \frac{\partial h_{t-1}}{\partial h_{t-2}} \cdots \frac{\partial h_{k+1}}{\partial h_k}.$$

Due to such recursive computation of the gradients, it is quite common to encounter problems such as exploding or vanishing gradients.[11] For example, if $\frac{\partial h_t}{\partial h_{t-1}} \propto R$, then $\frac{\partial h_t}{\partial h_k} \propto R^{t-k}$. This means that if R is large the gradient may explode, instead if R is small the gradient may vanish. Since the ability to learn long time dependencies requires good gradient flow through the network, vanishing or exploding gradients can significantly affect the training of RNNs.

Long short-term memory (LSTM) networks solve this issue by adopting an architecture with an *additive* gradient structure instead of multiplicative, making gradient flow easier. This architecture allows learning of dependencies over hundreds of time steps.

[11] For further discussion on this issue see Y. Bengio, P. Simard, and P. Frasconi, "Learning Long-term Dependencies with Gradient Descent is Difficult," *IEEE Transactions on Neural Networks*, vol. 5(2), pp. 157–166, 1994.

9.5.1 NARX networks

The nonlinear autoregressive network with exogenous inputs (NARX) is a type of recurrent neural network based on a nonlinear version of the ARX model that is commonly used in time-series modeling. The defining equation for the NARX model is

$$y(t) = f(y(t-1), y(t-2), \ldots, y(t-n_y), u(t-1), u(t-2), \ldots, u(t-n_u)),$$

where the next value of the dependent output signal $y(t)$ depends on previous values of the output signal itself and previous values of an independent (exogenous) input signal. We can implement a NARX model by using a feedforward neural network to approximate the function f. A diagram of the resulting network is shown in Figure 9.12. We can consider the output $\hat{y}(t)$ of the NARX network to be an

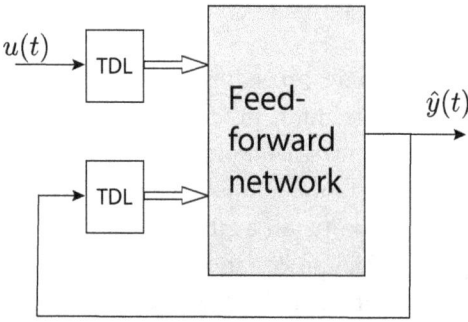

Figure 9.12 NARX network. TDL indicates a Tapped Delay Line (see Figure 9.13).

estimate of the output of some nonlinear dynamic system that we are trying to model.

The output is fed back to the input of the feedforward neural network as part of the standard NARX architecture. Since the true output $y(t)$ is available during the training of the network, we may create a "series-parallel" architecture in which the true output is used instead of feeding back the estimated output. The diagram of such architecture is represented in Figure 9.14. This has two advantages. The first is that the input to the feedforward network is more accurate. The second is that the resulting network has a purely feedforward architecture, and static backpropagation can be used for training.

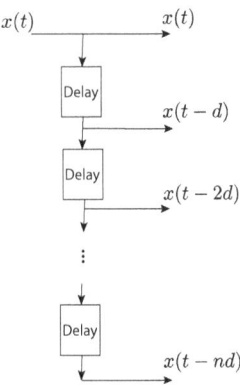

Figure 9.13 A tapped delay line outputs the input signal at arbitrary delay length values.

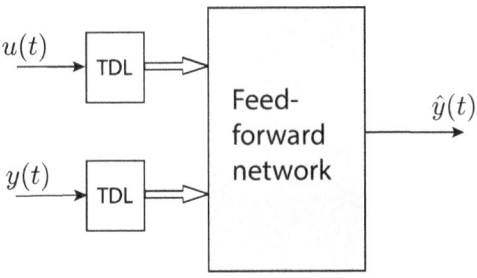

Figure 9.14 NARX network with series-parallel architecture.

Example 9.3 (*GDP time series*) We consider the US GDP quarterly time series shown in Figure 9.15. We build a two-layer NARX network in order to predict the GDP value at a new quarter given current and past values of GDP and other input signals.[12]

[12] Data files and code for this example are available in the supplementary material in the file package ex_GDP.zip in the online resources.

Figure 9.15 GDP quarterly time series.

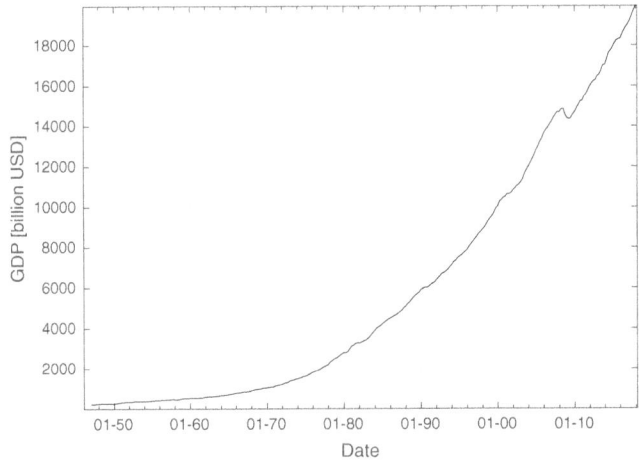

We consider as input signals the following data: personal consumption expenditures on durable goods, real disposable personal income, government consumption expenditures and investment, and imports of goods and services. These quantities are shown in Figure 9.16.

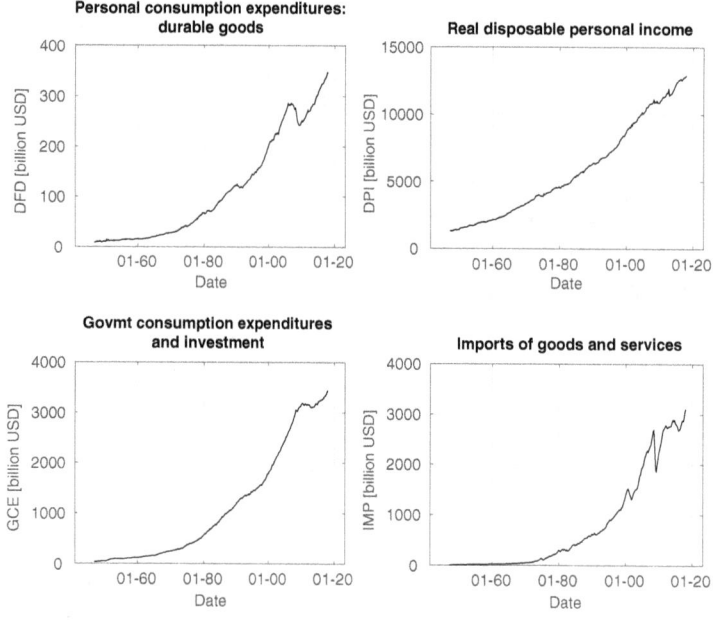

Figure 9.16 Input signals of the NARX network.

In order to evaluate the performance of the NARX network, we compare the estimated GDP with respect to its corresponding true value. Figure 9.17 shows the predicted and true GDP, whilst Figure 9.18 shows the relative prediction error. The mean relative prediction error on the validation set is 0.74%.

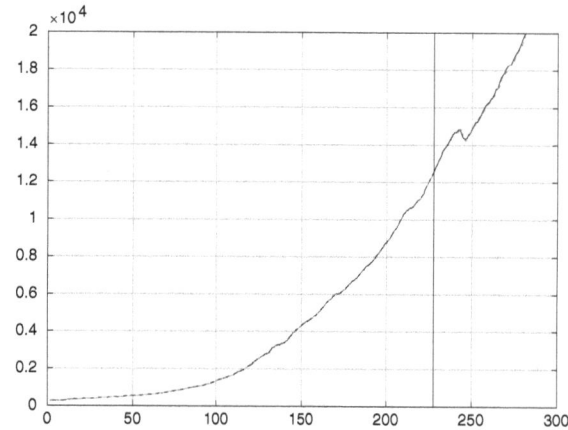

Figure 9.17 Predicted GDP (dashed curve) and true GDP (solid curve). The vertical line separates the training set (on the left) from the validation set (on the right).

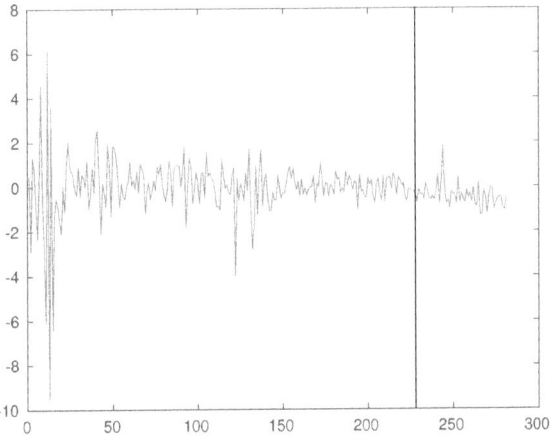

Figure 9.18 Relative prediction error. The vertical line separates the training set (on the left) from the validation set (on the right).

9.6 Selected modern deep learning approaches

In this last section we briefly overview some recent deep learning approaches that have led to significant advances in this field.

9.6.1 Residual neural networks

Residual neural networks (ResNets) use skip connections to jump over some layers and connect shallow layers directly with deeper layers. More specifically, a ResNet is obtained by stacking many residual blocks. The architecture of a residual block is shown in Figure 9.19. The rationale behind this type of architecture is that, instead of learning the direct mapping $f(x)$, the network learns the residual mapping $h(x) = f(x) - x$. This allows the so-called degradation problems that affect very deep neural networks to be avoided. Indeed, it has been observed that when one increases the depth of standard feedforward networks, the performance saturates and then degrades rapidly.[13] Instead, the introduction of residual blocks allows one to build deeper networks and reach breakthrough results, especially in the field of computer vision.

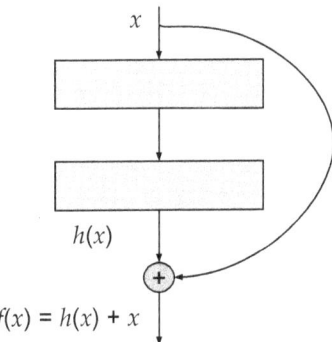

Figure 9.19 Residual block.

[13] For further details see K. He, X. Zhang, S. Ren, and J. Sun, "Deep Residual Learning for Image Recognition," *IEEE Conference on Computer Vision and Pattern Recognition (CVPR)*, pp. 700–778, 2016.

9.6.2 Sequence to sequence models

Sequence to Sequence (Seq2Seq) models are a special type of RNNs that are focused on converting sequences from one domain to another. They are commonly used, for instance, for machine translation and question answering. The architecture of Seq2Seq models is usually subdivided in two blocks: an encoder and a decoder. The encoder takes the input sequence and extracts a vector of hidden features that summarize the sequence information. Then, the decoder takes as input the hidden feature vector produced by the encoder and generates the output sequence. Both the encoder and decoder are RNNs, usually implemented using LSTM architectures.[14]

9.6.3 Attention mechanism

The attention mechanism was first proposed in the context of Seq2Seq models for natural language processing,[15] but then it has been widely used in many other applications. This mechanism helps the model to focus on the most relevant parts of the input sequence. If we consider a Seq2Seq model where the encoder is a RNN, instead of encoding the whole sentence in a single hidden state, we compute a hidden state for each word of the sentence. Then, at the decoder an output word $d^{(i)}$ is obtained using the hidden states produced by the encoder and an auxiliary context vector. The latter is produced by the attention mechanism, which assigns a weight $\omega^{(i,j)}$ for each encoder hidden state $h(j)$ and computes the context vector as $\sum_j \omega^{(i,j)} h(j)$. Higher weights will be assigned to the hidden representations of the most relevant parts of the input sequence. A schematic representation of the attention mechanism is shown in Figure 9.20.

[14] For further details see I. Sutskever, O. Vinyals, and Q. V. Le, "Sequence to Sequence Learning with Neural Networks," *Advances in Neural Information Processing Systems*, vol. 27, 2014, and K. Cho, B. van Merrienboer, C. Gulcehre, D. Bahdanau, F. Bougares, H. Schwenk, and Y. Bengio, "Learning Phrase Representations using RNN Encoder–Decoder for Statistical Machine Translation," *Conference on Empirical Methods in Natural Language Processing (EMNLP)*, pp. 1724–1734, 2014.

[15] See D. Bahdanau, K. Cho, and Y. Bengio, "Neural Machine Translation by Jointly Learning to Align and Translate," *International Conference on Learning Representations (ICLR)*, 2015.

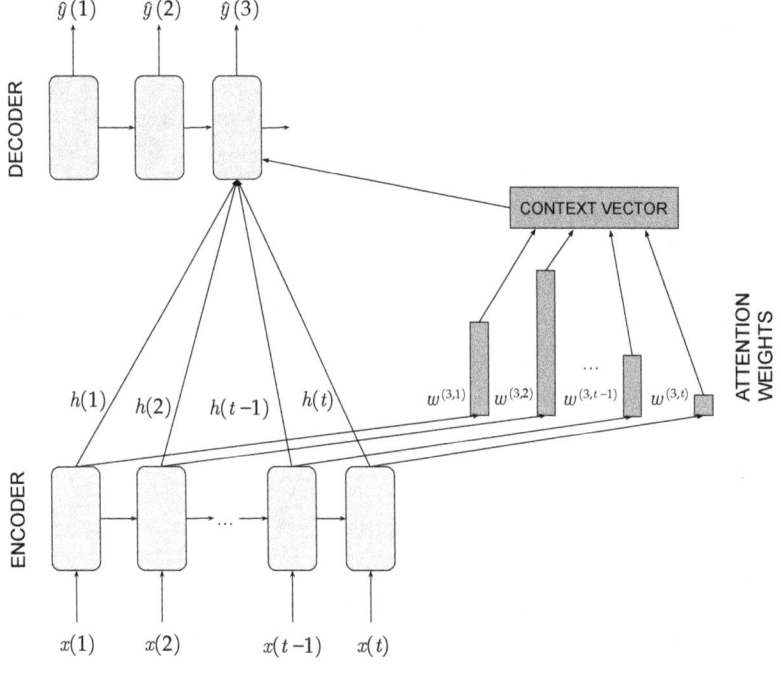

Figure 9.20 Attention mechanism.

The use of attention highly improved the performance of Seq2Seq models. However, they still have a strong limitation: they cannot process inputs (i.e., words) in parallel. This strongly increases the training and inference time, especially when we need to process very long sequences. Transformers were proposed in order to solve this issue. They are a combination of feed-forward neural networks and attention modules. The advantage of feed-forward neural networks is that they can work in parallel, since they are able to process each word of the input sequence at the same time. The attention module instead is fundamental in order to understand the dependencies among the words that compose the sequence. As Seq2Seq models, transformers are composed by an encoder and a decoder.[16] The encoder consists of two layers: a self-attention module followed by a feed-forward neural network. The role of the self-attention layer is to take into account all the other words of the sequence as it computes a hidden representation of a specific word and to focus only on the parts of the input sequence that are most relevant for that word. Then, the attention vectors are processed by the feed-forward NN to produce the output of the encoder. This output is then taken as input by the decoder, which has a structure similar to the one of the encoder with the addition of an encoder–decoder attention module that helps the decoder focus on the relevant parts of the input sequence. A schematic representation of the transformer is shown in Figure 9.21.

[16] Modern transformers usually have multiple encoders and decoders, for simplicity we consider only the case with just one encoder and one decoder.

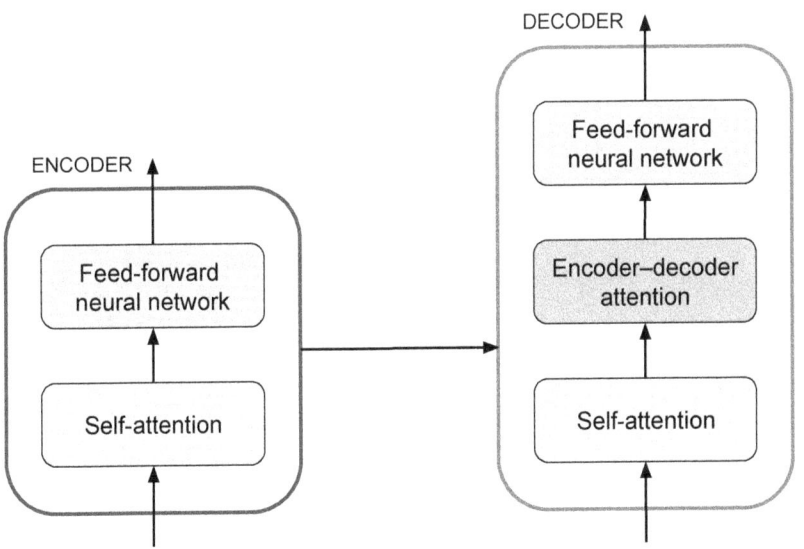

Figure 9.21 A schematic representation of the transformer.

The transformer is the key element that allowed recent deep learning models to reach outstanding results in the field of natural language processing.

9.6.4 Generative Adversarial Networks

Generative learning is a class of statistical models that focus on learning the underlying patterns of the data points in a given dataset in order to be able to generate new data instances of such dataset. Generative adversarial networks (GANs) are a type of generative models whose goal is to learn the distribution of a given dataset by being able to sample from it. A GAN is composed of two neural networks, a generator and a discriminator. The generator learns to generate new data samples, while the discriminator learns to classify the data samples as either real (i.e., data samples from the original dataset) or fake (i.e., generated by the generator). These two networks are trained together in a zero-sum game where the generator generates a batch of samples that are fed to the discriminator, along with real samples, and classified as real or fake. Typically, the input of the generator network is randomly sampled from a predefined distribution (e.g., a multivariate normal distribution), which is called latent space.[17] Generative adversarial networks have displayed impressive results, in particular in the context of image generation where they showed they were capable of generating images that were almost natural looking.[18]

[17] For further details see I. Goodfellow et al., "Generative Adversarial Networks," *Advances in Neural Information Processing Systems*, vol. 27, 2014.

[18] See for instance A. Brock, J. Donahue, and K. Symonyan, "Large Scale GAN Training for High Fidelity Natural Image Synthesis," *International Conference on Learning Representations (ICLR)*, 2019.

9.7 Exercises

Exercise 9.1 *Derivative of the output of a neural network*
As explained in Section 9.3.1, at each iteration of the SGD algorithm the weights of the network are updated by taking a step in the opposite direction of the gradient of the loss function. In order to compute the gradient we can use the backpropagation algorithm, as illustrated in Section 9.3.1. In this exercise, we will see a practical example of how to apply backpropagation in order to compute the partial derivative of the loss function with respect to a weight of the network.

Let $y = \phi(x)$ be the output of a network. Consider a data set $\mathcal{D} = \{(x^{(i)}, y_i)\}_{i=1}^m$, where $y_i \in \mathbb{R}$, and define the cross-entropy error function for binary classification:

$$\mathcal{L}(\theta, \mathcal{D}) = \sum_{i=1}^m y_i \log(\phi(x^{(i)})) + (1 - y_i) \log(1 - \phi(x^{(i)})).$$

Let the output of the network be given by a sigmoid function, as discussed in Section 9.2.2:

$$\phi(x) = \sigma(z) = \frac{1}{1 + \exp(-z)},$$

where $z = w^{(d)} h^{(d-1)} + b^{(d)}$ and $w^{(d)} \in \mathbb{R}^{1, F^{(d-1)}}$. Compute the derivative of the cross-entropy loss $\mathcal{L}$ with respect to $w^{(d)}$. Using the chain rule, you should first compute the derivative of the cross-entropy loss with respect to $\phi(x)$, then the derivative of the sigmoid function with respect to $w^{(d)}$.

Exercise 9.2 *New York Stock Exchange price forecasting*
In this exercise, we will perform stock price prediction using recurrent neural networks. We will use the data set contained in the file `price-split-adjusted.csv`[19] in the online resources. This data set contains the daily stock price of the New York Stock Exchange. For each stock, the data set contains the daily opening and closing price, the daily minimum and maximum price, and the daily stock volume. Most data spans are from 2010 to end 2016; for companies new on the stock market the time range is shorter.

[19] This data set is also available at www.kaggle.com/datasets/dgawlik/nyse.

1. Load the data set and choose a specific stock. Drop the symbol and volume features. Normalize the features such that they are all in the range between 0 and 1.[20]

2. Create all the possible sequences of length 20. Subdivide the obtained data set into a training set, a validation set, and a test set (the size of the training set should be 80% of the entire data set, the

[20] In Python, you may use for this purpose the function `MinMaxScaler` from the package `sklearn`.

size of the validation set and test set should be 10% of the original data set).

3. Train a basic RNN[21] that takes the values of the first 19 days of a sequence and predicts the values of the last day of the sequence. Use a RNN with two layers and 200 features. Set the number of epochs equal to 100.

4. Use the validation set to monitor the performance during training, then evaluate the trained model on the test set.

5. Repeat points 3 and 4 using a LSTM[22] instead than a basic RNN. Use the same settings defined in point 3. Compare the results.

[21] To solve this exercise you might use the Python library Keras, which contains numerous implementations of commonly used neural-network building blocks. In Keras, you can use the functions SimpleRNNCell and StackedRNNCells.

[22] In Keras, you may use the functions LSTMCell and StackedRNNCells.

Exercise 9.3 *Credit card fraud detection with neural networks*
In this exercise, we consider the problem of credit card fraud detection already discussed in Exercises 7.3, 7.5, and 8.1. We are now interested in addressing this problem using a neural network.

1. Load the data set contained in the file creditcard.csv in the online resources. Discard the information regarding the time of the transactions, since it is uninformative for our purpose. Subdivide the data set into a training set, a validation set and a test set. The size of the training set should be 60% of the entire data set, and the size of the test set should be 25%. The remaining 15% of the data set is used as validation set.

2. Rescale the features to have zero mean and variance equal to one.[23]

3. Train a fully connected network using the training set. Use a network with four fully connected layers[24] with ReLU activations (except the last layer) and 256 features. After each layer, except the last one, apply batch normalization and then dropout[25] with dropout probability equal to 30%. Set the number of epochs to 300.

4. Use the validation set to monitor the performance during training, then evaluate the trained model on the test set.

5. Compare the obtained results with those of Exercises 7.3, 7.5, and 8.1.

[23] In Python, you can use the function StandardScaler from the package sklearn.

[24] In Keras, you can use the functions Dense and Sequential.

[25] In Keras, you can use the functions Dropout and BatchNormalization.

Exercise 9.4 *Classification of fashion products*
Fashion-MNIST[26] is a data set of Zalando's article images, consisting of a training set of 60,000 examples and a test set of 10,000 examples. Each example is a grayscale image of dimension 28 × 28, associated with a label from 10 classes (namely: T-shirt, Trouser, Pullover, Dress,

[26] This data set is available at www.kaggle.com/datasets/zalando-research/fashionmnist.

Coat, Sandal, Shirt, Sneaker, Bag, and Ankle boot). We are now interested in performing classification of these images using a convolutional neural networks.

1. Load the training set and test set contained in the files fashion-mnist_train.csv and fashion-mnist_test.csv in the online resources, respectively. In the data sets the images are stored as vectors of length 784, reshape them as matrices 28×28. Randomly sample 20% of the images of the data set from the training set and use them as a validation set during the training process.

2. Train a convolutional neural network using the training set. Use a network with three convolutional layers[27] followed by two fully connected layers. Use ReLU activation functions. Set the kernel size of the convolutional layers equal to 3×3. After the first two convolutional layers, apply max pooling[28] using a neighborhood window of size 2×2. Use the following setting for the number of features: $F^{(1)} = 32$, $F^{(2)} = 64$, $F^{(3)} = 128$, and $F^{(4)} = 128$. Set the number of epochs equal to 50.

3. Use the validation set to monitor the performance during training. During training, compare the accuracy of the model on the training set with the one on the validation set. Is the model overfitting the training set?

4. In order to reduce the risk of overfitting, you can apply dropout after each layer (except the last one). In the first two convolutional layers, apply dropout after max pooling. Use a dropout probability equal to 25%. Train the model and evaluate its performance with respect to the ones obtained using the model defined in point 2.

5. Evaluate the performance of the model on the test set.

[27] In Keras, you can use the function Conv2D.

[28] In Keras, you can use the function MaxPooling2D.

10
Optimization tools

OPTIMIZATION METHODS play a fundamental role in machine learning. Indeed, if one may say that Bayesian statistics is one of the methodological foundations of machine learning, optimization techniques and linear algebra are its engine. Statistics helped us formulating a learning problem in an appropriate way, but it is then numerical optimization that allows us to actually solve the problem in practice. The field of optimization is vast, so here we limit ourselves to an overview of the key concepts and issues, with a focus on convexity. In machine learning, the goal is often to find the optimal values of model parameters that minimize a certain objective function. Convex optimization provides a framework to define and solve these optimization problems efficiently. The convexity of the objective function ensures that there are no local minima, guaranteeing that the *global* minimum can be found. Convex optimization enables efficient and global solution for many regularized loss minimization problems arising in data science, which include the Lasso, Ridge regression, SVMs, Logistic models, and so on. Other machine learning models, such as, prominently, neural networks, are instead intrinsically nonconvex but still they are solved numerically in an approximate sense using numerical optimization techniques such as the gradient algorithm.

Optimization forms an essential part of many of the tools already discussed in previous chapters of this book. For instance, Euclidean projections and data scores were defined by means of a minimization of a distance, eigenvalues of symmetric matrices are related to the extrema of certain quadratic functions, least squares involve the minimization of a quadratic norm, and maximum likelihood estimation is based on solving certain maximization problems. But what is optimization? Optimization is a field of applied mathematics also known as "mathematical programming," which is concerned with the modeling and solving of problems involving the minimization or maximization of an objective function of one or several variables, possibly under *constraints* on the variables. Optimization is also a *language* that allows us to describe precisely how a *decision* should be made. It includes as special cases machine learning problems (the decision may be about what parameters to use in a classification rule), financial decision problems (e.g., how to allocate resources in a portfolio of assets), and a wide range of design problems in engineering and applied science.

10.1 Optimization problems in standard form

An optimization problem in standard form is typically stated as

$$p^* \doteq \min_x f_0(x), \text{ s.t.: } f_i(x) \leq 0, \ i = 1, \ldots, m, \quad (10.1)$$

where

- $x \in \mathbb{R}^n$ is the *decision variable*;
- $f_0 \colon \mathbb{R}^n \to \mathbb{R}$ is the *objective* (or, *cost*) function;
- $f_i \colon \mathbb{R}^n \to \mathbb{R}, i = 1, \ldots, m$, represent the *constraints*;
- p^* is the *optimal value* of the problem.

The above problem is often is referred to as a "mathematical program" for historical reasons.

A point $x \in \mathbb{R}^n$ such that $f_i(x) \leq 0, i = 1, \ldots, m$, is called *feasible* for problem (10.1). The set

$$\mathcal{X}_{\text{opt}} \doteq \{x \in \mathbb{R}^n : x \text{ is feasible, and } f_0(x) = p^*\}$$

is called the *optimal set*, or the set of the optimal solutions of problem (10.1).

Penalization In some cases we deal with problems in which we aim at minimizing simultaneously two or more competing objectives, say $g_1(x)$ and $g_2(x)$. Such *multi-objective* problems may be converted into a family of standard single-objective problems by assigning suitable weights or preference levels to the individual objectives, thus creating an objective of the form $f_0(x) = \lambda_1 g_1(x) + \lambda_2 g_2(x)$, where $\lambda_1, \lambda_2 \geq 0$ represent the weights that reflect the desired tradeoffs between the two objectives. This is the typical case that arises in the context of regularized loss minimization, see, for example, (6.10) and (7.7).

Penalization is also used for approximately transforming the constrained problem (10.1) into an *unconstrained* one obtained by augmenting the objective with the constraint functions as penalties, that is,

$$g(\lambda) \doteq \min_x f_0(x) + \sum_{i=1}^{m} \lambda_i f_i(x), \qquad (10.2)$$

where $\lambda = (\lambda_1, \ldots, \lambda_m) \geq 0$ is a vector of weights, also known as *Lagrange multipliers*. For any choice of $\lambda \geq 0$ it is easy to show that

$$g(\lambda) \leq p^*,$$

so problem (10.2) is a *relaxation* of the original constrained problem (10.1). In certain cases it is possible to prove that there exist some special $\lambda^* \geq 0$ for which it actually holds that $g(\lambda^*) = p^*$. When this is the case, knowing this λ^* may allow for solving the constrained problem (10.1) by actually solving the unconstrained problem (10.2) with $\lambda = \lambda^*$.

10.1.1 Hard vs. "easy" problems

The difficulty of the numerical solution of problem (10.1) lies in the nature and shape of the objective function f_0 and of the constraint functions $f_1, \ldots, f_m$. Figure 10.1 shows a generic objective function f_0 of one variable.

One source of difficulty is the presence of local optima in the objective function, since many minimization algorithms tend to remain trapped in these local minima and fail to reach the global minimum, which is usually the solution we are really seeking. In the case of constrained problems, algorithms may also fail to find a feasible point, even if one actually exists. These issues do not occur if the objective function and constraints have a special structure called *convexity*, which essentially rules out the presence of local optima.

Nonconvex optimization problems include Boolean or integer optimization, in which all or some of the variables are constrained to be Boolean or integers,[1] cardinality-constrained problems, in which we seek to bound the number of nonzero elements in a vector variable,[2] and general nonlinear programming problems, which usually involve nonconvex objective functions and constraints. This latter class includes as special cases many machine learning problems (e.g., neural nets).

Not all nonconvex problems, however, are hard to solve. For example, the variance maximization problem encountered in PCA is nonconvex, yet its solution is "easy" to compute via SVD. Moreover, in certain unconstrained contexts (e.g., ARIMA or neural networks

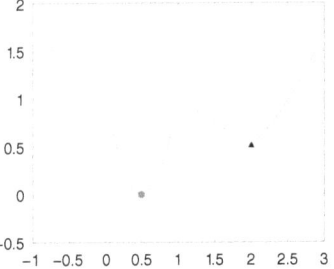

Figure 10.1 An objective function f_0 with a local minimum (triangle) and a global minimum (circle).

[1] Convex optimization can be sometimes used for getting good approximations or relaxations for this class of problems.
[2] Convex optimization can be used for getting good approximations, via ℓ_1 norm penalties for example.

models) nonconvexity may not be a an insurmountable issue, since local minima are usually "good enough." In the presence of constraints, however, nonconvexity usually is a serious issue, since algorithms may be unable to find a feasible point, even when one exists, so no usable solution may be returned to the user.

10.2 Convexity

10.2.1 Convex sets and convex functions

A set $X \subseteq \mathbb{R}^n$ is said to be convex if

$$x_1 \in X, x_2 \in X, \lambda \in [0,1] \quad \Rightarrow \quad \lambda x_1 + (1-\lambda)x_2 \in X.$$

In words, a set X is convex if the line segment joining any two points in X lies entirely in X, see Figure 10.2. A function $f\colon \mathbb{R}^n \to \mathbb{R}$ is *convex* if $\text{dom } f \subseteq \mathbb{R}^n$ is a convex set,[3] and

$$f(\lambda x_1 + (1-\lambda x_2)) \leq \lambda f(x_1) + (1-\lambda)f(x_2), \ \forall\, x_1, x_2 \in \mathbb{R}^n, \ \forall \lambda \in [0,1]. \tag{10.3}$$

We say f is *concave* if $-f$ is convex.

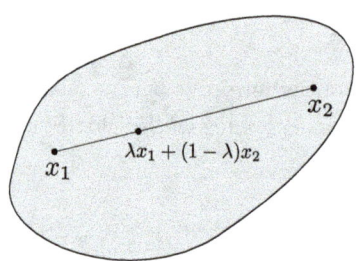

Figure 10.2 In a convex set the line segment joining any two point lies in the set.

[3] The domain $\text{dom } f$ of a function f is the set of points over which the function is well defined and such that $-\infty < f < \infty$. For example, for $f(x) = \|x\|_2$ we have $\text{dom } f = \mathbb{R}^n$, and for $f(x) = \log x$, for scalar x, we have $\text{dom } f = \{x\colon x > 0\}$.

Figure 10.3 The graph of a convex function is always below the chord.

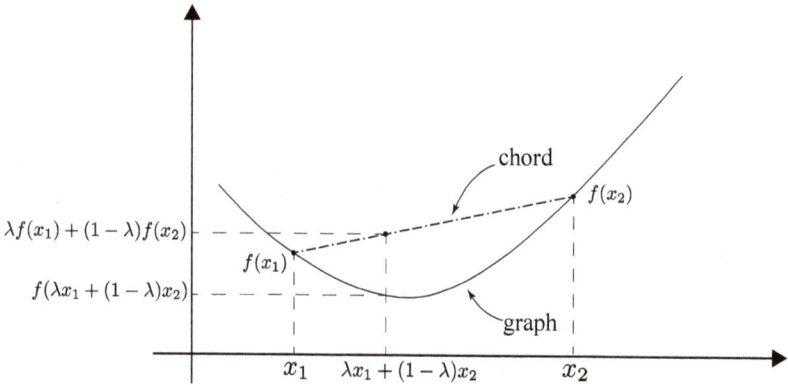

Convex sets arise naturally from sub-level sets of convex functions. In particular, it can be proved that for any convex function $f\colon \mathbb{R}^n \to \mathbb{R}$ and any scalar α the set

$$X_\alpha \doteq \{x \in \mathbb{R}^n \colon f(x) \leq \alpha\}$$

is convex. This fact is particularly relevant in the optimization context where *constraints* are posed on the decision variables in the form of functional inequalities like $f_i(x) \leq 0, \ i = 1,\ldots,m$. If the functions

$f_i(x)$ are convex, then each constraint defines a convex set $X_i = \{x: f_i(x) \leq 0\}$, and it is easy to prove that the intersection of all such sets

$$X = \{x: f_i(x) \leq 0, i = 1, \ldots, m\} = \bigcap_{i=1,\ldots,m} X_i$$

is also convex (the intersection of convex sets is always a convex set). The following are examples of convex functions:

- an affine (i.e., linear plus constant) function $f(x) = c^\top x + d$ is both convex and concave;

- the Euclidean norm $f(x) = \|x\|_2$ and the ℓ_1 norm $f(x) = \|x\|_1$ are convex (in fact, any norm is convex);

- the logistic loss function $f(x) = \log(1 + \exp(-x))$, for scalar x, is convex;

- the hinge loss function $f(x) = \max(0, 1 - x)$, for scalar x, is convex.

Proving convexity by directly checking if the definition (10.3) holds may be difficult in practice. For this reason, we often use a combination of *convexity rules*, such as:

- the nonnegative linear combination $f(x) = \sum_{i=1}^{m} a_i f_i(x)$ of convex functions $f_1, \ldots, f_m$ with nonnegative coefficients $a_i \geq 0$, $i = 1, \ldots, m$, is convex;

- the point-wise maximum $f(x) = \max_{i=1,\ldots,m} f_i(x)$ of convex functions $f_1, \ldots, f_m$ is convex (this rule extends to point-wise maximum over infinite and dense sets of functions);

- the composition $f(x) = g(h(x))$ of a convex function g with an affine map h is convex;

- for scalar x, the composition $f(x) = g(h(x))$ of a convex function g with an increasing convex function h is convex;

- a quadratic function $f(x) = x^\top A x + b^\top x + d$ is convex if and only if A is a positive semidefinite matrix, see Appendix A.1.3;

- a twice differentiable function $f(x)$ is convex if and only if its Hessian matrix (i.e., the matrix of its second-order derivatives with respect to all variables in $x \in \mathbb{R}^n$) is positive semidefinite.

Examples The cost function (7.14) appearing in the linear SVM problem is convex in the variable w since it is written as

$$f(w) = \|w\|_2^2 + c \sum_{i=1}^{m} \max(0, 1 - z_i(w)), \quad c \geq 0,$$

where $z_i(w) = y_i(w^\top x^{(i)} + b)$ is affine in w, and $\max(0, 1 - z_i)$ is convex. Therefore by the composition with affine function rule $\max(0, 1 - z_i(w))$ is convex in w, and by the nonnegative linear combination rule $c \sum_{i=1}^{m} \max(0, 1 - z_i(w))$ is convex too. Finally, this last function added to the convex quadratic function $\|w\|_2^2 = w^\top w$ yields overall a convex function.

The function

$$f(x) = x_1^2 - x_1 x_2 + 2x_2^2 - 3x_1 - 1.5x_2 = x^\top A x + c^\top x,$$

where

$$c^\top = [-3, -1.5], \text{ and } A = \begin{bmatrix} 1 & -1/2 \\ -1/2 & 2 \end{bmatrix}$$

is a quadratic function, hence its convexity depends on whether A is PSD or not. This characteristic can be checked numerically by computing the eigenvalues of A, which result to be $\lambda_1 = 2.2071$, $\lambda_2 = 0.7929$: both eigenvalues are nonnegative, thus $f(x)$ is convex. Alternatively, one may observe that A can be expressed as the sum of two PSD matrices

$$A = \frac{1}{2} \begin{bmatrix} 1 \\ -1 \end{bmatrix} \begin{bmatrix} 1 \\ -1 \end{bmatrix}^\top + \begin{bmatrix} 1/2 & 0 \\ 0 & 3/2 \end{bmatrix},$$

hence the quadratic form $x^\top A x$ is the sum of squares

$$x^\top A x = \frac{1}{2}(x_1 - x_2)^2 + \frac{1}{2}x_1^2 + \frac{3}{2}x_2^2 \geq 0.$$

The function $s_k(x)$ representing the sum of the k-largest components of an n-dimensional vector is convex (here, $k \leq n$), since

$$s_k(x) = \max_u u^\top x : u_i \in \{0, 1\}, \ i = 1, \ldots, n, \ \sum_{i=1}^{n} u_i = k.$$

For instance, with $n = 3$, $k = 2$, we have $s_2(x) = \max(x_1 + x_2, x_2 + x_3, x_3 + x_1)$, where each piece in the max is linear, hence convex. It can also be proved[4] that $s_k(x)$ can be explicitly represented as

$$s_k(x) = \min_t kt + \sum_{i=1}^{n} \max(0, x_i - t).$$

[4] See, for example, Sec. 9.3.1 of G. Calafiore and L. El Ghaoui, *Optimization Models*, Cambridge: Cambridge University Press, 2014.

The largest eigenvalue of a symmetric matrix X is a convex function in the matrix variable X, since

$$\lambda_{\max}(X) = \max_{z:\ z^\top z = 1} z^\top X z.$$

Here, $z^\top X z$ is linear (hence convex) in X, and the function we are interested in is obtained as the point-wise maximum (over the "parameter" z) of convex functions.

10.2.2 Convex problems

The problem in standard form

$$p^* \doteq \min_x f_0(x),\ \text{s.t.:}\ f_i(x) \leq 0,\ i=1,\ldots,m,\ Cx = d, \quad (10.4)$$

is convex if all functions $f_0, f_1, \ldots, f_m$ are convex. Here $C \in \mathbb{R}^{p,n}$, $d \in \mathbb{R}^p$ are given, and $x \in \mathbb{R}^n$ is the decision variable. Note that only *affine equality* constraints are allowed, and that one cannot replace $\leq$ signs by $\geq$ signs, without destroying convexity (unless also f_i is replaced by $-f_i$).

The set of vectors x that are *feasible*, that is, satisfy all the constraints, is called the feasible set. The feasible set of a convex optimization problem is a convex set. Sometimes the optimization problem arises more naturally in the form of a maximization problem:

$$\max_x f_0(x),\ \text{s.t.:}\ f_i(x) \leq 0,\ i=1,\ldots,m,$$
$$Cx = d,$$

which is convex if the function f_0 is concave, while the constraint functions functions $f_1, \ldots, f_m$ are all convex.

Convex problems enjoy the desirable property that any point that is locally optimal is also globally optimal, so for these problems one does not have the issue of numerical algorithms being trapped in local minima. Further, most classes of convex problems are solvable "efficiently" via suitable numerical algorithms. By efficiently we informally mean that the time needed for solving the problem on a computer grows gracefully (e.g., polynomially, and not exponentially or by combinatorial explosion) with the problem size.

Most supervised learning problems we have encountered so far in classification and regression share the common structure

$$\min_{w,b}\ \mathcal{L}(w,b) + \lambda \psi(w), \quad (10.5)$$

where $\psi(w)$ is typically a norm (e.g., the Euclidean, or the ℓ_1 norm), and $\mathcal{L}$ is a convex loss function, in particular:

- $\mathcal{L}(w,b) = \sum_{i=1}^{m}(w^\top x^{(i)} + b - y_i)^2$ in least-squares regression and its regularized versions (e.g., Lasso);

- $\mathcal{L}(w,b) = \sum_{i=1}^{m}|w^\top x^{(i)} + b - y_i|$ in ℓ_1 regression and its regularized versions;

- $\mathcal{L}(w,b) = \sum_{i=1}^{m}\max(0, 1 - y_i(w^\top x^{(i)} + b))$ in the linear SVM classifier;

- $\mathcal{L}(w,b) = \sum_{i=1}^{m}\log(1 + \exp(-y_i(w^\top x^{(i)} + b)))$ in the logistic classifier.

Training any of the above models thus amounts to solving some convex optimization problem. We next illustrate briefly some specialized classes of convex optimization problems that are specifically coded into widely available numerical solvers.

10.3 Convex problem classes

When one specifies the type of convex functions involved in the generic problem formulation (10.4) one obtains specific classes of convex problems. We next explore the most popular of such classes, in a hierarchical way from the most specialized class to the more general one.

10.3.1 Linear programs

Linear programming forms a class of convex problems obtained when all the functions $f_0, f_1, \ldots, f_m$ in (10.4) are linear or affine. That is, $f_0(x) = c^\top x$, and $f_i(x) = a_i^\top x - b_i$, $i = 1, \ldots, m$. Stacking all row vectors $a_i^\top$ and scalars b_i, $i = 1, \ldots, m$, into a matrix $A \in \mathbb{R}^{m,n}$ and vector $b \in \mathbb{R}^m$ respectively, we obtain the linear program (LP) representation in standard form

$$p^* \doteq \min_{x} c^\top x, \text{ s.t.: } Ax \leq b, \; Cx = d, \qquad (10.6)$$

where the vector inequality $Ax \leq b$ is to be intended element-wise. Linear programs are historically[5] the first class of convex problems to be extensively studied and implemented into efficient numerical solvers. Still to date LP models are a staple of large-scale industrial optimization and the efficiency and reliability of their ad-hoc computer solvers makes it possible to tackle real-world problems with millions of variables and constraints.

[5] George Dantzig is considered (in the West) to be the father of linear programming and of the Simplex algorithm for its solution developed in the 1950s, see, for example, J. K. Lenstra, A. H. G. Rinnooy Kan, and A. Schrijver (eds.), *History of Mathematical Programming: A Collection of Personal Reminiscences*, Amsterdam: Elsevier Science Publishers, 1991. Actually, a linear programming formulation and its solution was independently developed by the mathematician and economist Leonid Kantorovich in the USSR in 1939.

10.3 CONVEX PROBLEM CLASSES

Example 10.1 (*Sparse SVM*) Despite the outward simplicity of the LP model, many problems of practical relevance can be cast within this model. For instance, a sparse SVM training problem of the form

$$\min_{w,b} \sum_{i=1}^{m} \max(0, 1 - y_i(w^\top x^{(i)} + b)) + \lambda \|w\|_1, \quad \lambda \geq 0, \quad (10.7)$$

although apparently nonlinear, can be recast in LP format by applying a series of suitable "tricks," as described next. First, we observe that

$$\max(0, z) = \min_{s} s, \quad \text{s.t.: } s \geq 0, \ s \geq z,$$

hence

$$\sum_{i=1}^{m} \max(0, z_i) = \min_{s} \sum_{i=1}^{m} s_i, \quad \text{s.t.: } s_i \geq 0, \ s_i \geq z_i, \ i = 1, \ldots, m.$$

Similarly, we write

$$\|w\|_1 = \sum_{i=1}^{n} \max(w_i, -w_i)$$

$$= \min_{v} \sum_{i=1}^{n} v_i, \quad \text{s.t.: } v_i \geq w_i, \ v_i \geq -w_i, \ i = 1, \ldots, n.$$

Problem (10.7) can then be cast equivalently as

$$\min_{w,b,s,v} \sum_{i=1}^{m} s_i + \lambda \sum_{i=1}^{n} v_i,$$
$$\text{s.t.: } s_i \geq 0, \ s_i \geq 1 - y_i(w^\top x^{(i)} + b), \ i = 1, \ldots, m,$$
$$v_i \geq w_i, \ v_i \geq -w_i, \quad i = 1, \ldots, n,$$

which is an LP in standard form in the variables (w, b, s, v).

Example 10.2 (*A short-term financing problem*) A company faces the following net cash flow requirements:

Month	Jan	Feb	Mar	Apr	May	Jun
Net cash flow (in K$)	−150	−100	200	−200	50	300

and has the following available sources of funds:

- line of credit (max 100K, interest rate 1% per month);
- in any of the first three months it can issue 90-day commercial paper bearing a total interest of 2% for the three-month period;
- excess funds can be invested at 0.3% per month.

We denote with:

- x_i, the balance on the credit line for month $i = 1, 2, 3, 4, 5$;
- y_i, the amount of commercial paper issued ($i = 1, 2, 3$);
- z_i, the excess funds for month $i = 1, 2, 3, 4, 5$;
- z_6, the company's wealth in June.

The financial decision problem we consider is to maximize the company's final wealth, while respecting constraints and cash flow requirements, that is:

$$\text{maximize } z_6 \text{ subject to } \begin{cases} \text{Bounds on variables,} \\ \text{Cash-flow balance equations.} \end{cases}$$

Problem constraints include:

- non-negativity: $x_i \geq 0, i = 1, \ldots, 5$; $z_i \geq 0, i = 1, \ldots, 6$; $y_i \geq 0, i = 1, 2, 3$;

- upper bounds on x_is: $x_i \leq 100, i = 1, \ldots, 5$;

- cash flow balance equations.

This setting leads to the linear programming formulation

$$\begin{aligned}
\max_{x,y,z} \ & z_6, \\
\text{s.t.} \quad & x_1 + y_1 - z_1 = 150, \\
& x_2 + y_2 - 1.01 x_1 + 1.003 z_1 - z_2 = 100, \\
& x_3 + y_3 - 1.01 x_2 + 1.003 z_2 - z_3 = -200, \\
& x_4 - 1.02 y_1 - 1.01 x_3 + 1.003 z_3 - z_4 = 200, \\
& x_5 - 1.02 y_2 - 1.01 x_4 + 1.003 z_4 - z_5 = -50, \\
& -1.02 y_3 - 1.01 x_5 + 1.003 z_5 - z_6 = -300, \\
& 100 \geq x_i \geq 0, \quad i = 1, \ldots, 5, \\
& y_i \geq 0, \quad i = 1, 2, 3, \\
& z_i \geq 0, \quad i = 1, \ldots, 6.
\end{aligned}$$

Compactly:

$$\max_{x,y,z} z_6 : A \begin{bmatrix} x \\ y \\ z \end{bmatrix} = b, \ 0 \leq x \leq 100, \ x \geq 0, \ z \geq 0,$$

for appropriate vector $b \in \mathbb{R}^6$ and matrix $A \in \mathbb{R}^{6,14}$. Letting $v = (x, y, z)$, we may also observe that in this specific problem we can replace the equality constraints $Av = b$ with the inequality constraints $Av \geq b$, and yet equality will hold at the optimum. This is due to the fact that if an optimal solution v^* existed such that $Av^* > b$, then it would be possible to increase some of the z variables while remaining feasible: the increase in excess funds will eventually result in a higher value of the objective z_6, which would then contradict optimality. Our problem can thus be recast equivalently as

$$\max_{x,y,z} z_6 : Av \geq b, \ 0 \leq x \leq 100, \ x \geq 0, \ z \geq 0.$$

An optimal solution is found via a numerical LP solver, resulting in the following optimal investment strategy:[6]

[6] See the file short_term_financing.m in the online resources.

Month	x	y	z
1	0.00	150	0.00
2	50.98	49.02	0.00
3	0.00	203.43	351.94
4	0.00	–	0.00
5	0.00	–	0.00
6	–	–	**92.4969**

10.3.2 *Quadratic programs*

Quadratic programming involves the minimization of a convex quadratic function under linear or affine constraints. The objective f_0 in (10.4) is thus of the form

$$f_0(x) = x^\top Q x + c^\top x,$$

with $Q \succeq 0$, while $f_i(x) = a_i^\top x - b_i$, $i = 1, \ldots, m$. A quadratic program (QP) in standard form has the same type of constraints as an LP, but a different (and quadratic) objective:

$$p^* \doteq \min_x x^\top Q x + c^\top x, \text{ s.t.: } Ax \leq b, \; Cx = d, \qquad (10.8)$$

Clearly, the QP problem class includes the LP class, which is obtained by setting $Q = 0$.

Example 10.3 (*Mean/variance portfolio optimization*) Quadratic programs are relevant in the classical mean/variance portfolio theory, where one aims at minimizing the investment risk (variance) while guaranteeing a pre-specified level of expected return. With the notation introduced in Section 3.3, denoting with $\mu \in \mathbb{R}^n$ the vector of expected returns of a given collection of n assets, with $\Sigma \in \mathbb{S}_+^n$ the corresponding covariance matrix, and with $x \in \mathbb{R}^n$ the vector of portfolio allocations in the various assets, with budget conventionally set to one, that is, $\sum_i x_i = 1$, we have that the portfolio expected return is $\mu^\top x$ and the corresponding variance is $x^\top \Sigma x$. The problem of determining the portfolio allocation that yields the minimal risk while guaranteeing an expected return no smaller than a given γ value is cast as a convex QP:

$$\begin{aligned}
\min_x \quad & x^\top \Sigma x, \\
\text{s.t.:} \quad & x \geq 0, \\
& \sum_{i=1}^n x_i = 1, \\
& \mu^\top x \geq \gamma,
\end{aligned}$$

where we also imposed a *no-short-selling* constraint $x \geq 0$ on the portfolio positions. Notice that the above problem is indeed a convex QP since the covariance matrix Σ is always a positive semidefinite matrix.

Example 10.4 (*Elastic Net regression*) A variation on the Lasso regression problem entails the addition of an ℓ_2 regularization term to the usual Lasso objective, thus resulting in a training problem of the form

$$\min_w \|X^\top w - y\|_2^2 + \gamma_1 \|w\|_1 + \gamma_2 \|w\|_2^2, \quad \gamma_1, \gamma_2 \geq 0.$$

Expanding the squared norms, we have equivalently

$$\min_w w^\top X X^\top w - 2y^\top X^\top w + y^\top y + \gamma_1 \|w\|_1 + \gamma_2 w^\top w$$
$$= \min_w w^\top (XX^\top + \gamma_2 I) w - 2y^\top X^\top w + y^\top y + \gamma_1 \|w\|_1$$
$$= \min_{w,s} w^\top (XX^\top + \gamma_2 I) w - 2y^\top X^\top w + y^\top y + \gamma_1 \mathbf{1}^\top s,$$
$$\text{s.t.: } s \geq w, \; s \geq -w.$$

Since $XX^\top \succeq 0$, also $XX^\top + \gamma_2 I \succeq 0$, and hence the above objective function is a convex quadratic function and the overall Elastic Net regression problem is cast as a convex QP.

10.3.3 Second-order cone programs

The set
$$\mathcal{Q}^n \doteq \{x \in \mathbb{R}^n, y \in \mathbb{R} : \|x\|_2 \leq y\}$$

is a convex set known as the quadratic cone, or second-order cone (SOC). Second-order cone programs (SOCP) are convex programs expressed in standard form as

$$p^* \doteq \min_x c_0^\top x, \text{ s.t.: } \|A_i x + b_i\|_2 \leq c_i^\top x + d_i, \; i = 1, \ldots, m, \; Fx = g, \tag{10.9}$$

An SOC constraint is easily recognizable by the structure

$$\|\text{affine map of } x\|_2 \leq \text{affine function of } x.$$

This type of constraint is able to express a variety of convex conditions, including affine and (convex) quadratic as special cases, see Figure 10.4.

In particular, an affine constraint $q^\top x \leq d$ is trivially converted into an SOC constraint

$$\|Ax + b\|_2 \leq c^\top x + d \tag{10.10}$$

by taking $A = 0$, $b = 0$, $c = -q$. Similarly, the convex quadratic constraint

$$x^\top Q x + q^\top x + h \leq 0$$

can be proved to be equivalent to the SOC constraint (10.10) by taking L to be any matrix such that $Q = L^\top L$, and posing

$$A = \begin{bmatrix} q^\top x / 2 \\ L \end{bmatrix}, \; b = \begin{bmatrix} (1+h)/2 \\ L \end{bmatrix}, \; c = -q^\top/2, \; d = (1-h)/2,$$

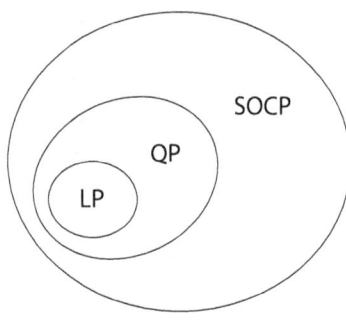

Figure 10.4 Hierarchies of convex problem families.

which results in

$$\left\| \begin{matrix} (1+q^\top x + h)/2 \\ Lx \end{matrix} \right\|_2 \leq (1 - q^\top x - h)/2.$$

The set

$$\mathcal{Q}_r^n \doteq \{x \in \mathbb{R}^n, y, z \in \mathbb{R} : \|x\|_2^2 \leq 2yz, \, y, z \geq 0\}$$

is a convex set known as the *rotated* second-order cone. The so-called *hyperbolic* constraint on the variables $x \in \mathbb{R}^n$, $y, z \in \mathbb{R}$,

$$x^\top x \leq 2yz, \quad y \geq 0, z \geq 0, \tag{10.11}$$

appearing in the definition of the rotated second-order cone can be cast in the form of a plain SOC constraint as

$$\left\| \begin{matrix} \sqrt{2}x \\ (y-z) \end{matrix} \right\|_2 \leq (y+z). \tag{10.12}$$

The set

$$\mathcal{P}_\alpha^n \doteq \{x \in \mathbb{R}^n, y, z \in \mathbb{R} : \|x\|_2 \leq y^\alpha z^{1-\alpha}, \, y, z \geq 0\}, \tag{10.13}$$

for $0 < \alpha < 1$, is also a convex set known as the *power cone*.

Example 10.5 (*Linear chance constraints*) Second-order cone programs are also of key importance for dealing with deterministic or stochastic uncertainty in linear programming. For instance, a probability constraint (also known a a *chance constraint*) of the form

$$\text{Prob}_a\{a^\top x \leq b\} \geq p, \tag{10.14}$$

where a is a normal (Gaussian) random n-dimensional vector and $p \in (0.5, 1)$ is a given level of probability, can be converted into the explicit SOC constraint

$$\bar{a}^\top x \leq b - \Phi^{-1}(p)\|\Sigma^{1/2} x\|_2,$$

where $\bar{a}$ is the expected value of a, $\Sigma^{1/2}$ is the matrix square root of its covariance matrix Σ, and Φ^{-1} is the inverse cumulative distribution function of a standard normal random variable.[7]

[7] For a derivation of this result see for instance Section 10.2.5 of G. Calafiore and L. El Ghaoui, *Optimization Models*, Cambridge: Cambridge University Press, 2014.

Example 10.6 (*Extractive summarization via group Lasso*) A problem arising in the context of extractive summarization of textual information takes the form

$$\min_{W=[w^{(1)}\cdots w^{(m)}]} \|X - XW^\top\|_F : \sum_{j=1}^m \|w^{(j)}\|_2 \leq \kappa,$$

where $X \in \mathbb{R}^{n,m}$ is a given data matrix, $W \in \mathbb{R}^{m,m}$ is the matrix variable, and κ is a given positive scalar. This problem can be cast as an SOCP by first introducing an epigraphic reformulation with a slack variable t as

$$\min_{W=[w^{(1)}\cdots w^{(m)}]} t : \|X - XW^\top\|_F \leq t, \sum_{j=1}^m \|w^{(j)}\|_2 \leq \kappa,$$

and then introducing slack variable for each of the terms $\|w^{(j)}\|_2$, obtaining

$$\min_{W,s,t} t,$$
$$\text{s.t.: } \|X - XW^\top\|_F \leq t,$$
$$\sum_{j=1}^m s_j \leq \kappa,$$
$$\|w^{(j)}\|_2 \leq s_j,$$

which is an SOCP in the variables $(w^{(1)},\ldots,w^{(m)},s,t)$.

10.3.4 Semidefinite programs

Semidefinite programs (SDPs) involve minimizing a linear objective subject to linear or affine constraint, and one or more constraints of the form

$$F(x) \succeq 0, \text{ with } F(x) \doteq F_0 + \sum_{i=1}^n x_i F_i,$$

and where $F_0,\ldots,F_n$ are given symmetric matrices. The constraint on the symmetric matrix $F(x)$ to be positive semidefinite, known as a linear matrix inequality (LMI), defines a convex set on the x variable. Convexity can be proved by considering that the constraint $F(x) \succeq 0$ is equivalent to a nonnegativity constraint on the minimal eigenvalue of $F(x)$:

$$f(x) \doteq \lambda_{\min}(F(x)) = \min_{z:\, z^\top z = 1} z^\top F(x) z \geq 0.$$

For any fixed z, the function $x \to z^\top F(x)z$ is affine in x, hence concave. The point-wise minimum of concave functions is concave, thus $f(x)$ is concave, hence $-f(x)$ is convex and the constraint $-f(x) \leq 0$ defines a convex set in x since it is the sub-level set of a convex function. An SDP in standard form,

$$p^* \doteq \min_x c^\top x, \text{ s.t.: } F(x) \succeq 0, \qquad (10.15)$$

is a convex optimization problem that can be solved efficiently for medium-sized problem instances (i.e., number of variables and dimension of the F_i matrices in the order of thousands). Semidefinite programs can model quite sophisticated problems and, in particular, they include as special cases LPs, QPs, and SOCPs, see Figure 10.5.

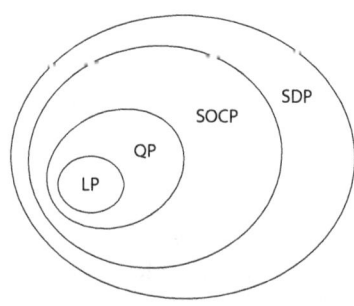

Figure 10.5 Hierarchies of convex problem families.

It can indeed be verified that an SOC constraint

$$\|Ax+b\|_2 \leq c^\top x + d,$$

is equivalent to the LMI constraint

$$\begin{bmatrix} c^\top x + d & (Ax+b)^\top \\ (Ax+b) & (c^\top x + d)I \end{bmatrix} \succeq 0.$$

Interfaces and numerical solvers for SDPs are available for most computing platforms, such as cvxpy for Python, or CVX for MATLAB®.

Example 10.7 (*Worst-case risk with partial covariance information*) We reconsider a classical mean/variance portfolio optimization setting, as described in Example 10.3. We here assume that the information about the covariance matrix Σ is incomplete. More precisely, we assume that some entries of Σ are known exactly, for other entries only the sign is known, and yet other entries are completely unknown. For example,

$$\Sigma \in \mathcal{S}, \quad \mathcal{S} \doteq \left\{ S \in \mathbb{S}^4_+ : S = \begin{bmatrix} 0.2 & + & + & ? \\ + & 0.1 & - & - \\ + & - & 0.3 & + \\ ? & - & + & 0.1 \end{bmatrix} \right\}.$$

As the risk of a given portfolio w is given by $\sigma^2 = w^\top \Sigma w$, we are interested in computing the worst-case risk of this portfolio, that is,

$$\sigma^2_{\text{wc}}(w) = \max_{\Sigma \in \mathcal{S}} w^\top \Sigma w.$$

Note that in the above problem the variable is Σ (a positive semidefinite matrix), and not w, which is assumed here to be a given and fixed portfolio vector. Computing the worst-case risk can be cast as an SDP by introducing variables $x = (x_1, \ldots, x_6)$ as placeholders for the unknown entries of Σ, and then solving

$$\sigma^2_{\text{wc}} = \max_x w^\top \Sigma(x) w,$$

$$\text{s.t.:} \quad \Sigma(x) \doteq \begin{bmatrix} 0.2 & x_1 & x_2 & x_3 \\ x_1 & 0.1 & x_4 & x_5 \\ x_2 & x_4 & 0.3 & x_6 \\ x_3 & x_5 & x_6 & 0.1 \end{bmatrix} \succeq 0,$$

$$x_1 \geq 0, \ x_2 \geq 0, \ x_6 \geq 0,$$
$$x_4 \leq 0, \ x_5 \leq 0.$$

The important aspect here is that $\Sigma(x)$ is linear in the variables x, and that the requirement that $\Sigma(x)$ be a covariance matrix is captured via the LMI constraint $\Sigma(x) \succeq 0$.

10.4 Robust optimization

The convex program (10.4) provides an effective model for describing many problems arising in decision making, statistics, finance, and so on. Moreover, it can typically be solved via efficient numerical algorithms. However, in practice, the data used for problem description (i.e., the functions f_0 and f_i, $i = 1, \ldots, m$) are imprecisely known, since they are affected by *uncertainty*. Such uncertainty may derive from errors in the estimation of the problem parameters, from fluctuations in time of the involved quantities (e.g., we use some parameters at the time of problem solution, but then the values of the parameters have changed at the future time of implementation of the solution), from incomplete knowledge, and so on. Solving an optimization problem on "nominal" data may give a nominally optimal solution which then performs very poorly on the actual perturbed data, or even turn out being infeasible in practice.

Example 10.8 (*Drug production – part 1*) A company produces two kinds of drugs, Drug I and Drug II, containing a specific active agent A, which is extracted from raw materials purchased on the market. There are two kinds of raw material, Raw I and Raw II, which can be used as sources of the active agent. The related production, cost, and resource data are given in Tables 10.1–10.3. We denote by $x_{\text{DrugI}}, x_{\text{DrugII}}$ the amount respectively of Drug I and Drug II, per 1000 packs produced. Let $x_{\text{RawI}}, x_{\text{RawII}}$ denote the amounts (in Kg) of raw materials to be purchased. The goal is to find the production plan which maximizes the profit for the company.

Parameter	Drug I	Drug II
Selling price (USD) per 1000 packs	6,500	7,100
Content of agent A (grams) per 1000 packs	0.500	0.600
Manpower required (hrs) per 1000 packs	90.0	100.0
Equipment required (hrs) per 1000 packs	40.0	50.0
Operational costs (USD) per 1000 packs	700	800

Table 10.1 Drug production data.

Material	Purch. price (USD/Kg)	Agent content (g/Kg)
Raw I	100.00	0.01
Raw II	199.90	0.02

Table 10.2 Contents of raw materials.

Budget (USD)	Manpw. (hrs)	Equip. (hrs)	Storage cap. (Kg)
100,000	2,000	800	1,000

Table 10.3 Resources.

According to the problem description, the objective to be maximized in this problem has the form

$$f_0(x) = f_{\text{income}}(x) - f_{\text{costs}}(x),$$

where

$$f_{\text{costs}}(x) = 100 x_{\text{RawI}} + 199.90 x_{\text{RawII}} + 700 x_{\text{DrugI}} + 800 x_{\text{DrugII}}$$

represents the purchasing and operational costs, and

$$f_{\text{income}}(x) = 6500 x_{\text{DrugI}} + 7100 x_{\text{DrugII}}$$

represents the income from selling the drugs.

Further, we have a total of five inequality constraints, and additional sign constraints on the variables:

- Balance of active agent:

$$0.01 x_{\text{RawI}} + 0.02 x_{\text{RawII}} - 0.5 x_{\text{DrugI}} - 0.6 x_{\text{DrugII}} \geq 0.$$

This constraint says that the amount of raw material must be enough to produce the drugs.

- Storage constraint:

$$x_{\text{RawI}} + x_{\text{RawII}} \leq 1{,}000.$$

This constraint says that the capacity of storage for the raw materials is limited.

- Manpower constraint:

$$90.0 x_{\text{DrugI}} + 100.0 x_{\text{DrugII}} \leq 2{,}000,$$

which expresses the fact that the resources in manpower are limited: we cannot allocate more than 2000 hours to the project.

- Equipment constraint:

$$40.0 x_{\text{DrugI}} + 50.0 x_{\text{DrugII}} \leq 800.$$

This says that the resources in equipment are limited.

- Budget constraint:

$$100.0 x_{\text{RawI}} + 199.90 x_{\text{RawII}} + 700 x_{\text{DrugI}} + 800 x_{\text{DrugII}} \leq 100{,}000.$$

This limits the total budget.

- Sign constraints:

$$x_{\text{RawI}} \geq 0, \ x_{\text{RawII}} \geq 0, \ x_{\text{DrugI}} \geq 0, \ x_{\text{DrugII}} \geq 0.$$

Solving this problem,[8] which is an LP, we obtain the optimal value $p^* = 14{,}085.13$ and a corresponding optimal solution:

$$x^*_{\text{RawI}} = 0, \ x^*_{\text{RawII}} = 438.789, \ x^*_{\text{DrugI}} = 17.552, \ x^*_{\text{DrugII}} = 0.$$

Note that both the budget and the balance constraints are active (i.e., they hold with equality at the optimum), which means that the production process utilizes the entire budget and the full amount of active agent

[8] See code in drugproduction.m in the online resources.

contained in the raw materials. This solution promises the company a quite respectable profit of about 14%.

Suppose that the management came up to the above optimal plan, which was put into production. But soon after market conditions changed, so that the purchase price of Raw II raised to 210 USD/Kg, while also the operational costs increased by 10% for both Drug I and Drug II. This unanticipated change made the devised optimal solution no longer optimal, and actually not even feasible, since this solution would now violate the budget constraint. A robust optimization approach would account right from the design phase of the possible fluctuations in prices and other data, and devise an optimal solution which is able to withstand such variations in the worst case.

From a "nominal" convex optimization problem of the form

$$p^* \doteq \min_x f_0(x), \qquad (10.16)$$
$$\text{s.t.:} \quad f_i(x) \leq 0, \; i = 1, \ldots, m,$$

we then move to a "robust" version of the problem of the form

$$p^*_{\text{wc}} \doteq \min_x \max_{u \in \mathcal{U}} f_0(x, u), \qquad (10.17)$$
$$\text{s.t.:} \quad f_i(x, u) \leq 0, \; \forall u \in \mathcal{U}, \; i = 1, \ldots, m,$$

where we have now made explicit the dependence of the objective and constraint functions on the uncertainty u, and we have modeled the range of possible uncertainty variation via the uncertainty set $\mathcal{U}$. Formally, problem (10.17) inherits the convexity of problem (10.16).[9] An explicit reformulation of (10.17) as a plain convex program, however, is possible only in some cases with special structure. We shall discuss next some of these cases, which are of interest in a financial context.

10.4.1 Robust LP

The robust counterpart of a nominal LP

$$p^* \doteq \min_x \hat{c}^\top x, \qquad (10.18)$$
$$\text{s.t.:} \quad \hat{a}_i^\top x \leq b_i, \; i = 1, \ldots, m,$$

is a program of the form

$$p^*_{\text{wc}} \doteq \min_x \max_{u \in \mathcal{U}_0} (\hat{c} + u_0)^\top x, \qquad (10.19)$$
$$\text{s.t.:} \quad (\hat{a}_i + u_i)^\top x \leq b_i, \; \forall u_i \in \mathcal{U}_i, \; i = 1, \ldots, m,$$

where $\hat{c} \in \mathbb{R}^n$ and $\hat{a}_i \in \mathbb{R}^n$, $i = 1, \ldots, m$, are the nominal data vectors, and u_i, $i = 0, 1, \ldots, m$ are the uncertainty terms. Observe that

[9] Indeed, from the max rule, if $f_0(x, u)$ is convex in x for any fixed u, then the point-wise maximum function $\bar{f}_0(x) \doteq \max_{u \in \mathcal{U}} f_0(x, u)$ is also convex. Similarly, if $f_i(x, u)$ is convex in x for any fixed u, for $i = 1, \ldots, m$, then the feasible set of (10.17) is the intersection of convex sets $X_u \doteq \{f_i(x, u) \leq 0, \; i = 1, \ldots, m\}$ as u ranges over $\mathcal{U}$, and hence it is convex.

the robust inequality constraints in (10.19) can be rewritten equivalently as
$$\max_{u_i \in \mathcal{U}_i} (\hat{a}_i + u_i)^\top x \leq b_i, \quad i = 1, \ldots, m,$$
so an explicit reformulation of the robust LP problem is obtained if we are able to express in explicit form the point-wise maximum functions
$$\bar{f}_i(x) \doteq \max_{u_i \in \mathcal{U}_i} (\hat{a}_i + u_i)^\top x = \hat{a}_i^\top x + \max_{u_i \in \mathcal{U}_i} u_i^\top x, \quad i = 1, \ldots, m, \quad (10.20)$$
as well as the worst-case objective function $\bar{f}_0(x) \doteq \max_{u_i \in \mathcal{U}_i} (\hat{c}_i + u_i)^\top x$, which as exactly the same structure. We next consider a few relevant cases in which such explicit formulation is possible.

Scenario uncertainty The simplest form of uncertainty description arises when each set $\mathcal{U}_i$ is a discrete collection of possible uncertainty scenarios, that is, $\mathcal{U}_i = \left\{ u_i^{(1)}, \ldots, u_i^{(q_i)} \right\}$. In this case, problem (10.19) can be solved by considering directly all scenarios:
$$p^*_{\text{wc}} \doteq \min_{x,t} t, \quad (10.21)$$
$$\text{s.t.: } \hat{c}^\top x + u_0^{(k)\top} x \leq t, \quad k = 1, \ldots, q_0,$$
$$\hat{a}_i^\top x + u_i^{(k)\top} x \leq b_i, \quad k = 1, \ldots, q_i, \quad i = 1, \ldots, m.$$

In this case, the robust version of the problem is still an LP, with a larger number of constraints with respect to the original problem.

Remark 10.1 (*Convex inequalities at scenarios*) An interesting observation about scenario-based robustness is that it gives a much larger uncertainty coverage than what may be imagined at first sight. Indeed, the fact that a linear inequality $(\hat{a} + u)^\top x \leq b$ is satisfied simultaneously for a finite number of scenarios $u \in \{u^{(1)}, \ldots, u^{(q)}\}$ implies that it is satisfied for all (densely infinite) points u that are in the convex hull of the given scenarios! This property is a consequence of the fact that the left-hand side of the inequality is a convex function of the uncertainties, and can hence be generalized as follows: suppose that a function $f(x, u)$ is convex in u for any given x, and let $f(x, u^{(i)}) \leq b$ for all given $u^{(i)}$, $i = 1, \ldots, q$. Then, defining $\mathcal{P}$ as the polytope generated as the convex hull of the $u^{(i)}$s, that is, $\mathcal{P} \doteq \text{co}\{u^{(1)}, \ldots, u^{(q)}\}$, it holds that $f(x, u) \leq b$ for all $u \in \mathcal{P}$. Indeed, any point $u \in \mathcal{P}$ can be written as a convex combination of the points generating the convex hull, that is, $u = \sum_{i=1}^{q} \theta_i u^{(i)}$, for some $\theta_1, \ldots, \theta_q \geq 0$, $\sum_{i=1}^{q} \theta_i = 1$. Then, Jensen's inequality for convex functions prescribes that
$$f(x, u) = f(x, \sum_{i=1}^{q} \theta_i u^{(i)}) \leq \sum_{i=1}^{q} \theta_i f(x, u^{(i)}) \leq \sum_{i=1}^{q} \theta_i b = b.$$

Interval uncertainty In the interval uncertainty case we assume that $\mathcal{U}_i$ is a box (or hyperrectangle). Without loss of generality, we can assume that the box is symmetric and centered in zero, and therefore is characterized as

$$\mathcal{U}_i = \{u \in \mathbb{R}^n : |u| \leq \rho^{(i)}\},$$

where $\rho^{(i)} \in \mathbb{R}_+^n$ is a vector of given interval bounds. Considering (10.20), we have that

$$\max_{\zeta \in \mathcal{U}_i} \zeta^\top x = \max_{|\zeta| \leq \rho^{(i)}} \zeta^\top x = \sum_{k=1}^n \max_{|\zeta_k| \leq \rho_k^{(i)}} \zeta_k x_k$$
$$= \sum_{k=1}^n \rho_k^{(i)} |x_k| = \rho^{(i)\top} |x|.$$

Hence, the robust LP (10.19) is rewritten explicitly as

$$p_{\text{wc}}^* \doteq \min_x \hat{c}^\top x + \rho^{(0)\top} |x|, \tag{10.22}$$
$$\text{s.t.:} \quad \hat{a}_i^\top x + \rho^{(i)\top} |x| \leq b_i, \quad i = 1, \ldots, m.$$

The above problem can be formulated as a standard LP by introducing a slack vector variable $z \in \mathbb{R}^n$ and writing

$$p_{\text{wc}}^* \doteq \min_{x,z} \hat{c}^\top x + \rho^{(0)\top} z, \tag{10.23}$$
$$\text{s.t.:} \quad \hat{a}_i^\top x + \rho^{(i)\top} z \leq b_i, \quad i = 1, \ldots, m,$$
$$-z \leq x \leq z.$$

Example 10.9 (*Drug production – part 2*) We reconsider Example 10.8. Instead of considering only the nominal setting, we now anticipate possible fluctuations in prices by assuming a 10% interval uncertainty around the nominal values of the material purchase prices, and 5% interval uncertainty around the nominal values of the operational costs. Solving the robust version of the problem[10] we find a worst-case optimal production strategy

$$x_{\text{RawI}}^* = 0, \ x_{\text{RawII}}^* = 401.14, \ x_{\text{DrugI}}^* = 16.05, \ x_{\text{DrugII}}^* = 0.$$

This type of strategy gives a nominal profit of about 14%, and a guaranteed worst-case profit of about 4.3%, while insuring that the production strategy and budget constraints remain feasible for all admissible instances of the uncertainty.

[10] See file drugproduction_robust.m in the online resources.

Ellipsoidal or spherical uncertainty In the ellipsoidal uncertainty case we assume that $\mathcal{U}_i$ is an ellipsoid centered in zero, characterized as

$$\mathcal{U}_i = \{u \in \mathbb{R}^n : u = W_i z, \|z\|_2 \leq 1\},$$

where $W_i \in \mathbb{R}^{n,p_i}$, $p_i \leq n$, is a given matrix describing the shape of the ellipsoid. The spherical case is a special case, obtained for $W_i = r_i I_n$, where $r_i \geq 0$ is the radius. We have that

$$\max_{u \in \mathcal{U}_i} u^\top x = \max_{\|z\|_2 \leq 1} z^\top (W_i^\top x) = \|W_i^\top x\|_2.$$

Hence, the robust LP (10.19) is in this case rewritten explicitly as

$$p_{wc}^* \doteq \min_x \hat{c}^\top x + \|W_0^\top x\|_2, \qquad (10.24)$$
$$\text{s.t.:} \quad \hat{a}_i^\top x + \|W_i^\top x\|_2 \leq b_i, \ i = 1, \ldots, m.$$

In this case, the robust version of the original LP is no longer an LP: it is an SOCP.

10.4.2 Robust QP

The robust counterpart of a nominal QP

$$p^* \doteq \min_x x^\top \hat{Q} x + \hat{c}^\top x, \qquad (10.25)$$
$$\text{s.t.:} \quad \hat{a}_i^\top x \leq b_i, \ i = 1, \ldots, m,$$

$\hat{Q} \geq 0$, is a program of the form

$$p_{wc}^* \doteq \min_x \max_{u \in \mathcal{U}_0, \Delta \in \mathcal{D}} x^\top (\hat{Q} + \Delta) x + (\hat{c} + u_0)^\top x,$$
$$\text{s.t.:} \quad (\hat{a}_i + u_i)^\top x \leq b_i, \ \forall u_i \in \mathcal{U}_i, \ i = 1, \ldots, m,$$

where $\mathcal{D}$ is the uncertainty set for the matrix perturbation Δ. All the uncertainties in the linear terms can be treated as discussed in the previous section, hence we shall next concentrate only on the quadratic term, assuming all linear uncertainty terms $u_0, u_1, \ldots, u_m$ to be zero. We thus consider henceforth the problem

$$p_{wc}^* \doteq \min_x \max_{\Delta \in \mathcal{D}} x^\top (\hat{Q} + \Delta) x + \hat{c}^\top x, \qquad (10.26)$$
$$\text{s.t.:} \quad \hat{a}_i^\top x \leq b_i, \ i = 1, \ldots, m.$$

Scenario uncertainty In the scenario uncertainty case we assume that $\mathcal{D} = \{\Delta^{(1)}, \ldots, \Delta^q\}$, where $\Delta^{(i)} \in \mathbb{S}^n$, are such that $Q + \Delta^{(i)} \succeq 0$, for $i = 1, \ldots, q$. In this case problem (10.26) is easily converted into a convex program with quadratic constraints:

$$p_{wc}^* \doteq \min_{x,t} t + \hat{c}^\top x, \qquad (10.27)$$
$$\text{s.t.:} \quad \hat{a}_i^\top x \leq b_i, \ i = 1, \ldots, m,$$
$$x^\top (\hat{Q} + \Delta^{(i)}) x \leq t, \ i = 1, \ldots, q.$$

Observe that, since Remark 10.1 holds also in this case, problem (10.27) offers robustness not only for the considered scenarios $\Delta^{(1)}, \ldots, \Delta^q$ but actually for all Δ in the convex hull of these scenarios.

Norm-bounded uncertainty In the norm-bounded uncertainty case we assume that $\mathcal{D} = \{\Delta \in \mathbb{S}^n : \hat{Q} + \Delta \succeq 0, \|B^{-\top}\Delta B^{-1}\| \leq \gamma\}$, where the matrix norm is either the Frobenius norm or the maximum singular value norm, and B is a given invertible matrix. The key point is to evaluate $\max_{\Delta \in \mathcal{D}} x^\top \Delta x$. Letting $\tilde{\Delta} \doteq B^{-\top}\Delta B^{-1}$, we have that the condition $\Delta \in \mathcal{D}$ translates to condition $\tilde{\Delta} \in \tilde{\mathcal{D}}$, where $\tilde{\mathcal{D}} = \{\tilde{\Delta} \in \mathbb{S}^n : \hat{Q} + B^\top \tilde{\Delta} B \succeq 0, \|\tilde{\Delta}\| \leq \gamma\}$. Therefore, we have that

$$\max_{\Delta \in \mathcal{D}} x^\top \Delta x = \max_{\tilde{\Delta} \in \tilde{\mathcal{D}}} x^\top B^\top \tilde{\Delta} B x$$

$$[\text{letting } v \doteq Bx] = \max_{\tilde{\Delta} \in \tilde{\mathcal{D}}} v^\top \tilde{\Delta} v$$

$$\leq \max_{\tilde{\Delta} : \|\tilde{\Delta}\| \leq \gamma} v^\top \tilde{\Delta} v$$

$$\leq \max_{\tilde{\Delta} : \|\tilde{\Delta}\| \leq \gamma} \lambda_{\max}(\tilde{\Delta}) \|v\|_2^2$$

$$\leq \gamma \|v\|_2^2,$$

where the first inequality in the above chain follows from the fact that we enlarged the feasibility set of the maximization problem by dropping the condition $\hat{Q} + B^\top \tilde{\Delta} B \succeq 0$, the second inequality follows from Rayleigh's theorem for symmetric matrices, and the third inequality follows from the fact that $\lambda_{\max}(\tilde{\Delta}) \leq \|\tilde{\Delta}\|$. Now, we observe that the matrix

$$\tilde{\Delta}^* = \gamma \frac{vv^\top}{\|v\|_2^2}$$

is symmetric, positive semidefinite (hence $\hat{Q} + B^\top \tilde{\Delta} B \succeq 0$), has norm $\|\tilde{\Delta}^*\| = \gamma$, and attains the upper bound, that is, $v^\top \tilde{\Delta}^* v = \gamma \|v\|_2^2$. This implies that all the inequalities in the previous chain are actually equalities, therefore $\max_{\Delta \in \mathcal{D}} x^\top \Delta x = \gamma \|Bx\|_2^2$. We can hence write the explicit version of problem (10.26) as

$$p_{\text{wc}}^* \doteq \min_x x^\top (\hat{Q} + \gamma B^\top B) x + \hat{c}^\top x, \qquad (10.28)$$
$$\text{s.t.:} \quad \hat{a}_i^\top x \leq b_i, \quad i = 1, \ldots, m,$$

which is still a convex QP.

10.4.3 Robust least squares

In nominal least squares we solve

$$\min_{x \in \mathbb{R}^n} \|\hat{A}x - \hat{b}\|_2,$$

for a nominal matrix $\hat{A} \in \mathbb{R}^{m,n}$ and vector $\hat{b} \in \mathbb{R}^m$. We next consider a robust version in which the $\hat{A}$ matrix is subject to uncertainty:

$$p^*_{\text{wc}} = \min_{x \in \mathbb{R}^n} \max_{\Delta \in \mathcal{D}} \|(\hat{A} + \Delta)x - \hat{b}\|_2.$$

The scenario uncertainty case can be treated straightforwardly, similar to the previously discussed cases. Further, since $\|(\hat{A} + \Delta)x - \hat{b}\|_2$ is convex in Δ for any given x, the scenario robustness gives coverage for all the uncertainties in the convex hull of the scenarios.

Norm-bounded uncertainty If $\mathcal{D} = \{\Delta : \|\Delta\| \leq \gamma\}$, then we have that

$$\begin{aligned}
\max_{\Delta \in \mathcal{D}} \|(\hat{A} + \Delta)x - \hat{b}\|_2 &= \max_{\Delta : \|\Delta\| \leq \gamma} \|(\hat{A}x - \hat{b}) + \Delta x\|_2 \\
&\leq \max_{\Delta : \|\Delta\| \leq \gamma} \|\hat{A}x - \hat{b}\|_2 + \|\Delta x\|_2 \\
&\leq \max_{\Delta : \|\Delta\| \leq \gamma} \|\hat{A}x - \hat{b}\|_2 + \|\Delta\| \|x\|_2 \\
&= \|\hat{A}x - \hat{b}\|_2 + \gamma \|x\|_2, \quad (10.29)
\end{aligned}$$

where the first inequality follows from the triangle inequality for norms, and the second inequality follows from the definition of matrix norm. Further, we observe that matrix

$$\Delta^* = \gamma \frac{(\hat{A}x - \hat{b})x^\top}{\|\hat{A}x - \hat{b}\|_2 \|x\|_2}$$

is such that $\|\Delta^*\| = \gamma$ and $\|(\hat{A} + \Delta^*)x - \hat{b}\|_2 = \|\hat{A}x - \hat{b}\|_2 + \gamma \|x\|_2$, so all inequalities leading to (10.29) are actually equalities. Therefore, the robust least-squares problem, under norm-bounded uncertainty, is written explicitly as

$$\begin{aligned}
p^*_{\text{wc}} &= \min_{x \in \mathbb{R}^n} \max_{\Delta \in \mathcal{D}} \|(\hat{A} + \Delta)x - \hat{b}\|_2 \\
&= \min_{x \in \mathbb{R}^n} \|\hat{A}x - \hat{b}\|_2 + \gamma \|x\|_2.
\end{aligned}$$

We observe that robustness led to the addition of a ℓ_2-type regularization term to the original objective.

Spherical uncertainty on the columns of $\hat{A}$ We next consider the case where $\mathcal{D} = \{\Delta = [\delta_1, \ldots, \delta_n] : \|\delta_i\|_2 \leq r_i, i = 1, \ldots, n\}$, for given uncertainty radii $r_i \geq 0$, $i = 1, \ldots, n$. Similar to the previous case, we have that

$$\max_{\Delta \in \mathcal{D}} \|(\hat{A}x - \hat{b}) + \Delta x\|_2 = \max_{\|\delta_i\|_2 \leq r_i, i=1,\ldots,n} \|(\hat{A}x - \hat{b}) + \sum_{i=1}^{n} \delta_i x_i\|_2$$

$$\leq \max_{\|\delta_i\|_2 \leq r_i, i=1,\ldots,n} \|\hat{A}x - \hat{b}\|_2 + \|\sum_{i=1}^{n} \delta_i x_i\|_2$$

$$\leq \max_{\|\delta_i\|_2 \leq r_i, i=1,\ldots,n} \|\hat{A}x - \hat{b}\|_2 + \sum_{i=1}^{n} \|\delta_i x_i\|_2$$

$$= \max_{\|\delta_i\|_2 \leq r_i, i=1,\ldots,n} \|\hat{A}x - \hat{b}\|_2 + \sum_{i=1}^{n} \|\delta_i\|_2 |x_i|$$

$$= \|\hat{A}x - \hat{b}\|_2 + \sum_{i=1}^{n} r_i |x_i|. \tag{10.30}$$

Further, let

$$v \doteq \frac{\hat{A}x - \hat{b}}{\|\hat{A}x - \hat{b}\|_2}, \quad \delta_i^* \doteq r_i v \operatorname{sgn}(x_i), \ i = 1, \ldots, n.$$

Then, $\Delta^* x = \sum_{i=1}^{n} \delta_i^* x_i = v(\sum_{i=1}^{n} r_i |x_i|)$, and one can easily check that $\|(\hat{A}x - \hat{b}) + \Delta^* x\|_2 = \|\hat{A}x - \hat{b}\|_2 + \sum_{i=1}^{n} r_i |x_i|$, so equality actually holds in the previous chain of inequalities, and thus we have in this case that

$$p_{\text{wc}}^* = \min_{x \in \mathbb{R}^n} \max_{\Delta \in \mathcal{D}} \|(\hat{A} + \Delta)x - \hat{b}\|_2$$

$$= \min_{x \in \mathbb{R}^n} \|\hat{A}x - \hat{b}\|_2 + \sum_{i=1}^{n} r_i |x_i|.$$

We see that robustness led to the addition of a weighted ℓ_1-type regularization term. In particular, for $r_i = r$ for all i, we obtain the so-called square-root lasso model:

$$\min_{x \in \mathbb{R}^n} \|\hat{A}x - \hat{b}\|_2 + r\|x\|_1.$$

10.5 Exercises

Exercise 10.1 *Convexity*
Are the following functions convex, concave, or neither? Explain.

- $f(x) = x_1^2 - x_2^2$
- $f(x) = x_1^2 + (2x_1 - x_2)^2$
- $f(x) = \max(0, x_1 + 2x_2) + \max(x_3 - x_1, x_3 - x_2)$
- $f(x) = \sum_{i=1}^n \max(x_i, -x_i)$

Exercise 10.2 *Gradient descent*
Let $f: \mathbb{R}^n \to \mathbb{R}$ be a differentiable function, and let $x^{(0)} \in \mathbb{R}^n$ be a given initial point in the domain of f. The gradient descent algorithm for finding a stationary point of f works by repeating iterations of the form
$$x^{(k+1)} = x^{(k)} - s_k \nabla f(x^{(k)}), \quad k = 0, 1, \ldots, \qquad (10.31)$$
where $s_k > 0$ is a stepsize, which is typically determined at each iteration by considering the restriction of f along the direction of the gradient: $\phi_k(s) = f(x_k - s_k \nabla f(x^{(k)}))$. A suitable s_k is computed either by minimizing $\phi_k(s)$ over $s > 0$, an approach called *exact line search*, or by just determining a value s_k that guarantees a sufficient objective decrease, that is, such that
$$\phi_k(s_k) \le \phi_k(0) - s_k \alpha \|\nabla f(x_k)\|_2^2, \qquad (10.32)$$
for some given $\alpha \in (0,1)$.

1. Let the standard assumption of Lipschitz continuous gradient hold for f, that is, there exist $L > 0$ such that
$$\|\nabla f(x) - \nabla f(y)\|_2 \le L\|x - y\|_2, \quad \forall x, y \in \mathrm{dom}\, f.$$
Prove that in this case f can be approximated around x_k via a global upper approximation given by a convex quadratic function $h_k(x)$, that is,
$$f(x) \le h_k(x), \quad h_k(x) \doteq f(x_k) + \nabla f(x_k)^\top (x - x_k) + \frac{L}{2}\|x - x_k\|_2^2, \quad \forall x.$$

2. Show that under the Lipschitz continuous gradient hypothesis condition (10.32) is satisfied at each step by the constant stepsize
$$s_k = \bar{s} \doteq \frac{2}{L}(1-\alpha).$$

270   10 OPTIMIZATION TOOLS

Exercise 10.3 *Gradient descent algorithm for Ridge regression*
Consider a Ridge regression problem of the form

$$\min_x \frac{1}{2}\|Ax - b\|_2^2 + \frac{\lambda}{2}\|x\|_2^2, \qquad (10.33)$$

for given $A \in \mathbb{R}^{m,n}$, $b \in \mathbb{R}^m$ and $\lambda > 0$. Given a initial point $x^{(0)} \in \mathbb{R}^n$ (e.g., $x^{(0)} = 0$), we want to solve this problem using the simple gradient descent algorithm in (10.31). Consider the notation introduced in Exercise 10.2.

1. Has the objective function in (10.33) a Lipschitz continuous gradient? In the positive case, determine explicitly a value for the Lipschitz constant L.

2. Find explicitly the optimal stepsize s_k which solves the exact line-search problem for each step of the gradient algorithm.

Exercise 10.4 *Proximal gradient algorithm*
The gradient algorithm requires the function f to be minimized to be differentiable, and it may fail to work on nondifferentiable objectives. We next consider the situation in which $f(x) = f_0(x) + h(x)$, where $f_0(x)$ is convex and differentiable, while $h(x)$ is convex but possibly nondifferentiable. In this case, a class of algorithms called *proximal gradient* algorithms works via modified gradient steps as

$$x^{(k+1)} = \operatorname{prox}_{sh}\left(x^{(k)} - s\nabla f_0(x^{(k)})\right), \quad k = 0, 1, \ldots, \qquad (10.34)$$

where $\operatorname{prox}_g(x)$ denotes the *proximal mapping* of g, defined as

$$\operatorname{prox}_g(x) \doteq \arg\min_z \left(g(z) - \frac{1}{2}\|z - x\|_2^2\right).$$

1. Suppose $h(x)$ is the indicator function of a convex set $\mathcal{X}$, that is, $h(x) = 0$ if $x \in \mathcal{X}$ and $h(x) = \infty$ otherwise. Show that in this case $\operatorname{prox}_h(x)$ gives the Euclidean projection of x onto $\mathcal{X}$. Explain the functioning of a proximal gradient algorithm for solving the constrained optimization problem $\min_x f_0(x)$ subject to $x \in \mathcal{X}$.

2. Let $h(x)$ denote the indicator function of $\mathcal{X} = \{x \in \mathbb{R}^n : x \geq 0\}$. Compute explicitly $\operatorname{prox}_h(x)$. Describe the functioning of a proximal gradient algorithm for solving a nonnegative-constrained least-squares problem of the form $\min_x \|Ax - b\|_2^2$ subject to $x \geq 0$.

3. Let $\mathcal{X} = \{x \in \mathbb{R}^n : a^\top x \leq b\}$, for given $a \in \mathbb{R}^n$, $a \neq 0$, and $b \in \mathbb{R}$. Express explicitly $\operatorname{prox}_h(x)$, where h is the indicator function of $\mathcal{X}$.

Exercise 10.5 *Diversification*

You have $12,000 to invest at the beginning of the year, and three different funds from which to choose. The municipal bond fund has a 7% yearly return, the local bank's certificates of deposit (CDs) have an 8% return, and a high-risk account has an expected (hoped-for) 12% return. To minimize risk, you decide not to invest any more than $2000 in the high-risk account. For tax reasons, you need to invest at least three times as much in the municipal bonds as in the bank CDs. Denote by x_B, x_{CD}, x_{HR} the amounts (in thousands) invested in bonds, CDs, and high-risk account, respectively. Assuming the year-end yields are as expected, what are the optimal investment amounts for each fund? Discuss the problem formulation and its solution.

11
Mean/variance portfolio optimization

IN THIS CHAPTER we discuss the classical Markowitz's portfolio allocation model, whose success was dictated both by its conceptual simplicity in capturing the fundamental tradeoff between risk and expected return, and by the computational effectiveness of the ensuing convex quadratic programming models. The fundamental idea in the Markowitz's mean/variance approach to portfolio design is to model the assets' returns as stationary random variables, with known expected returns and covariance matrix. By mixing together the assets in a *portfolio*, the investor is able to synthesize many different combinations of expected return and related risk for their investment. Optimal portfolios are mixes that achieve best-possible tradeoffs between risk and expected return. The Markowitz's model has been the workhorse of professional portfolio management for years, despite the fact that investors knew its limitations, in particular its sensitivity to the input parameters, that is, to the covariance matrix, and especially to the expected returns. After the 2008 subprime financial crisis the Markowitz model has been further criticized, but never fully abandoned to date. Indeed, it appears that the key issues are mostly not related to the model itself, but to the difficulty of determining reliable values for its parameters. Recent research thus concentrated on one side on reliable estimation of the model parameters, and on the other side on introducing *robustness* to parameter uncertainty in the model. In this chapter we introduce the elements of the classical mean/variance portfolio design approach, and discuss extensions of the basic model that include transaction costs, market impact, and alternative measures of risk beyond the variance. Chapter 12 will then discuss some variations and advances beyond the mean/variance model.

11.1 Prices and returns

Denote with $a_1, \ldots, a_n$, a collection of n assets, and with $p_i(k)$ the market price of a_i at time $k\Delta$, where k is an integer and Δ is a fixed period of time representing the *data frequency*, for example, one month, one day, one minute. If not specified otherwise, we usually consider daily price series.

The *simple return* of an investment in asset i over the kth period (from $(k-1)\Delta$ to $k\Delta$), is

$$r_i(k) \doteq \frac{p_i(k) - p_i(k-1)}{p_i(k-1)}, \quad i = 1, \ldots, n; \; k = 1, 2, \ldots.$$

The gross return, or *gain*, is defined as

$$g_i(k) \doteq 1 + r_i(k) = \frac{p_i(k)}{p_i(k-1)}, \quad i = 1, \ldots, n; \; k = 1, 2, \ldots.$$

The *log-return* is defined as

$$\rho_i(k) \doteq \log(1 + r_i(k)), \quad i = 1, \ldots, n; \; k = 1, 2, \ldots.$$

For high-frequency data the returns $r_i(k)$ are typically very small, so we have that

$$\rho_i(k) = \log(1 + r_i(k)) \simeq r_i(k)$$

hence simple returns and log-returns are often used interchangeably, for daily or shorter data frequencies.

Observe that the return $r_i(k)$ over the period $[(k-1)\Delta, k\Delta]$ is only known at the end of this time interval, that is, at time $k\Delta$. At the beginning of each time interval, instead, the value of the return is unknown. To model this uncertainty, it is typically assumed that the a-priori values of the returns $r_i(k)$ at time $k-1$ are described by random variables. In particular, in the single-period portfolio optimization setting, it is usually assumed that these random variables are stationary (i.e., their statistical properties do not depend on time), independent in time and identically distributed. In other words, the predicted value of the return vector $r(k)$ is assumed to be a random vector $r = [r_1 \ldots r_n]^\top$ with some given probability distribution and hence given expected value

$$\hat{r} \doteq \mathrm{E}\{r\}$$

and covariance matrix

$$\Sigma \doteq \mathrm{E}\{(r - \hat{r})(r - \hat{r})^\top\}.$$

In this chapter, we shall work under these classical assumptions, that is, at each period k, the return vector $r(k)$ is represented by a random vector r whose probability distribution remains constant in time, and $r(1), r(2), \ldots$ are mutually independent (i.i.d. returns hypothesis). These assumptions are of course debatable and perhaps unrealistic in practice, especially in multi-stage decision settings or over long periods of time, where it is plausible that market conditions, and hence the returns' distribution parameters, vary in consequence of external events and macro-economic conditions. Nevertheless, the i.i.d. assumption is still widely accepted as a working hypothesis for the purpose of single-period portfolio optimization, once we understand its limitations, and use it in the appropriate context and practical purpose.

A *portfolio* of assets $a_1, \ldots, a_n$ is defined by a vector $w(k) \in \mathbb{R}^n$ whose entry $w_i(k)$, $i = 1, \ldots, n$, describes the (signed) amount of money invested in asset a_i at time k, where $w_i(k) > 0$ denotes a regular "long" position and $w_i(k) < 0$ denotes a "short" position[1] on asset i at time k. Due to the above stationarity assumptions, in single-period portfolio analysis and design we can drop the dependency on time k, and hence assume that the portfolio vector $w \in \mathbb{R}^n$ contains the money amounts w_i invested in the various assets at the beginning of any time period Δ, again with $w_i > 0$ for long positions and $w_i < 0$ for short positions. Then, we look at the value of the portfolio components at the end of that period, which are given by

$$w_i^+ = w_i + r_i w_i, \quad i = 1, \ldots, n,$$

where $r_1, \ldots, r_n$ are the assets' (random) returns over the considered period. For each position, the difference between the final value and the initial one is

$$\delta_i \doteq w_i^+ - w_i = r_i w_i, \quad i = 1, \ldots, n.$$

Observe that if $w_i > 0$ (long position on the ith asset), then a positive gain $r_i w_i$ is experienced on the ith position if the corresponding return $r_i > 0$. On the other hand, if $w_i < 0$ (short position on the ith asset), then a positive gain $r_i w_i$ is experienced on the ith position if the corresponding return $r_i < 0$. Short positions thus permit to "bet" on an asset's negative return, while regular long positions permit to "bet" on an asset's positive return. The total wealth variation at the end of the investment period is therefore

$$\delta = \sum_{i=1}^n \delta_i = \sum_{i=1}^n r_i w_i = r^\top w.$$

The total initial capital invested in both long and short positions is denoted with W_{ini}, and the overall return of the investment is

$$\rho \doteq \frac{\delta}{W_{\text{ini}}}.$$

[1] See Remark 11.1 for a discussion on how negative positions can be set in practice.

There are two different views about how W_{ini} is computed in the presence of short positions. The original Markowitz's view was that $W_{\text{ini}} \doteq \sum_{i=1}^{n} |w_i|$, since both long and short positions imply putting capital at stake, either in the form of invested cash (for long positions) or in the form of borrowed asset shares (for short positions). The second view is that $W_{\text{ini}} \doteq \sum_{i=1}^{n} w_i$, and this view reflects the way in which hedge funds and modern brokerage accounts operate: if a client has one dollar on the trading account, they may invest two dollars long on an asset and one dollar short on another, and the balance on the account would indicate that one dollar was invested. This second view thus focuses on the *cash budget* of the investment, and it is also the one taken by classical authors such as R. C. Merton.[2] Further, most of the current literature on portfolio optimization assumes $W_{\text{ini}} \doteq \sum_{i=1}^{n} w_i$, so we shall conform with this view.

Since r is random, both the wealth change δ and the portfolio return ρ are random variables, and it holds that

$$\bar{\delta} \doteq \mathrm{E}\{\delta\} = \hat{r}^\top w, \tag{11.1}$$

$$\sigma_\delta^2 \doteq \mathrm{E}\{(\delta - \bar{\delta})^2\} = w^\top \Sigma w, \tag{11.2}$$

$$\bar{\rho} \doteq \mathrm{E}\{\rho\} = \frac{1}{W_{\text{ini}}} \hat{r}^\top w, \tag{11.3}$$

$$\sigma_\rho^2 \doteq \mathrm{E}\{(\rho - \bar{\rho})^2\} = \frac{1}{W_{\text{ini}}^2} w^\top \Sigma w. \tag{11.4}$$

[2] See, for example, R. C. Merton, "An Analytic Derivation of the Efficient Portfolio Frontier," *Journal of Financial and Quantitative Analysis*, vol. 7(4), 1972.

Without loss of generality we can impose the normalization condition that $W_{\text{ini}} = 1$, so that $\delta = \rho = r^\top w$. Since $W_{\text{ini}} = \mathbf{1}^\top w$, the condition $W_{\text{ini}} = 1$ represents a simple linear equality constraint[3] on the portfolio vector w. Under the normalization condition that $W_{\text{ini}} = 1$, equations (11.3)–(11.4) simplify to

$$\bar{\rho} \doteq \mathrm{E}\{\rho\} = \hat{r}^\top w, \tag{11.5}$$

$$\sigma_\rho^2 \doteq \mathrm{E}\{(\rho - \bar{\rho})^2\} = w^\top \Sigma w. \tag{11.6}$$

[3] This is actually one additional reason why $W_{\text{ini}} \doteq \sum_{i=1}^{n} w_i$ is generally preferred over $W_{\text{ini}} \doteq \sum_{i=1}^{n} |w_i|$. In fact, while the constraint $\sum_{i=1}^{n} w_i = 1$ is linear in w and hence convex, the constraint $\sum_{i=1}^{n} |w_i| = 1$ is a complicated nonconvex equality constraint which may introduce difficulties in the numerical treatment of the ensuing portfolio optimization problems.

Remark 11.1 (*How short positions are set*) Uninitiated investors are often puzzled as to how short positions are set in practice, that is, how can one hold a *negative amount* of an asset. While regular long positions are set by just buying on the market, usually through a broker, a certain number of shares of an asset at the current price, short positions are set via a financial "trick," allowed by the broker. More precisely, if a client wants to open a short position on an asset a_i, they can do so by borrowing from the broker a number x of shares of that asset and by immediately selling it on the market at the current price $p_i(k)$ per share, thus receiving $xp_i(k)$ dollars from the operation. Since this money is formally received, and not spent, by the client, we set $w_i = -xp_i(k)$. The broker allows such operation after checking that the client has enough funds available to cover possible losses, and by charging a small fee. Then, at the end of the investment period, that is, at time $k+1$, the client must return the x borrowed shares to the broker. In order to do so, since the client does *not* own such shares (they were sold immediately after being borrowed from the broker, remember?), they must buy the x shares on the market, at the current price $p_i(k+1)$, thus spending $xp_i(k+1)$ dollars, and then return the x shares to the broker. Overall, the client received $xp_i(k)$ at the beginning of the period, and spent $xp_i(k+1)$ at the end of the period. So, if the asset value declined over the period, that is, if $p_i(k+1) < p_i(k)$, the client gains $x(p_i(k) - p_i(k+1))$ currency units, while if the asset value raised over the period, that is, if $p_i(k) < p_i(k+1)$, the client looses $x(p_i(k+1) - p_i(k))$ currency units.

In practice, brokers usually do not impose that the debt in assets' shares is fulfilled and closed at the end of the trading period. Simply, the gains/losses are marked to the market and posted in the form of corresponding value increases/decreases on the client's account positions. In this way, short positions can be held for several trading periods, analogously to long positions. In current trading platforms all the described mechanism is made completely transparent to the user, who can simply enter in a computer form a selling order for shares that they do not own in order to open a short position. Such a position can be left open for the desired time, and eventually closed simply via a buy order.

11.2 Optimal risk/return tradeoffs

A key problem we are interested in is to design portfolios that provide some optimal tradeoff between expected return $\bar{\rho}$ and associated risk σ_ρ^2. Clearly, any rational investor likes to have a "large" expected return, so making $\bar{\rho}$ large is one of the goals. However, investors are also usually *risk averse*, in the sense that they prefer sure investments over volatile and uncertain ones. Therefore, making risk σ_ρ^2 small is another goal in portfolio design. The issue here is that, in practice, these two goals are competing against each other since, typically, investment opportunities with large expected return come with high associated risk, and investments with low risk yield low returns. Looking at one of these objectives alone while neglecting the other yields to "extreme" optimal portfolios, as shown next. The simplest approach one can take towards finding portfolios that provide a tradeoff between expected return and risk (portfolio variance) is to consider an objective function that blends together the objectives in (11.8) (maximize expected return) and in (11.9) (minimize risk). This can be done by considering a problem with a mixed objective of the form

$$\max_w \hat{r}^\top w - \gamma w^\top \Sigma w, \quad \text{s.t.:} \ w \in \mathcal{W}, \tag{11.7}$$

where $\gamma > 0$ is a tradeoff (or *risk aversion*) parameter: $\gamma \to 0$ will tend to provide portfolios that maximize the return and neglect the associated risk, while as γ increases the risk term will give more and more weight, so yielding portfolios that are progressively more averse to risk. In (11.7) the set $\mathcal{W}$ represents the set of admissible portfolios. In the standard case, we simply have $\mathcal{W} = \{w \in \mathbb{R}^n : \sum_{i=1}^n w_i = 1\}$, representing the normalization on the initial cash budget. Various types of other constraints can be inserted in $\mathcal{W}$, for instance we may limit the maximum exposure in each individual assets by adding constraints of the type $|w_i| \leq u_i$, where u_i is the upper bound on the exposure in the ith asset, or we can forbid the short selling on an asset by imposing $w_i \geq 0$. More complex constraints related to diversification requirements can also be introduced, and also the problem can be adapted to take into account transaction costs, market impact, uncertainty in the returns and covariances, and so on. Some of these advanced features are discussed in subsequent sections.

Example 11.1 (*Extreme "optimal" portfolios*) Consider first the problem of maximizing the expected portfolio return $\bar{\rho}$, under no requirement on risk. Assuming a unit initial budget, and imposing additionally that no shorting is allowed, the problem becomes

$$\max_{w} \hat{r}^\top w, \quad \text{s.t.:} \quad \sum_{i=1}^n w_i = 1, \; w_i \geq 0, \; i = 1, \ldots, n. \tag{11.8}$$

Direct inspection of this problem suggests that the optimal solution is obtained by finding an index $i_{\max}$ such that $i_{\max} \in \arg\max_{i=1,\ldots,n} \hat{r}_i$ and then setting $w_{i_{\max}} = 1$ and $w_i = 0$ for all $i \neq i_{\max}$. In words, this means spotting the asset $a_{i_{\max}}$ that has the largest return, and investing the whole budget long on that asset. Clearly, such strategy corresponds to the "putting all eggs in one basket" approach, which is known to perform poorly in terms of associated risk. Indeed, the investment risk is ignored altogether in problem (11.8).

The other extreme case arises when we look for a portfolio with minimum risk, with no attention towards the return:

$$\sigma_{\min}^2 = \min_{w} w^\top \Sigma w, \quad \text{s.t.:} \quad \mathbf{1}^\top w = 1. \tag{11.9}$$

This problem is a convex QP, with one linear equality constraint. Assuming Σ is invertible, it turns out that this problem has a closed-form solution which can be obtained from a Lagrangian duality approach, by solving the equivalent problem

$$\sigma_{\min}^2 = \max_{\lambda} \min_{w} \; w^\top \Sigma w + \lambda(\mathbf{1}^\top w - 1)$$
$$= \max_{\lambda} - \left(\frac{\lambda^2}{4} \mathbf{1}^\top \Sigma^{-1} \mathbf{1} + \lambda \right)$$
$$= - \min_{\lambda} \frac{\lambda^2}{4} \mathbf{1}^\top \Sigma^{-1} \mathbf{1} + \lambda$$
$$= (\mathbf{1}^\top \Sigma^{-1} \mathbf{1})^{-1},$$

which is attained for $\lambda = -2(\mathbf{1}^\top \Sigma^{-1} \mathbf{1})^{-1}$ and $w = -(\lambda/2)\Sigma^{-1}\mathbf{1}$, that is, for the optimal portfolio

$$w = \frac{\Sigma^{-1}\mathbf{1}}{\mathbf{1}^\top \Sigma^{-1} \mathbf{1}}. \tag{11.10}$$

In practice, as we shall see next, an investor is seldom interested in such extreme situations (maximum return, with no control on risk, or minimum risk, with no control on expected return) but rather in finding optimal tradeoffs between expected return and associated risk.

11.2.1 The efficient frontier

Let $w(\gamma)$ denote an optimal solution of problem (11.7) for a given $\gamma > 0$, and let

$$\bar{\rho}_\gamma \doteq \hat{r}^\top w(\gamma), \quad \sigma_\gamma^2 \doteq w(\gamma)^\top \Sigma w(\gamma).$$

The *efficient frontier* is the curve obtained by the parametric plot of $(\sigma_\gamma^2, \bar{\rho}_\gamma)$ in the risk/return plane as γ varies from 0 to ∞. In practice, the frontier curve is approximated numerically by solving problem (11.7) repeatedly for a suitable number of gridded values of γ, see Figure 11.1. The portfolios $w(\gamma)$ are called *efficient portfolios*. Any feasible portfolio $w \in \mathcal{W}$ is represented in the risk/return plane by a point of coordinates $(\sigma_\rho^2(w), \bar{\rho}(w))$, where $\sigma_\rho^2(w) = w^\top \Sigma w$ is the risk and $\bar{\rho}(w)$ is the expected return of the portfolio. We say that a portfolio $w^{(a)} \in \mathcal{W}$ is *dominated* if there exists a $w^{(b)} \in \mathcal{W}$ which is better than $w^{(a)}$ both in terms of return and risk, that is, such that $\sigma_\rho^2(w^{(b)}) < \sigma_\rho^2(w^{(a)})$ and $\bar{\rho}(w^{(b)}) > \bar{\rho}(w^{(a)})$. We say $w^{(a)} \in \mathcal{W}$ is *efficient* if it is not dominated, that is, if there exists no other portfolio in $\mathcal{W}$ that is better than $w^{(a)}$ in terms of both risk and return.

With reference to Figure 11.1, observe that all portfolios in the thicker segment of the frontier, such as $w^{(c)}$, yield both smaller risk and higher return than $w^{(b)}$: it would then be irrational (as long as be believe in our theoretical setting) to prefer $w^{(b)}$ over any of these portfolios. Whether to choose portfolio $w^{(a)}$ or $w^{(c)}$ is instead a matter of the investor's preference: both portfolios are efficient and no one

11 MEAN/VARIANCE PORTFOLIO OPTIMIZATION

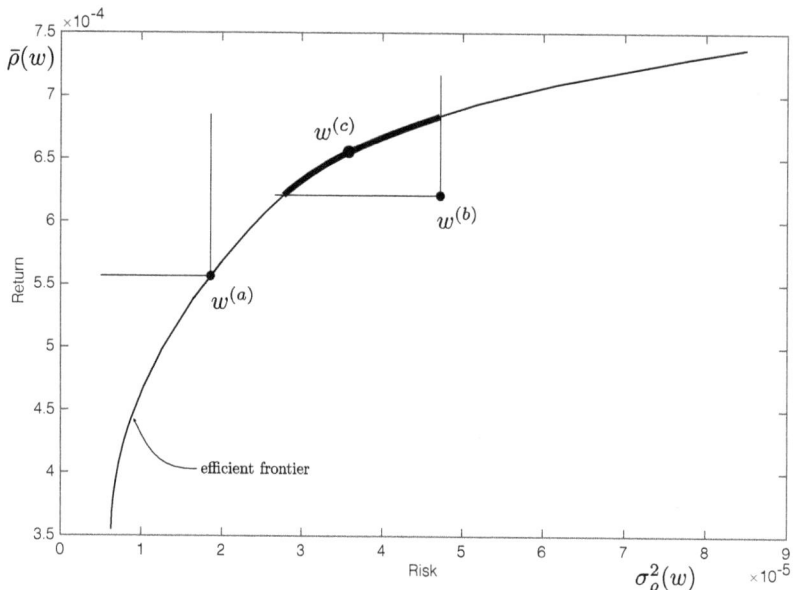

Figure 11.1 Example of an efficient frontier plot for long-only portfolios. Portfolio $w^{(a)}$ and $w^{(c)}$ are both efficient. Portfolio $w^{(b)}$ is dominated.

is univocally "better" than the other, for example, $w^{(c)}$ has higher return than $w^{(a)}$, but it also has a higher risk, so one rational but risk averse investor may choose $w^{(a)}$ while another rational but more adventurous investor may choose $w^{(c)}$, and both are "optimal" choices.

All portfolios generated by the solution of (11.7), for any $\gamma > 0$, are efficient. This fact is easily proved by contradiction, for suppose that for some $\gamma > 0$ the portfolio $w(\gamma)$ was not efficient: then there would exist another portfolio $w^* \in \mathcal{W}$ such that $\sigma_\rho^2(w^*) < \sigma_\gamma^2$ and $\bar{\rho}(w^*) > \bar{\rho}_\gamma$. But then multiplying the first inequality by $-\gamma$ and summing up the two previous inequalities we would obtain that

$$\bar{\rho}(w^*) - \gamma \sigma_\rho^2(w^*) > \bar{\rho}_\gamma - \gamma \sigma_\gamma^2,$$

which would mean that w^* has a larger objective value than $w(\gamma)$ in (11.7), and this is not possible since $w(\gamma)$ attains the global optimum of this objective. As a consequence of the previous fact, all feasible and dominated portfolios map to points in the region strictly below the efficient frontier curve, while efficient portfolios map to the boundary of that region, that is, on the curve itself (the efficient frontier).

Remark 11.2 (*Variance and standard deviation*) In the previous discussion we measured the risk of a portfolio w of assets according to the variance $\sigma_\rho^2 = w^\top \Sigma w$ of its random return. Since $\sigma_\rho^2 \geq 0$, we can equivalently use the square root of the variance (i.e., the standard deviation) as a measure of risk. In such setting, the risk/return plane has the standard deviation

$\sigma_\rho(w)$ in the horizontal axis and the expected return on the vertical axis. This choice may be preferable in some cases since both expected return and risk are expressed in the same units (e.g., dollars), instead of dollars and "squared" dollars as in the variance case. In the optimization context, however, the variance is often preferred to the standard deviation, since its expression is a simple quadratic form $w^\top \Sigma w$, which was traditionally easier to manipulate than $\sqrt{w^\top \Sigma w}$. This argument is actually no longer valid for modern SOCP solvers, which can deal very effectively with the standard deviation in the form $\|L^\top w\|_2$, where $\Sigma = LL^\top$ is a Cholesky decomposition of the returns' covariance matrix.

Remark 11.3 (*Alternative constructions on the efficient frontier*) By looking at Figure 11.1 and recalling that no portfolio can map to a point in the risk/return plane *above* the efficient frontier, we can find a point on the frontier by fixing a lower bound value ϱ for the expected return and solving

$$\min_{w \in \mathcal{W}} w^\top \Sigma w, \quad \text{s.t.:} \quad \hat{r}^\top w \geq \varrho. \tag{11.11}$$

It can be immediately verified that any optimal solution $w(\varrho)$ to this problem is efficient (proceed again by contradiction: if a portfolio $\tilde{w}$ exists with larger return and smaller risk such $\tilde{w}$ would be feasible for (11.11) and yield a better objective that the optimal solution $w(\varrho)$, which is impossible). The efficient frontier, or a portion of it, can then be approximated numerically by selecting a range $[\varrho_{\min}, \varrho_{\max}]$ for ρ, and then solving (11.11) repeatedly for several gridded values of ϱ in this interval.

By an analogous reasoning, we can also approach the problem by fixing an upper bound level v for the risk, and then maximizing the expected return of the portfolio under the constraint that the risk does not exceed v, that is,

$$\max_{w \in \mathcal{W}} \hat{r}^\top w, \quad \text{s.t.:} \quad w^\top \Sigma w \leq v. \tag{11.12}$$

Also this problem yields optimal solutions $w(v)$ which are efficient portfolios. A portion of the efficient frontier can in this case be approximated numerically by selecting a range $[v_{\min}, v_{\max}]$ for v, and then solving (11.12) repeatedly for several gridded values of v in this interval.

Equations (11.7), (11.11), and (11.12) are three essentially equivalent versions of the optimal mean/variance-tradeoff portfolio optimization problem.

11.2.2 Closed-form solutions for the optimal tradeoff

Equation (11.7) defining optimal tradeoff portfolios is a convex optimization problem, whenever $\mathcal{W}$ is convex. Such a problem can only be solved numerically via appropriate solvers, in general. There are, however, two simple particular cases in which the solution can be obtained in closed form. Namely, the cases when $\mathcal{W} = \mathbb{R}^n$ and when

$\mathcal{W} = \{w : \mathbf{1}^\top w = 1\}$. For both cases, we shall assume that Σ is positive definite, hence invertible.

Optimal tradeoff in the unconstrained case We consider the problem

$$\max_w \hat{r}^\top w - \gamma w^\top \Sigma w. \tag{11.13}$$

The optimum in this case can be simply characterized by computing the gradient of the objective function and setting it to zero:

$$\nabla_w(\hat{r}^\top w - \gamma w^\top \Sigma w) = \hat{r} - 2\gamma \Sigma w = 0,$$

which yields the optimal solution

$$w(\gamma) = \frac{1}{2\gamma} \Sigma^{-1} \hat{r}. \tag{11.14}$$

We observe that, in this setting, the only effect of the γ parameter is to reduce the overall exposure of the portfolio, that is, increasing γ decreases risk by reducing the overall investment in proportion to γ^{-1}. The efficient frontier in this case also takes an explicit form, since

$$\bar{\delta}(\gamma) \doteq \hat{r}^\top w(\gamma) = \frac{1}{2\gamma} \hat{r}^\top \Sigma^{-1} \hat{r},$$

$$\sigma_\delta^2(\gamma) \doteq w(\gamma)^\top \Sigma w(\gamma) = \frac{1}{4\gamma^2} \hat{r}^\top \Sigma^{-1} \hat{r},$$

from which we see that

$$\bar{\delta}(\gamma) = c\sqrt{\sigma_\delta^2(\gamma)}, \quad \text{where } c = \sqrt{\hat{r}^\top \Sigma^{-1} \hat{r}},$$

hence the parametric curve $(\bar{\delta}(\gamma), \sigma_\delta^2(\gamma))$ is a branch of square-root function. It has to be observed, however, that in this case the quantity $\bar{\delta}(\gamma) = \hat{r}^\top w(\gamma)$ represents the absolute wealth variation, and not the relative return of the portfolio, which is instead given by

$$\bar{\rho}_\gamma = \frac{\hat{r}^\top w(\gamma)}{W_{\text{ini}}(\gamma)},$$

where $W_{\text{ini}}(\gamma)$ is the total initial capital invested in portfolio $w(\gamma)$. If we consider

$$W_{\text{ini}} = \mathbf{1}^\top w(\gamma) = \frac{1}{2\gamma} a, \quad a \doteq \mathbf{1}^\top \Sigma^{-1} \hat{r},$$

we have for the actual relative expected return and variance

$$\bar{\rho}_\gamma \doteq \frac{\hat{r}^\top w(\gamma)}{W_{\text{ini}}(\gamma)} = \frac{1}{a} \hat{r}^\top \Sigma^{-1} \hat{r},$$

$$\sigma_\gamma^2 \doteq \frac{w(\gamma)^\top \Sigma w(\gamma)}{W_{\text{ini}}^2} = \frac{1}{a^2} \hat{r}^\top \Sigma^{-1} \hat{r}.$$

The point here is that the portfolio expected return and variance do not depend on γ, hence equation (11.13) is not really producing a frontier curve (in terms of actual expected relative return and variance) but rather a single point: the relative portfolio composition remains constant with γ, only the total invested amount is proportional to γ^{-1}.

Optimal tradeoff for $\mathcal{W} = \{w \colon \mathbf{1}^\top w = 1\}$ We next consider the problem
$$\max_{w} \hat{r}^\top w - \gamma w^\top \Sigma w, \quad \text{s.t.:} \ \mathbf{1}^\top w = 1. \qquad (11.15)$$

The solution in this case can be obtained by solving the dual formulation of the problem
$$\min_{\lambda} \max_{w} \hat{r}^\top w - \gamma w^\top \Sigma w + \lambda(\mathbf{1}^\top w - 1). \qquad (11.16)$$

The inner maximization yields
$$\begin{aligned} g(\lambda) &= \max_{w} \hat{r}^\top w - \gamma w^\top \Sigma w + \lambda(\mathbf{1}^\top w - 1) \\ &= \frac{1}{4\gamma}(\hat{r} + \lambda \mathbf{1})^\top \Sigma^{-1}(\hat{r} + \lambda \mathbf{1}) - \lambda, \end{aligned}$$

where the maximum over w is attained at
$$w^*(\lambda) = \frac{1}{2\gamma}\Sigma^{-1}(\hat{r} + \lambda \mathbf{1}).$$

Then, $\min_\lambda g(\lambda)$ is attained at
$$\lambda^* = \frac{2\gamma - \mathbf{1}^\top \Sigma^{-1} \hat{r}}{\mathbf{1}^\top \Sigma^{-1} \mathbf{1}},$$

and substituting this value of λ in $w^*(\lambda)$ we finally obtain the optimal portfolio
$$\begin{aligned} w(\gamma) &= \frac{1}{2\gamma}\Sigma^{-1}\hat{r} + \left(1 - \frac{\mathbf{1}^\top \Sigma^{-1}\hat{r}}{2\gamma}\right) \frac{\Sigma^{-1}\mathbf{1}}{\mathbf{1}^\top \Sigma^{-1}\mathbf{1}} \\ &= w^{(a)}(\gamma) + \left(1 - \frac{\mathbf{1}^\top \Sigma^{-1}\hat{r}}{2\gamma}\right) w_{\text{minrisk}}, \qquad (11.17) \end{aligned}$$

where $w^{(a)}(\gamma)$ is the optimal solution (11.14) of the unconstrained problem (11.13), and w_{minrisk} is the minimum risk portfolio (11.10) that solves (11.9).

Remark 11.4 (*Optimal tradeoff under the $\|w\|_1 = 1$ constraint*) The problem
$$\max_{w} \hat{r}^\top w - \gamma w^\top \Sigma w, \quad \text{s.t.:} \ \|w\|_1 = 1 \qquad (11.18)$$

does not admit a closed-form solution, in general. Actually, the nonlinear equality constraint $\|w\|_1 = 1$ makes this problem nonconvex and hard to solve numerically. There exist, however, a range of values of γ for which the problem can be recast equivalently in the form of a convex quadratic problem and hence solved efficiently. In particular, if $\gamma > 0$ is such that

$$\gamma < \frac{1}{2}\|\Sigma^{-1}\hat{r}\|_1, \qquad (11.19)$$

then problem (11.18) is equivalent to

$$\max_w \hat{r}^\top w - \gamma w^\top \Sigma w, \quad \text{s.t.:} \ \|w\|_1 \leq 1, \qquad (11.20)$$

in the sense that the two problems have the same optimal solutions. This fact is proved by observing that the gradient of the objective $\hat{r}^\top w - \gamma w^\top \Sigma w$ is zero for $w = \Sigma^{-1}\hat{r}/(2\gamma)$, so for the point of zero gradient it holds that $\|w\|_1 = \|\Sigma^{-1}\hat{r}\|_1/(2\gamma)$ and under condition (11.19) we have $\|w\|_1 > 1$, which means that the gradient of the objective is nonzero in the feasible set $\{w\colon \|w\|_1 \leq 1\}$. The fact that the gradient of the objective is nonzero over the feasible set of (11.20) means that any optimal solution for this problem will be on the boundary of the feasible set,[4] that is, the constraint $\|w\|_1 \leq 1$ will be active at optimum, so the solution will be such that $\|w\|_1 = 1$, and hence it will be optimal also for the original equality-constrained problem (11.18).

The inequality formulation (11.20) of the problem has the clear advantage of being convex, and hence amenable to efficient numerical solution. Moreover, by solving (11.20) in the regime of γ values specified in (11.19), we actually solve (11.18). Also, one may argue that values of $\gamma > \|\Sigma^{-1}\hat{r}\|_1/2$ should be avoided since they tend to over-weight the risk term in the objective and hence to produce portfolios that, in absence of the equality constraint, would shrink towards zero according to (11.14) by reducing the total invested amount in proportion to γ^{-1}.

[4] This fact can be proved by contradiction: suppose there exist an optimal solution w^* that is in the interior of the feasible set, that is, such that $\|w^*\|_1 < 1$. Then there would exist an open neighborhood of w^* also within the interior of the feasible set and we could move infinitesimally away from w^* in the direction of the gradient while remaining feasible and increasing the objective value: this would contradict the fact that w^* was the global optimum of the problem, which meant that no other feasible solution could yield a better objective value.

Example 11.2 Consider a collection of $n = 5$ assets whose random return vector r is characterized via its expected value $\hat{r}$ and covariance matrix Σ, with

$$\hat{r} = \begin{bmatrix} 2 \\ 4 \\ 4 \\ 3 \\ 6 \end{bmatrix}, \quad \Sigma = \begin{bmatrix} 3 & -1 & 1 & -1 & -1 \\ -1 & 4 & 1 & 1 & 1 \\ 1 & 1 & 9 & -1 & 1 \\ -1 & 1 & -1 & 16 & 1 \\ -1 & 1 & 1 & 1 & 21 \end{bmatrix}.$$

These n individual assets can be represented in the risk (std. dev.)/return plane as n points of coordinates $(\sigma_i, \hat{r}_i)$, where $\sigma_i \doteq \sqrt{\Sigma_{ii}}$, $i = 1, \ldots, n$, as depicted by asterisks in Figure 11.2. Assuming the set of admissible portfolios to be

$$\mathcal{W} = \{w \in \mathbb{R}^n \colon w \geq 0, \ \sum_{i=1}^n w_i = 1\},$$

we can construct by points the curve of the efficient frontier by solving problem (11.7) repeatedly for many values of γ in an interval $[\gamma_{\min}, \gamma_{\max}]$.

11.2 OPTIMAL RISK/RETURN TRADEOFFS

In particular, we solved the quadratic program

$$\max_{w} \hat{r}^\top w - \gamma_k w^\top \Sigma w, \quad \text{s.t.:} \quad w \geq 0, \quad \mathbf{1}^\top w = 1 \quad (11.21)$$

for $m = 40$ log-spaced values of γ_k between 10^{-3} and 10^3. The resulting curve is shown in Figure 11.2. Small values of γ_k lead to portfolios in the upper-right region of the curve (high return, high risk), while large values of γ_k lead to portfolios in the lower-left region of the curve (low return, low risk). Due to the presence of the $w \geq 0$ constraint, the solution of (11.21) cannot be found in closed form, in general. Instead, it required the solution of a sequence of convex quadratic problems (one for each γ_k value).

It can be observed that the portfolios on the efficient frontiers have risk/return characteristics that are *better* than those of the individual component assets. For instance, for $\gamma_k = 10^3$ we obtain the leftmost portfolio on the curve, with composition $w = [0.4999, 0.3498, 0.0245, 0.0732, 0.0525]$, expected return 3.032 and risk $\sigma_w = 1.0243$. Hence, the risk of such portfolio is better (lower) than the risk of the component asset with lower risk (asset 1), while also having a better (higher) expected return. Appropriate mixtures w of the individual assets, therefore, really create new financial objects, with enhanced risk/return profiles with respect to the original individual components.

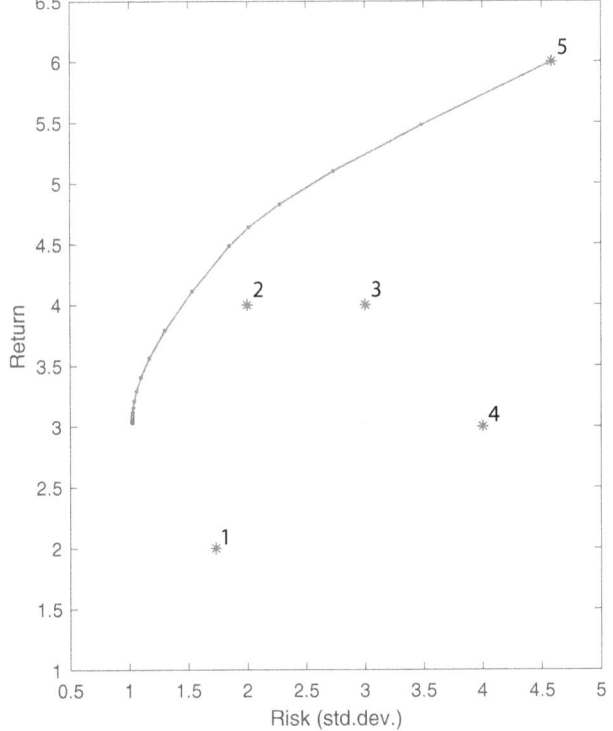

Figure 11.2 Five assets (asterisks) and the efficient frontier generated by optimal portfolios $w \in \mathcal{W}$ combining these assets; see code file ex_efffront.m in the online resources.

11.2.3 Design with factor covariance models

We discussed earlier in Section 3.5.3 that factor models for the covariance matrix may bring substantial computational advantages when solving a plain unconstrained portfolio tradeoff problem of the form (11.13). This advantage is actually retained also in the general constrained case. Assume that the covariance matrix Σ has a factor model of the form

$$\Sigma = LL^\top + D,$$

with $D \doteq \mathrm{diag}\,(\sigma_1^2,\ldots,\sigma_n^2)$ and $L \in \mathbb{R}^{n,k}$, with $k \ll n$. Then, the objective of the portfolio design problem (11.7) is rewritten as

$$\hat{r}^\top w - \gamma w^\top (D + LL^\top) w = \hat{r}^\top w - \gamma \|L^\top w\|_2^2 - \gamma \sum_{i=1}^n \sigma_i^2 w_i^2.$$

Problem (11.7) can then be rewritten as

$$\max_{w,z} \quad \hat{r}^\top w - \gamma \|z\|_2^2 - \gamma \sum_{i=1}^n \sigma_i^2 w_i^2, \tag{11.22}$$
$$\text{s.t.:} \quad w \in \mathcal{W},$$
$$\quad z = L^\top w.$$

In the above formulation, we added equality constraints $z = L^\top w$ in order to make appear a quadratic term $z^\top z$ with diagonal kernel matrix (the identity, in this case) in the objective. In general, it is advantageous to introduce fictitious equality constraints if they render *sparse* the kernel matrix of the quadratic terms in the objective. Indicatively, the QP formulation with full kernel matrix requires a numerical effort that scales approximately as n^3, while the formulation with diagonal kernel matrix scales linearly in n and cubically in k: when $k \ll n$ this may yield dramatic computational advantages.

11.3 Additional criteria and constraints

In this section we discuss common criteria and constraints that may be added to the stylized problem (11.7) in order to adapt it to a more realistic setting. These criteria may include costs that are added to the objective function of (11.7) with suitable tradeoff coefficients, or explicit requirements on the portfolio composition that are added to the constraint set $\mathcal{W}$.

11.3.1 Linear transaction costs

One aspect that should be considered in practice is that whenever we modify our invested position, either by increasing the exposure or by decreasing it, we may incur additional transaction costs to be paid to the broker. Suppose that our initial position is described by a portfolio vector $w^{(0)}$, while the new position we want to take is described by w, so that the portfolio variation is described by vector $d \doteq w - w^{(0)}$. In the *linear* transaction cost model we assume that the cost for varying the ith position is given by

$$f(d_i) = \begin{cases} c_+ d_i & \text{if } d_i \geq 0, \\ c_- d_i & \text{if } d_i < 0, \end{cases}$$

where $c_+, c_- > 0$ are the unit costs for a long or a short transaction, respectively. Note that we can write $f(d_i) = \max(c_+ d_i, -c_- d_i)$, which shows that the individual transaction cost is a convex function in d_i, since it is the max of two linear functions. The total cost of the transactions is

$$\text{TC}(w) = \sum_{i=1}^n f(d_i) = \sum_{i=1}^n \max(c_+(w_i - w_i^{(0)}), -c_-(w_i - w_i^{(0)})),$$

which is a convex function in w. The transaction cost can be added as a penalty in the objective function, thus yielding a problem of the form

$$\max_{w \in \mathcal{W}} \hat{r}^\top w - \gamma w^\top \Sigma w - \nu \text{TC}(w), \qquad (11.23)$$

where $\nu \geq 0$ is a penalty parameter. Alternatively, the transaction costs may be considered as a constraint; for instance

$$\max_{w \in \mathcal{W}} \hat{r}^\top w - \gamma w^\top \Sigma w, \qquad (11.24)$$
$$\text{s.t.: } \text{TC}(w) \leq \epsilon(\mathbf{1}^\top w)$$

imposes the constraint that the transaction costs do not exceed a given fraction $\epsilon \in (0,1)$ of the total invested capital. Note that both problem (11.23) and problem (11.24) can be cast as convex QPs by the suitable addition of slack variables. In particular, we rewrite (11.23) as

$$\max_{w \in \mathcal{W}, y \in \mathbb{R}^n} \hat{r}^\top w - \gamma w^\top \Sigma w - \nu \mathbf{1}^\top y,$$
$$\text{s.t.: } c_+(w - w^{(0)}) \leq y,$$
$$-c_-(w - w^{(0)}) \leq y.$$

Similarly, we rewrite (11.24) as

$$\max_{w \in \mathcal{W}, y \in \mathbb{R}^n} \hat{r}^\top w - \gamma w^\top \Sigma w,$$
$$\text{s.t.: } \mathbf{1}^\top y \leq \epsilon(\mathbf{1}^\top w),$$
$$c_+(w - w^{(0)}) \leq y,$$
$$-c_-(w - w^{(0)}) \leq y.$$

Observe also that in the case when $c_+ = c_- = c$ the linear transaction cost function is simply expressed as

$$\mathrm{TC}(w) = c\|w - w^{(0)}\|_1, \quad \text{for } c_+ = c_- = c.$$

This latter expression is also often used to measure the level of *turnover* or overall change in composition from portfolio w_0 to portfolio w.

As discussed above, linear transaction costs lead to convex problem formulations. Unfortunately, however, more complex or structured cost models cannot in general be represented via convex programs. For instance, fixed costs or fixed-plus-linear costs cannot be accommodated into a convex programming framework and are thus harder to deal with computationally. Nevertheless, modern optimization platforms, such as Mosek or Gurobi, are capable of dealing with such nonconvex mixed-integer optimization problems with reasonable ease, at least when the problem size is moderate.

11.3.2 Short-selling constraints

Constraints on the maximum short exposure on security i can be simply set as $w_i \geq -s_i$, where $s_i \geq 0$ is the given limit. A constraint on the total short exposure of the portfolio can instead be set as

$$\sum_{i=1}^{n} \max(-w_i, 0) \leq S,$$

where S is the given total allowable short exposure. This is a convex constraint that can be easily recast in the form of a collection of linear inequalities, via the addition of a slack vector variable t:

$$\sum_{i=1}^{n} t_i \leq S, \quad t \geq 0, \ t + w \geq 0.$$

Also, we may consider leverage strategies in which short selling is used to produce cash that is employed for buying other securities. We can express such strategy using jointly the constraints

$$\mathbf{1}^\top w = 1, \quad \|w\|_1 \leq c, \tag{11.25}$$

for given $c \geq 1$, see Figure 11.3.

Indeed, by indicating with $w^+ = \max(w, 0)$ the vector with the positive part of w and with $w^- = \max(-w_i, 0)$ the negative part, we have that $w = w^+ - w^-$ and the constraints in (11.25) are rewritten as

$$\sum_{i=1}^{n} w_i^+ - \sum_{i=1}^{n} w_i^- = 1, \quad \sum_{i=1}^{n} w_i^+ + \sum_{i=1}^{n} w_i^- \leq c.$$

11.3 ADDITIONAL CRITERIA AND CONSTRAINTS

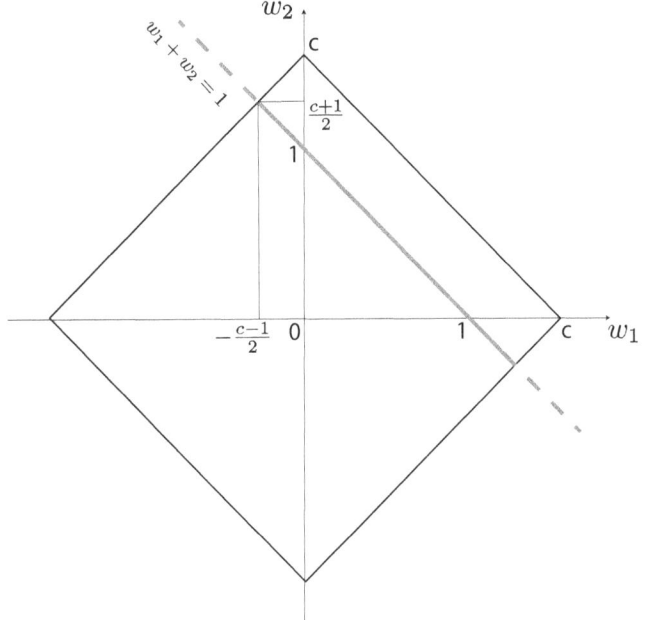

Figure 11.3 The thicker solid line shows the locus of portfolios satisfying $\mathbf{1}^\top w = 1$, $\|w\|_1 \leq c$, in a two-dimensional example.

Obtaining $\sum_{i=1}^{n} w_i^+$ from the first equation and substituting it into the second inequality, we obtain that

$$\sum_{i=1}^{n} w_i^- \leq \frac{c-1}{2}, \quad \sum_{i=1}^{n} w_i^+ \leq \frac{c+1}{2},$$

which shows that the constraints in (11.25) induce joint constraints on the total long and short exposures of the portfolio. For instance, for $c = 1$ we obtain a no short-selling (long only) strategy, and for $c = 1.6$ we obtain a strategy that allows at most 30% short exposure and 130% long exposure.

11.3.3 Market impact

When the traded amounts are small, for example, in the case of small private investors, we may assume that the security prices are independent of the amounts traded, that is, the individual buy/sell order has no impact on the price. For large trades such as the ones made by institutional or professional investors, however, the traded amount does have an adverse impact on the underlying security price. For this reason, large investors tend to slice a large order into orders of smaller size, and then the strategy with which these sub-orders are executed in time is a problem in its own right, known as the optimal execution problem. Next, we discuss how to take into account market impact in our single-stage portfolio design setting.

While there is no standard model for market impact, in practice it has been observed empirically that the expected market-impact cost (MIC) of a trade $d_i \doteq w_i - w_i^{(0)}$ may be approximately described by a *power law* of the form[5]

$$\text{MIC}(w_i) = a_i |d_i|^\beta = a_i |w_i - w_i^{(0)}|^\beta,$$

where β is typically assumed to be 3/2 (or sometimes 5/3), and $a_i \geq 0$ is a coefficient that should be estimated for a specific market and security. Approximately, one may assume $a_i \simeq \sigma_i v_i^{1-\beta}/\beta$, where σ_i is the volatility (estimated standard deviation) of the return of security i, and v_i is the average currency volume of trades in security i during the considered trading period. For a change in portfolio composition $d \doteq w - w^{(0)}$ the overall market-impact cost is therefore

$$\text{MIC}(w) = \sum_{i=1}^{n} a_i |w_i - w_i^{(0)}|^\beta.$$

[5] See, for example, B. Tóth, Y. Lemperiere, C. Deremble, J. de Lataillade, J. Kockelkoren, and J.-P. Bouchaud, "Anomalous Price Impact and the Critical Nature of Liquidity in Financial Markets." *Physical Review X*, vol. 1, 021006, 2011.

Function $\text{MIC}(w)$ is convex in w, for $\beta > 1$. In particular, we can express

$$\text{MIC}(w) = \min_u \sum_{i=1}^{n} a_i u_i, \quad \text{s.t.:} \ |w_i - w_i^{(0)}|^\beta \leq u_i, \ i = 1, \ldots, n,$$

where each constraint $|w_i - w_i^{(0)}|^\beta \leq u_i$ is equivalent to $|w_i - w_i^{(0)}| \leq u_i^{1/\beta}$, which in turn can be expressed as $(w_i - w_i^{(0)}, u_i, 1) \in \mathcal{P}_{1/\beta}^n$ where $\mathcal{P}_{1/\beta}^n$ is the power cone defined in (10.13). In advanced parsers/solvers for convex optimization, power cone constraints can be dealt with directly. For instance, in CVX a constraint of the form $|x|^b \leq u$ is expressed as pow_abs(x,b) <= u, which is convex in x for $b > 1$.

11.3.4 Diversification constraints

Diversification constraints describe requirements on limits in exposure in individual assets or groups of assets. The simplest form of diversification constraint is to require that no individual investment contains more than a certain percentage ϵ of the total available budget B, that is, $|w_i| \leq \epsilon B$, $i = 1, \ldots, n$. More generally, we can define one or several groups of assets defined by subsets J_k of the indices $\{i = 1, \ldots, n\}$, for $k = 1, \ldots, K$, and impose that no more than an ϵ_k fraction of the total budget is invested in the kth group J_k, which typically represents a sector (e.g., technology, retail, oil, industrial, financial, etc.), that is,

$$\sum_{j \in J_k} |w_j| \leq \epsilon_k B, \quad k = 1, \ldots, K.$$

These constraints can be easily converted into a collection of linear inequality constraints in w and slack vector variable u:

$$\sum_{j \in J_k} u_j \leq \epsilon_k B, \quad -u_j \leq w_j \leq u_j, \quad k = 1, \ldots, K.$$

A more complex type of diversification constraint arises when we want to impose that no group of m assets contains more than a fraction $1 - \epsilon$ of the total budget, that is,

$$\max_{J: \text{card}(J) = m} \sum_{j \in J} |w_j| \leq \epsilon B.$$

Clearly, the maximum over the subsets of assets of cardinality m is attained by the subset containing the m-largest elements of $|w|$. Denoting with $|w|_{[1]}, |w|_{[2]}, \ldots, |w|_{[n]}$ the elements of $|w|$ in decreasing order, the previous constraint can be written equivalently as

$$\sum_{j=1}^{m} |w|_{[j]} \leq \epsilon B. \tag{11.26}$$

Remark 11.5 (*Sum of k-largest elements of a vector*) For $x \in \mathbb{R}^n$, the function

$$s_k(x) = x_{[1]} + \cdots + x_{[k]},$$

representing the sum of the k-largest elements of x, is a convex function. This fact can be proved by observing that $s_k(x)$ can be expressed as the point-wise maximum of several linear functions, namely

$$s_k(x) = \max_{h \in \{0,1\}^n} h^\top x, \quad \text{s.t.: } \|h\|_0 = k, \tag{11.27}$$

where $\|\cdot\|_0$ is the cardinality function, which returns the number of nonzero entries of its vector argument. Here, the number of linear functions involved in the maximum is "n choose k," that is, $n!/(k!(n-k)!)$, which happens to become a very large number as n and k increase. So, while the representation in (11.27) is useful from a theoretical perspective for proving convexity, it is not really useful in practice. The following representation gives instead a practical expression for $s_k(x)$:

$$s_k(x) = \min_{\alpha} k\alpha + \sum_{i=1}^{n} (x_i - \alpha)_+, \tag{11.28}$$

where $(x_i - \alpha)_+ \doteq \max(0, x_i - \alpha)$. We next prove (11.28). First, we show that inequality holds in (11.28) since, for all α,

$$s_k(x) = \sum_{i=1}^{k} x_{[i]} = k\alpha + \sum_{i=1}^{k} (x_{[i]} - \alpha) \leq k\alpha + \sum_{i=1}^{k} (x_{[i]} - \alpha)_+$$

$$\leq k\alpha + \sum_{i=1}^{n} (x_{[i]} - \alpha)_+ = k\alpha + \sum_{i=1}^{n} (x_i - \alpha)_+.$$

Further, for $\alpha = x_{[k+1]}$ we have that $(x_{[i]} - x_{[k+1]}) \geq 0$ for $i \leq k$, and $(x_{[i]} - x_{[k+1]}) \leq 0$ for $i > k$, thus

$$kx_{[k+1]} + \sum_{i=1}^{n}(x_i - x_{[k+1]})_+ = kx_{[k+1]} + \sum_{i=1}^{n}(x_{[i]} - x_{[k+1]})_+$$
$$= kx_{[k+1]} + \sum_{i=1}^{k}(x_{[i]} - x_{[k+1]})$$
$$= \sum_{i=1}^{k} x_{[i]} = s_k(x),$$

hence the bound is attained by $\alpha = x_{[k+1]}$.

Function $s_k(x)$ is expressed via (11.28) in the CVX function sum_largest. Note also that the sum of the k smallest elements of x is simply expressible as

$$\underline{s}_k(x) \doteq x_{[n]} + \cdots + x_{[n-k+1]} = -s_k(-x),$$

and it is hence a concave function.

Using (11.28), the diversification constraint (11.26) is first restated using a slack vector variable u as

$$\sum_{j=1}^{m} u_{[j]} \leq \epsilon B, \quad -u \leq w \leq u,$$

and then we use (11.28) to express $s_m(u) = \sum_{j=1}^{m} u_{[j]}$, obtaining

$$\min_{\alpha} m\alpha + \sum_{i=1}^{n}(u_i - \alpha)_+ \leq \epsilon B, \quad -u \leq w \leq u.$$

Note that the first inequality is satisfied if there exists some α such that $m\alpha + \sum_{i=1}^{n}(x_i - \alpha)_+ \leq \epsilon B$, hence we have

$$m\alpha + \sum_{i=1}^{n}(u_i - \alpha)_+ \leq \epsilon B, \quad -u \leq w \leq u.$$

We can now introduce a further slack vector variable z, and restate the previous conditions as

$$m\alpha + \mathbf{1}^\top z \leq \epsilon B, \quad -u \leq w \leq u, \quad z \geq 0, \quad u - \alpha \mathbf{1} \leq z. \quad (11.29)$$

Overall, condition (11.26) on the n variables in w is equivalent to conditions (11.29) on the $3n + 1$ variables in (w, u, z, α).

11.4 Sharpe ratio optimization

The Sharpe ratio (SR)[6] is a measure of reward to risk, which quantifies the amount of expected return in excess of the risk-free rate per

[6] Named after William Sharpe, Nobel laureate in economics, 1990.

unit risk. If we assume that there is available an asset which yields a return r_f with *zero* risk, then all portfolios w with expected return $\hat{r}^\top w \leq r_f$ and positive risk are dominated by the risk-free asset. In other words, in order to accept a positive risk, an investor should expect a return larger than r_f. The SR of a portfolio w with random return $\rho(w) \doteq r^\top w$ is defined as

$$\mathrm{SR}(w) \doteq \frac{\mathrm{E}\{\rho(w)\} - r_f}{\sqrt{\mathrm{var}\{\rho(w)\}}} = \frac{\hat{r}^\top w - r_f}{\sqrt{w^\top \Sigma w}},$$

where $r_f \geq 0$ is the return of the *risk-free* asset, for instance, the return one gets from money deposited on a bank savings account. The higher the SR value, the better the investment, from a reward/risk perspective. Geometrically, we observe that portfolios with a given SR map to points lying on a straight line in the risk/return plane; this line crosses the vertical axis at r_f and its slope is the SR, see Figure 11.4.

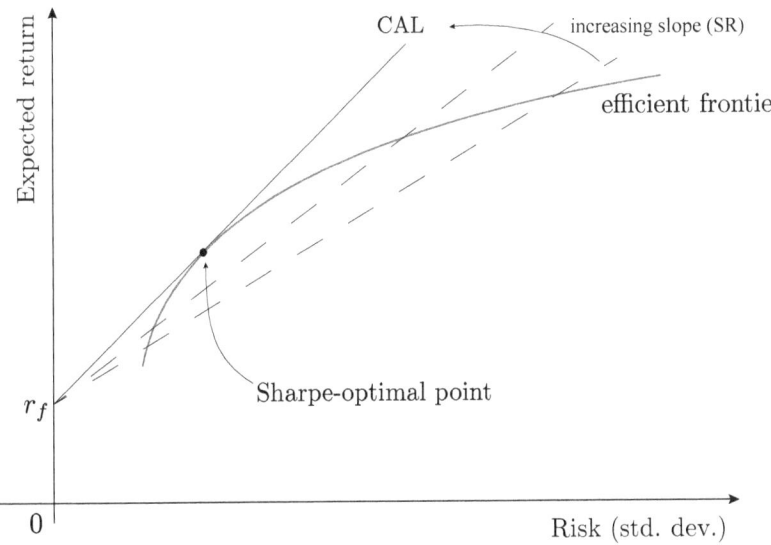

Figure 11.4 Portfolios with constant SR map on the risk/return plane to points on a line with slope SR passing through $(0, r_f)$.

The SR optimization problem amounts to finding a portfolio that maximizes the SR:

$$\max_{w \in \mathcal{W}} \quad \frac{\hat{r}^\top w - r_f}{\sqrt{w^\top \Sigma w}}, \qquad (11.30)$$
$$\text{s.t.:} \quad \hat{r}^\top w \geq r_f.$$

Assuming that the risk-free asset is not included among the assets considered in the portfolio, and under the condition that $\hat{r}^\top x \geq r_f$

and $\Sigma \succ 0$, the portfolio that maximizes the SR corresponds, geometrically, to the point of tangency to the efficient frontier of the line passing from the $(0, r_f)$ point, the so-called capital allocation line (CAL). In the formulation (11.30) the problem is *not* convex. However, we can obtain an equivalent convex formulation as follows. First, we introduce a slack variable $t \geq 0$ and rewrite the problem equivalently as

$$\max_{w \in \mathcal{W}, t} \frac{\hat{r}^\top w - r_f}{t},$$
$$\text{s.t.: } \hat{r}^\top w \geq r_f,$$
$$\sqrt{w^\top \Sigma w} \leq t.$$

Next, with the change of variables $z = w/t$, $\alpha = 1/t$, we obtain

$$\max_{z \in \alpha \mathcal{W}, \alpha \geq 0} \hat{r}^\top z - \alpha r_f,$$
$$\text{s.t.: } \hat{r}^\top z \geq \alpha r_f,$$
$$\sqrt{z^\top \Sigma z} \leq 1,$$

which is a convex formulation. This problem is typically cast as an SOCP by considering a Cholesky factorization $\Sigma = LL^\top$:

$$\max_{z \in \alpha \mathcal{W}, \alpha \geq 0} \hat{r}^\top z - \alpha r_f,$$
$$\text{s.t.: } \hat{r}^\top z \geq \alpha r_f,$$
$$\|L^\top z\|_2 \leq 1.$$

Observe that if, for instance, $\mathcal{W} = \{w: \mathbf{1}^\top w = 1, \|w\|_1 \leq c\}$, for given $c \geq 1$, then the above problem specifies to

$$\max_{z, \alpha} \hat{r}^\top z - \alpha r_f,$$
$$\text{s.t.: } \hat{r}^\top z \geq \alpha r_f,$$
$$\|L^\top z\|_2 \leq 1,$$
$$\mathbf{1}^\top z = \alpha, \quad \|z\|_1 \leq \alpha c, \quad \alpha \geq 0.$$

Example 11.3 We consider again the data in Example 11.2, and assume a risk-free rate $r_f = 3$ (say, this means 3% return). We have in this case $\Sigma = LL^\top$ with

$$L = \begin{bmatrix} 1.732 & 0 & 0 & 0 & 0 \\ -0.5774 & 1.915 & 0 & 0 & 0 \\ 0.5774 & 0.6963 & 2.86 & 0 & 0 \\ -0.5774 & 0.3482 & -0.3178 & 3.93 & 0 \\ -0.5774 & 0.3482 & 0.3814 & 0.1696 & 4.513 \end{bmatrix}.$$

11.5 ALTERNATIVE MEASURES OF RISK

To find the Sharpe-optimal portfolio in $\mathcal{W} = \{w \in \mathbb{R}^5 : w \geq 0, \mathbf{1}^\top w = 1\}$ we solve the SOCP

$$\max_{z,\alpha} \quad \hat{r}^\top z - \alpha r_f,$$
$$\text{s.t.:} \quad \hat{r}^\top z \geq \alpha r_f,$$
$$\|L^\top z\|_2 \leq 1,$$
$$\mathbf{1}^\top z = \alpha, \quad z \geq 0, \quad \alpha \geq 0,$$

which yields

$$z = [0.0, 0.2443, 0.0916, 0.0, 0.1595]^\top, \quad \alpha = 0.4953,$$

and hence the Sharpe-optimal portfolio

$$w^* = [0.0, 0.4932, 0.1849, 0.0, 0.3219]^\top,$$

whose return has expected value 4.6438 and standard deviation 2.0189, as shown by the small circle in Figure 11.5.

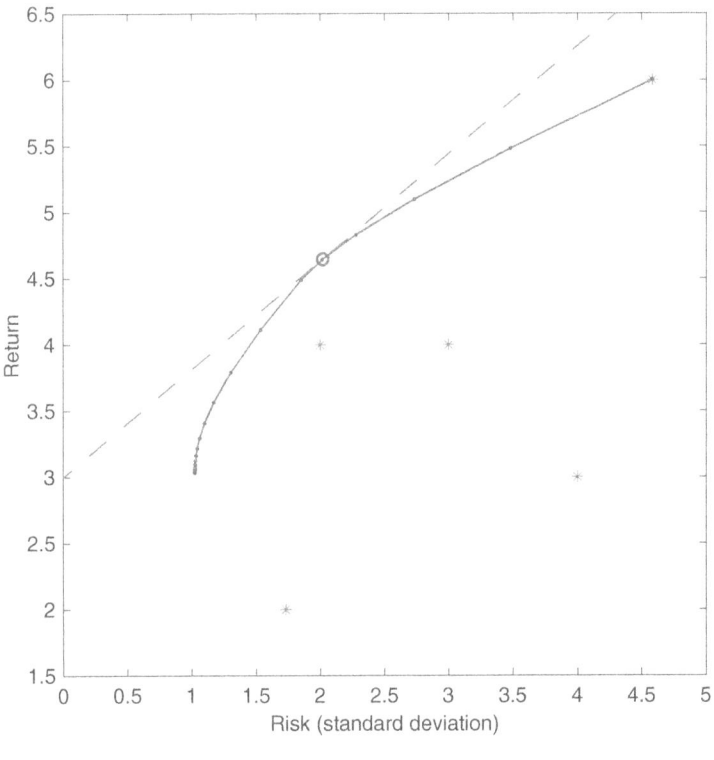

Figure 11.5 Five component assets (asterisks), their efficient frontier (solid line), and the capital asset line (CAL, dashed). The Sharpe-optimal portfolio corresponds to the point of tangency between the CAL and the efficient frontier, as shown by a circle; see code file ex_efffront.m in the online resources.

11.5 Alternative measures of risk

The variance $\sigma^2(w)$ of the portfolio return $\rho(w) \doteq r^\top w$ is surely not the only possible measure of riskiness of a portfolio. First, by looking at the definition

$$\sigma^2(w) \doteq \mathrm{E}\{(\rho(w) - \mathrm{E}\{\rho(w)\})^2\}$$

we see that $\sigma^2(w)$ is a *symmetric* measure of risk, in the sense that it counts equally the positive and the negative deviations of the portfolio return from its expectation. This may be misleading in the case when the underlying return distribution is skewed, or when the investor has asymmetric aversions to risk. Put simply, a long investor may dislike negative deviations of $\rho(w)$ from its mean, but have nothing against positive deviations. Second, the variance accounts for fluctuations around the mean via a square function, and this tends to over-emphasize large-but-rare deviations, that is, large deviations from the mean that happen with low probability may deeply impact the variance.

For the above reasons several other quantifications of risk, besides the variance, are frequently encountered in the financial literature and practice. We next review some of the most relevant risk measures for portfolio design.

11.5.1 Value-at-risk

Value-at-risk (VaR) is a risk measure that uses information about the whole return probability distribution, rather than only of the mean and covariance matrix. Let $F_\rho(x)$ be the cumulative distribution function of the portfolio return $\rho = r^\top w$:

$$F_\rho(x) \doteq \mathrm{Prob}\{\rho \leq x\}, \quad x \in \mathbb{R}.$$

The corresponding quantile function is defined as

$$Q_\rho(\alpha) \doteq \min\{x \colon F_\rho(x) \geq \alpha\}, \quad \alpha \in (0,1),$$

see Figure 11.6.

For given probability level $\alpha \in (0,1)$, the the $\mathrm{VaR}_\rho(\alpha)$ of the portfolio is defined as

$$\mathrm{VaR}_\rho(\alpha) \doteq -Q_\rho(\alpha).$$

In words, $\mathrm{VaR}_\rho(\alpha)$ represents that level of *loss* that is not exceeded with probability at least $1 - \alpha$. Observe that it holds that

$$Q_\rho(\alpha) \leq x \quad \Leftrightarrow \quad \alpha \leq F_\rho(x). \tag{11.31}$$

Further, when $F_\rho(x)$ is continuous and strictly monotone, we simply have that $Q_\rho(\alpha) = F_\rho^{-1}(\alpha)$, and it also holds that $Q_\rho(\alpha) \geq x \Leftrightarrow \alpha \geq F_\rho(x)$.

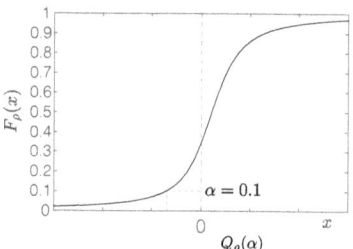

Figure 11.6 Cumulative distribution function and quantile.

Using VaR, risk can then be controlled by setting an a-priori loss probability level α (e.g., $\alpha = 0.05$ for 5% probability), and then imposing that

$$\text{VaR}_\rho(\alpha) \leq \ell, \qquad (11.32)$$

where $\ell > 0$ is the loss level we want to guarantee not to be exceeded (with probability $\geq 1 - \alpha$). One problem with VaR is that its computation requires exact knowledge of the return distribution. Further, even if one knows the distribution, computing VaR and imposing (11.32) explicitly is problematic from a computational point of view.

Normal return distribution A notable exception arises under the assumption that the returns distribution is normal with mean $\hat{r}$ and covariance Σ. In such case, since $F_\rho(x)$ is continuous and strictly monotone, we have that $\text{VaR}_\rho(\alpha) \leq \ell$ if and only if $\alpha \geq F_\rho(-\ell)$. Observe that

$$F_\rho(-\ell) \doteq \text{Prob}\{\rho \leq -\ell\} = \text{Prob}\{(\rho - \hat{\rho})/\sigma \leq (-\ell - \hat{\rho})/\sigma\}$$
$$= \Phi((-\ell - \hat{\rho})/\sigma),$$

where $\hat{\rho} = \hat{r}^\top w$ is the portfolio expected return, $\sigma^2 = w^\top \Sigma w$ is the portfolio variance, and Φ is the cumulative distribution function of a standard normal random variable (i.e., a normal variable with zero mean and unit variance). Therefore,

$$\text{VaR}_\rho(\alpha) \leq \ell \Leftrightarrow F_\rho(-\ell) \leq \alpha \Leftrightarrow \Phi((-\ell - \hat{\rho})/\sigma) \leq \alpha$$
$$\Leftrightarrow (-\ell - \hat{\rho})/\sigma \leq \Phi^{-1}(\alpha).$$

For $\alpha < 0.5$, we have that $\Phi^{-1}(\alpha) < 0$, and from symmetry of the standard normal distribution it holds that $\Phi^{-1}(\alpha) = -\Phi^{-1}(1 - \alpha)$, hence the last expression above is equivalent to

$$\ell + \hat{\rho} \geq \sigma \Phi^{-1}(1 - \alpha).$$

Letting $\Sigma = LL^\top$, we write $\sigma = \sqrt{w^\top \Sigma w} = \|L^\top w\|_2$, and hence the constraint (11.32) translates into the explicit SOC constraint

$$\Phi^{-1}(1 - \alpha)\|L^\top w\|_2 \leq \ell + \hat{r}^\top w, \quad \alpha < 0.5. \qquad (11.33)$$

The expression in (11.33) relies on the assumption of normality of the returns distribution. When the actual distribution is unknown the problem can be addressed via a *robust* VaR approach, in which we impose the constraint

$$\sup \text{VaR}_\rho(\alpha) \leq \ell, \qquad (11.34)$$

where the supremum is taken with respect to all return probability distributions having the given expected return $\hat{r}$ and covariance matrix Σ. It turns out that, via application of the Chebyshev bound, the robust constraint can be converted into an equivalent SOC constraint of the form[7]

$$\kappa_\alpha \|L^\top w\|_2 \leq \ell + \hat{r}^\top w, \qquad (11.35)$$

where

$$\kappa_\alpha \doteq \sqrt{\frac{1-\alpha}{\alpha}}. \qquad (11.36)$$

Observe that the constraints for the Normal case (11.33) and for the robust (distribution independent) case (11.35) have identical structure, the only difference being in the "safety coefficient" of the norm term, which is $\Phi^{-1}(1-\alpha)$ in the Normal case and $\kappa_\alpha > \Phi^{-1}(1-\alpha)$ in the robust case; see Figure 11.7 for a plot of these coefficients.

11.5.2 Expected shortfall (conditional VaR)

While VaR expresses the level of loss that can only be surpassed with probability no larger than α, it gives no indication of the amount of loss that is incurred when the adverse event actually happens. The *expected shortfall* (or *conditional* VaR; CVaR) provides instead precisely such indication, being defined as

$$\text{CVaR}_\rho(\alpha) \doteq -\mathbb{E}\{\rho \,|\, \rho \leq Q_\rho(\alpha)\}, \qquad (11.37)$$

where $\rho = r^\top w$ is the random return of the portfolio. In words, the expected shortfall $\text{CVaR}_\rho(\alpha)$ measures how large losses can be expected if the return of the portfolio drops below its α-quantile. The expected shortfall at level α is the average of the VaR for all levels below α, that is,

$$\text{CVaR}_\rho(\alpha) = \frac{1}{\alpha}\int_0^\alpha \text{VaR}_\rho(u)\,du.$$

Further, $\text{CVaR}_\rho(\alpha) \geq \text{VaR}_\rho(\alpha)$, and both the CVaR and the VaR are decreasing functions of α. For a normal return distribution, the expected shortfall has the explicit expression[8]

$$\text{CVaR}_\rho(\alpha) = \frac{\varphi(\Phi^{-1}(1-\alpha))}{\alpha}\sigma - \hat{\rho},$$

where φ is the standard normal probability density function, $\hat{\rho} = \hat{r}^\top w$ is the portfolio expected return, and $\sigma = \sqrt{w^\top \Sigma w} = \|L^\top w\|_2$. In the normal case, therefore, a constraint of the form

$$\text{CVaR}_\rho(\alpha) \leq \ell$$

[7] See L. El Ghaoui, M. Oks, and F. Oustry, "Worst-case Value-at-Risk and Robust Portfolio Optimization: A Conic Programming Approach," *Operations Research*, vol. 51(4), pp. 543–556, 2003.

[8] See D. Bertsimas, G. J. Lauprete, and A. Samarov, "Shortfall as a Risk Measure: Properties, Optimization and Applications," *Journal of Economic Dynamics and Control*, vol. 28(7), pp. 1353–1381, 2004.

is equivalent to the explicit SOC constraint

$$\frac{\varphi(\Phi^{-1}(1-\alpha))}{\alpha}\|L^\top w\|_2 \leq \ell + \hat{r}^\top w, \quad (11.38)$$

where a plot of the coefficient of the norm term is shown in Figure 11.7. Also, the worst-case expected shortfall constraint

$$\sup \mathrm{CVaR}_\rho(\alpha) \leq \ell,$$

where the sup is taken over all the return distributions having expected value $\hat{r}$ and covariance Σ, is equivalent to the SOC constraint

$$\kappa_\alpha \|L^\top w\|_2 \leq \ell + \hat{r}^\top w, \quad (11.39)$$

where κ_α is the same as given in (11.36).

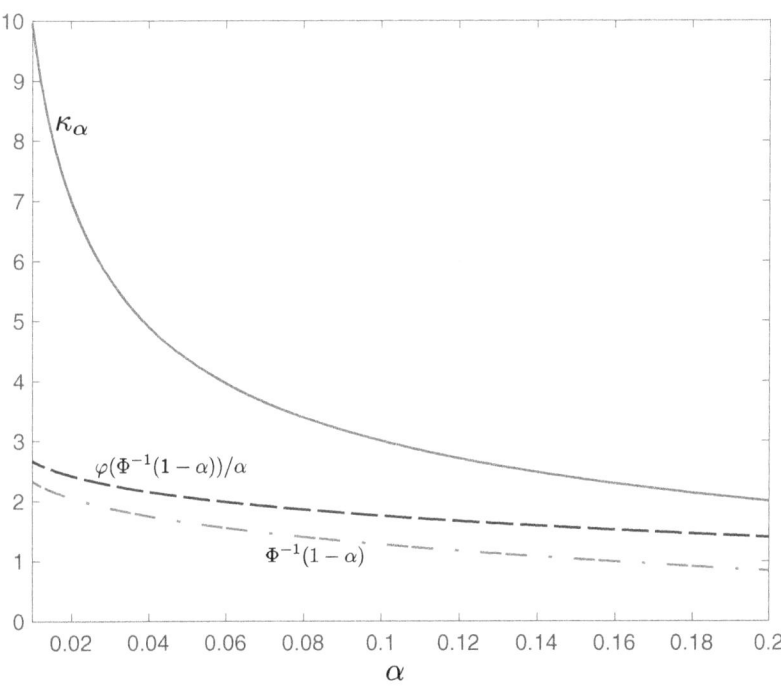

Figure 11.7 Safety coefficients for the Normal VaR constraint (11.33) (dash-dotted), the normal CVaR constraint (11.38) (dashed), and the robust VaR and CVaR constraints (11.35) and (11.39) (solid).

Conditional VaR is commonly assumed to be preferable to VaR as a measure of risk, since it possesses several properties which make it a *coherent* risk measure.[9] Moreover, the CVaR can be proved to be a convex function of the portfolio w, although such a function is not generally obtainable in explicit form for generic distributions of the returns.

It is worth to observe that the CVaR of a given portfolio w can be estimated in a natural way from samples or scenarios of the returns. If we have m samples $r^{(1)}, \ldots, r^{(m)}$ of the return vectors, and a given

[9] See P. Artzner, F. Delbaen, J. M. Eber, and D. Heath, "Coherent Measures of Risk," *Mathematical Finance*, vol. 9(3), pp. 203–228, 1999.

portfolio w, we can obtain samples of portfolio returns as $\rho_i \doteq w^\top r^{(i)}$, $i = 1, \ldots, m$. Let
$$\rho_{[1]} \leq \rho_{[2]} \leq \cdots \leq \rho_{[m]}$$
denote the ordered returns in increasing order. Then, we can estimate $\mathrm{CVaR}_\rho(\alpha)$ as
$$\widehat{\mathrm{CVaR}}_\rho(\alpha) = -\frac{1}{k} \sum_{j=1}^k \rho_{[j]},$$
where k is the largest integer $\leq \alpha m$. Note that the sum of the k smallest elements of a vector is a concave function, hence due to the minus sign above we see that $\widehat{\mathrm{CVaR}}_\rho(\alpha)$ is convex and, recalling Remark 11.5, we can express it via the sum-k-largest function s_k as
$$\widehat{\mathrm{CVaR}}_\rho(\alpha) = \frac{1}{k} s_k(-R^\top w),$$
where $R \doteq [r^{(1)} \cdots r^{(m)}]$.

11.5.3 Expected absolute variation and semi-variance

As we mentioned at the beginning of Section 11.5, there exist alternatives to the variance as a measure of variability around the mean. For example, the expected absolute variation (EAV), defined as
$$\mathrm{EAV}(w) \doteq \mathrm{E}\{|\rho(w) - \mathrm{E}\{\rho(w)\}|\},$$
measures the average of the absolute deviations of the portfolio returns from the mean. Compared to the variance, it has the advantage of being less sensitive to large deviations. On the other hand, the EAV cannot be determined in closed form, in general. The semi-variance
$$\sigma_-^2(w) \doteq \mathrm{E}\{\min(\rho(w) - \mathrm{E}\{\rho(w)\}, 0)^2\},$$
gives instead an asymmetric measure focused on the downside risk, since it accounts only for the fluctuations of $\rho(w)$ that fall below the expected value. Clearly, one may as well consider a semi-EAV measure, or any type of variation on the theme, as long as the introduced measure is justified by some practical requirement.

Finally, we observe that both the EAV and the semi-variance of a given portfolio w can be estimated from m samples $r^{(1)}, \ldots, r^{(m)}$ of the return vectors. Indeed, we estimate the expected return of the portfolio as
$$\hat{\rho}(w) = \frac{1}{m} \sum_{i=1}^m w^\top r^{(i)} = \frac{1}{m} \mathbf{1}^\top R^\top w,$$

and

$$\widehat{\mathrm{EAV}}(w) = \frac{1}{m} \sum_{i=1}^{m} |w^\top r^{(i)} - \hat{\rho}(w)|,$$

$$\hat{\sigma}_-^2(w) = \frac{1}{m} \sum_{i=1}^{m} \min(w^\top r^{(i)} - \hat{\rho}(w), 0)^2.$$

Both these empirical functions are convex in the portfolio vector w.

11.6 Portfolio optimization in practice

In practice, single-period portfolio optimization is applied in a rolling-horizon fashion, as schematized in Figure 11.8.

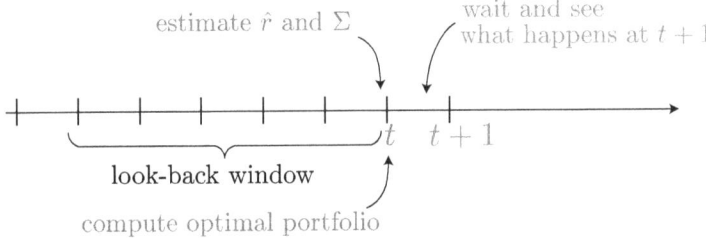

Figure 11.8 Rolling-horizon scheme for portfolio optimization.

Considering for simplicity a plain mean/variance approach, at the beginning of each time period t, we do the following:

1. Estimate the expected return vector $\hat{r}(t)$ and covariance matrix $\Sigma(t)$ of the returns on the basis of expert advice and/or observed return data on a given *look-back window* of T_{lb} periods before the current t.

2. Solve the single-period portfolio optimization problem and find $w(t|t)$.

3. Trade accordingly and wait for the end of the period to see the outcome of the investment, resulting in the portfolio value $w(t|t+1)$.

4. Let $t \leftarrow t+1$ and repeat from step 1, and so on until desired.

Backtesting refers to applying the above procedure to *past* market data and collecting performance metrics to asses "what would have happened in the past" if we applied our portfolio optimization technique and rolling-horizon strategy. In particular, on past data, at each t we can compute the optimal portfolio $w(t|t)$ and then observe the realized values of the components assets at $t+1$ and update the portfolio value as $w(t|t+1)$ and compute the realized return

$\rho(t) = (\mathbf{1}^\top w(t|t+1) - \mathbf{1}^\top w(t|t))/\mathbf{1}^\top w(t|t)$. This process is repeated over the entire backtesting data set and thus the sequence of hypothetically realized returns $\rho(t)$, $t = 1, 2, \ldots$ is generated. Typical metrics that can be computed in backtesting are, for instance:

- the total long and short exposure of $w(t|t)$, $t = 1, 2, \ldots$;

- the number of assets contained in the portfolio $w(t|t)$, $t = 1, 2, \ldots$;

- the total cost of transaction at each $t = 1, 2, \ldots$;

- the turnover $\|w(t|t+1) - w(t+1|t+1)\|_1$, $t = 1, 2, \ldots$;

- the expected return $\hat{r}(t)^\top w(t|t)$ and the realized return $\rho(t)$, $t = 1, 2, \ldots$;

- the estimated Sharpe ratio of the portfolio realized returns;

- the maximum drawdown (peak-to-trough decline).

From the series of realized returns $\rho(t)$, $t = 1, 2, \ldots$, it is also possible to evaluate the portfolio "alpha" and "beta." Beta refers to that part of portfolio return that can be attributed to the overall market (systematic risk). Alpha is the portion of a portfolio's return that cannot be attributed to market returns and is thus independent of them (idiosyncratic risk). The alpha and beta parameters can be estimated via linear regression, imposing that the observed portfolio returns are approximately given by

$$\rho(t) \simeq \alpha + \beta m(t), \quad t = 1, 2, \ldots,$$

where $m(t)$ is the return of an index that represents the overall market of reference, such as, for example, the Standard & Poor's 500 Index.

While backtesting is a widely accepted procedure to "validate" a proposed portfolio design idea, it must be stressed that plain backtesting may lead to overfitting and may give poor predictions of *future* performance of the proposed method. This is due to two main reasons: first, backtesting is based on a *single* realization of a stochastic process (the market is the stochastic process, and the single realization we are using is the observed reality that happened) and, second, care should be exerted not to use backtesting data to "tune" the parameters of the optimization problem, since this would make validation data spill-over to training. To overcome these issues, it is suggested to run the backtesting not only on the real data observed in the past, but also on variations and perturbations of this data.

Put simply, we generate many versions of possible realities, for example, by bootstrapping the historical return data (i.e., building a new return dataset by resampling with replacement from the original dataset), and we run a backtesting simulation on each of these synthetic realities, so producing empirical distributions of our portfolio's performance indices.

11.7 Exercises

Exercise 11.1 *Robust portfolio optimization*

In remark 11.3 we have seen that the efficient frontier can be computed by solving the quadratic programming (QP) problem (11.11):

$$\min_{w \in \mathcal{W}} \quad w^\top \Sigma w,$$
$$\text{s.t.:} \quad \hat{r}^\top w \geq \varrho.$$

Here, we consider $\mathcal{W} = \{w \in \mathbb{R}^n : \sum_{i=1}^n w_i = 1\}$. In practice, we usually use empirical mean and covariance matrix of historical asset return data to estimate $\hat{r}$ and Σ.

1. Use the data in file 77Stocks.mat from the online resources which contains log-return time series for $p = 77$ stocks over $n = 503$ days, from January 2007 to December 2008. Compute the sample mean vector $\hat{r}$ and covariance matrix Σ. Solve the Markowitz portfolio problem by sweeping the target return ϱ from 0 to $10\varrho^*$, where $\varrho^* \doteq \max(\hat{r}_1, \ldots, \hat{r}_n)$. Plot the efficient frontier on a 2D plane where the x-axis corresponds to the standard deviation of the portfolio and the y-axis to the expected return.

2. We now consider a robust version of portfolio optimization problem given by

$$\min_{w \in \mathcal{W}} \quad w^\top \Sigma w,$$
$$\text{s.t.:} \quad r^\top w \geq \varrho, \ \forall r \in \mathcal{U},$$

where $\mathcal{U} = \{r = +\alpha \Sigma^{1/2} v, \ \|v\|_2 \leq 1\}$ is an α-standard deviation uncertainty set for r; $\alpha \geq 0$ is a scalar parameter.

Show that this problem is equivalent to

$$\min_{w \in \mathcal{W}} \quad w^\top \Sigma w,$$
$$\text{s.t.:} \quad \alpha \|\Sigma^{1/2} w\|_2 \leq w^\top \hat{r} - \varrho. \quad (11.40)$$

3. Now let us plot the efficient frontiers by solving equation (11.40). Use the empirical mean and covariance matrix computed in the first point of the exercise from historical data. Sweep the parameter ϱ from 0 to $10\varrho^*$, solve the above portfolio optimziation problem for each ϱ and plot the efficient frontier. Let $\alpha = 0$, then we will obtain the nominal efficient frontier. Let $\alpha = 0.2$, then we will get a robust efficient frontier. Compare the two efficient frontiers.

4. Now let us take a look at the efficient frontiers based on the realized mean and covariance, rather than the expectations based on historical data. Let us denote an optimal nominal portfolio with target return ϱ and $\alpha = 0$ by $w^{(0)}(\varrho)$, and an optimal robust portfolio with target return ϱ and $\alpha = 0.2$ by $w^{(0.2)}(\varrho)$. We computed these optimal portfolios in the previous question using $\hat{r}$ and Σ, the empirical mean and covariance matrix of historical data. Now let us assume the actual return that the market eventually realizes is

$$r(\beta) = \hat{r} + \beta \Sigma^{1/2} v,$$

where v is a unit vector ($\|v\|_2 = 1$) that is given in the data file 77Stocks.mat, and β is a parameter such that $r(\beta)$ is on the boundary of the β-standard deviation uncertainty set of r. Let us assume the actual covariance that the market eventually realizes is still Σ. So, the actual return of a portfolio is given by $r(\beta)^\top x^{(\alpha)}(\varrho)$ and its standard deviation is $\sqrt{w^{(\alpha)}(\varrho)^\top \Sigma w^{(\alpha)}(\varrho)}$. Let $\beta = 0, 0.2, 0.5$, plot the efficient frontier for $w^{(0)}(\varrho)$ and $w^{(0.2)}(\varrho)$ by sweeping t from 0 to $10\varrho^*$ for each given β. How does the efficient frontier change with α and β? Briefly comment on the result.

Exercise 11.2 *Transaction costs via the reversed Huber function*
In Section 11.3, we have already discussed some common criteria and constraints that may be added in order to adapt the portfolio optimization problem to a realistic setting. In this exercise, we see how to use the reversed Huber function to model transaction costs and market impact. We consider the following portfolio optimization problem:

$$\max_{w \in \mathcal{W}} \hat{r}^\top w - \gamma w^\top \Sigma w - \nu \cdot T(w - w^{(0)}), \quad (11.41)$$

where $\mathcal{W} = \{w \in \mathbb{R}^n : w \geq 0, \sum_{i=1}^n w_i = 1\}$. The function $T(\cdot)$ represents transaction costs and market impact, and $w^{(0)} \in \mathbb{R}^n$ is the vector of initial positions. The function T has the form

$$T(w - w^{(0)}) = \sum_{i=1}^n B_M(w - w^{(0)}),$$

where the function $B_M(\cdot)$ is piece-wise linear for small inputs and quadratic for large inputs. In this way, we seek to capture the fact that transaction costs are dominant for smaller trades, while market impact kicks in for larger ones. Precisely, we define B_M to be the so-called "reversed Huber" function with cut-off parameter M: for a scalar z,

$$B_M(z) \triangleq \begin{cases} |z| & \text{if } |z| \leq M, \\ \frac{z^2 + M^2}{2M} & \text{otherwise.} \end{cases}$$

1. Prove that the reversed Huber function is equivalent to the following optimization problem:

$$B_M(z) = \min_{t,v} v + t + \frac{t^2}{2M}: \ |z| \leq v+t, \ v \leq M, \ t \geq 0. \quad (11.42)$$

Is the reversed Huber function convex in z?

2. For given $w, w^{(0)} \in \mathbb{R}^n$, show that

$$T(w - w^{(0)}) = \min_{t,v} \mathbf{1}^T(v+t) + \frac{1}{2M} t^T t: \ v \leq M\mathbf{1}, \ t \geq 0, \ |w - w^{(0)}| \leq v + t,$$

where, $v, t \in \mathbb{R}^n$, and $\mathbf{1} \in \mathbb{R}^n$ is the vector of ones.

3. Now re-formulate the optimization problem in equation (11.41) as a convex optimziation problem. Which category does this problem belong to (LP, QP, SOCP, etc.)?

4. Load the data from 77Stocks.mat, compute the mean $\hat{r}$ and the covariance matrix Σ of the assets. Initial position $w^{(0)}$ is given in the data file. Let $\nu = 5 \times 10^{-5}$. Draw the efficient frontiers by sweeping the parameter γ from 0 to 10 with increments of 1 for each $M = 0.01, 0.02, 0.03, 0.1$. Also plot the efficient frontier without considering the transaction costs and market impact. Briefly comment on the result.

Exercise 11.3 *Portfolio optimization with transaction costs*

An investor's Dollar position in n assets is described by vector $w^{(0)} \in \mathbb{R}^n$, where $w_i^{(0)}$ is the current Dollar exposure in asset i. The investor wishes to update their portfolio by conducting transactions on the market. Let $w \in \mathbb{R}^n$ be the updated portfolio, and let $d_i \doteq w_i - w_i^{(0)}$ denote the transacted amount in the ith asset.

Each transaction is subject to a cost: for the ith transaction, we pay a rate of $c_+ \geq 0$ per-dollar on the amount we buy, a rate of $c_- \geq 0$ per-dollar on the amount we sell while we hold a positive amount of the asset, and a rate $c_{\text{short}} \geq c_-$ per-dollar on the amount we sell short. The total cost of transactions is thus $TC(d) = \sum_i f(d_i)$, where $f(d_i) \geq 0$ is the cost of the ith transaction. This is a generalization of the case we treated in Section 11.3.1, where we considered $c_{\text{short}} = c_-$.

We assume that we have an estimate $\hat{r} \in \mathbb{R}^n$ of the expected return of the n assets for the considered investment horizon. We also assume that the covariance matrix Σ of the returns is known.

Letting $W_{\text{ini}} \doteq \sum_{i=1}^n w_i^{(0)}$ denote the initial total wealth, we observe that the net expected return of our investment at the end of the period is

$$\rho(d) = \frac{\hat{r}^T w - TC(d)}{W_{\text{ini}}} = \text{const.} + \frac{\hat{r}^T d - TC(d)}{W_{\text{ini}}},$$

and the variance of the return is

$$w^\top \Sigma w = (d + w^{(0)})^\top \Sigma (d + w^{(0)}).$$

The problem we consider is to maximize the net expected return at the end of the period, while guaranteeing that the risk remains below a given bound, that is, $w^\top \Sigma w \leq \bar{\sigma}^2$, where $\bar{\sigma} \geq 0$ is given:

$$\max_{d} \quad \hat{r}^\top d - TC(d), \qquad (11.43)$$
$$\text{s.t.: } (d + w^{(0)})^\top \Sigma (d + w^{(0)}) \leq \bar{\sigma}^2.$$

1. Justify the fact that problem (11.43) can be rewritten in the following epigraphic form by introducing a new slack variable η:

$$\max_{d,\eta} \quad \hat{r}^\top d - \eta, \qquad (11.44)$$
$$\text{s.t.: } (d + w^{(0)})^\top \Sigma (d + w^{(0)}) \leq \bar{\sigma}^2,$$
$$TC(d) \leq \eta.$$

2. Prove the fact that the condition $TC(d) \leq \eta$ can be replaced by the conditions $\sum_{i=1}^{n} u_i \leq \eta$, $f(d_i) \leq u_i$, $i = 1, \ldots, n$, where u_i are additional slack variables. Then, express each constraint $f(d_i) \leq u_i$ as an equivalent set of linear inequalities in d_i, u_i. *Hint:* Express the transaction cost $f(d_i)$ as the point-wise maximum of three affine functions.

3. Using the previous findings, show that problem (11.44) can be expressed explicitly as a SOCP. *Hint:* Use the fact that $\Sigma \succeq 0$ can be factored as $\Sigma = \Sigma^{1/2} \Sigma^{1/2}$.

Exercise 11.4 *Budget constraint and self-financing portfolio*
We consider again the portfolio optimization problem (11.44) discussed in the previous exercise; the same definitions and notation apply here. We observe that in the previous formulation we imposed no a-priori limit on the budget to be invested in the portfolio. Also, the previous formulation implicitly assumed the presence of a cash account that we can use to pay for the transaction costs and/or to increase the positions in the assets.

Here, we take a different perspective, and assume the following:

(i) The portfolio vectors $w^{(0)} \in \mathbb{R}^{n+1}$ and $w \in \mathbb{R}^{n+1}$ now include cash as an additional asset. We assume without loss of generality that the cash position is the $(n+1)$th element of the portfolio.

(ii) Cash is risk free, and it bears a deterministic return $r_f > 0$ at the end of the investment period (the risk-free rate).

(iii) The investor has no access to other funds while managing the portfolio. This means that if the initial total wealth is $W_{ini} = \sum_{i=1}^{n+1} w_i^{(0)}$ and we update the portfolio according to

$$w_i = w_i^{(0)} + d_i, \quad i = 1, \ldots, n+1,$$

where d_i is the vector of transacted amounts, then the total wealth in w must be equal to the initial wealth W_{ini}, minus the transaction expenses. In other words, the new holdings and the transaction costs must be paid using the assets initially present in the portfolio, and no external injection of funds is allowed. We say in such case that the portfolio is *self-financing*.

(iv) The cash transactions involve no cost. That is, the total transaction cost is

$$TC(d) = \sum_{i=1}^{n} f(d_i),$$

where we notice that the summation stops at $i = n$, that is, it assumes no cost for the cash transaction d_{n+1}.

(v) We let $\tilde{r}^\top \doteq [\hat{r}^\top \; r_f] \in \mathbb{R}^{n+1}$, where $\hat{r} \in \mathbb{R}^n$ is the vector containing the expected returns of the risky assets, and $r_f > 0$ is the risk-free rate of return. We also denote by $\Sigma \in \mathbb{R}^{n,n}$ the covariance matrix of the risky returns.

1. Justify the fact that the self-financing condition imposes a "budget constraint" on the problem, which takes the form

$$\sum_{i=1}^{n+1} d_i + TC(d) = 0.$$

Problem (11.44) thus translates to the following optimization problem

$$\max_{d,\eta} \quad \tilde{r}^\top d - \eta, \qquad (11.45)$$

$$\text{s.t.:} \; (d + w^{(0)})^\top \begin{bmatrix} \Sigma & 0 \\ 0 & 0 \end{bmatrix} (\delta + w^{(0)}) \leq \bar{\sigma}^2,$$

$$TC(d) \leq \eta,$$

$$\sum_{i=1}^{n+1} d_i + TC(d) = 0.$$

Discuss about the convexity of this problem formulation.

2. Consider the "budget constraint" $\sum_i d_i + TC(d) = 0$. Justify the fact that in this problem we can replace this equality constraint with an inequality constraint of the form $\sum_i d_i + TC(d) \leq 0$. Further, prove that $\sum_i d_i + TC(d) \leq 0$ if and only if $TC(d) \leq \eta$ and $\sum_i d_i + \eta \leq 0$.

3. Assuming the transaction cost structure considered in the previous exercise, express problem (11.45) in the form of an equivalent SOCP.

Exercise 11.5 *Median risk*
We consider a single-period portfolio optimization problem with n assets. We use past samples, consisting of single-period return vectors $r^{(1)}, \ldots, r^{(m)}$, where $r^{(k)} \in \mathbb{R}^n$ contains the returns of the assets from period $k-1$ to period k. We denote with $\hat{r} \doteq (1/m)(r^{(1)} + \cdots + r^{(m)})$ the vector of sample averages; it is an estimate of the expected return, based on the past samples.

As a measure of risk, we use the following quantity. Denote by $\rho_k(w)$ the return at time t (if we had held the position w at that time). Our risk measure is

$$\mathcal{R}_1(w) \doteq \frac{1}{m} \sum_{k=1}^{m} |\rho_k(w) - \hat{\rho}(w)|,$$

where $\hat{\rho}(w)$ is the portfolio's sample average return.

1. Show that $\mathcal{R}_1(w) = \|R^\top w\|_1$, with R an $n \times m$ matrix that you will determine. Is the risk measure $\mathcal{R}_1$ convex?

2. Show how to minimize the risk measure $\mathcal{R}_1$, subject to the condition that the sample average of the portfolio return is greater than a target μ, using linear programming. Make sure to put the problem in standard form, and define precisely the variables and constraints.

3. Comment on the qualitative difference between the resulting portfolio, and one that would use the more classical, variance-based risk measure, given by

$$\mathcal{R}_2(w) \doteq \frac{1}{m} \sum_{k=1}^{m} (\rho_k(w) - \hat{\rho}(w))^2.$$

12
Beyond the mean/variance model

IN THIS CHAPTER we explore several aspects in portfolio allocation that go beyond the classical single-period mean/variance model discussed in Chapter 11. In particular, Section 12.1 discusses an alternative approach to portfolio design based on the concept of risk budgets and risk parity. Section 12.2 discusses the problem of designing portfolios that have as goal the replication (tracking) of a given index, with focus on the problem of replicating the index using few assets (sparse tracking). In Section 12.3 we tackle the issue of uncertainty in the data (the expected returns and covariance matrix), and discuss how to make the portfolio resilient, or robust, to such variations. Finally, in Section 12.4 we introduce financial decision problem that span over multiple time periods, and discuss optimization-based techniques for designing effective recourse policies that make the future decisions react dynamically to uncertainties.

12.1 Risk contributions and risk budgets

Due to the difficulty of estimating expected returns reliably,[1] it is interesting to consider the construction of portfolios whose composition does not depend on the expected returns. One example of such portfolios is the minimum variance portfolio derived in (11.10). Another one is the co-called "$1/n$" portfolio, in which the portfolios weights are simply $w_i = 1/n$, $i = 1,\ldots,n$. A systematic approach for determining portfolio allocations that are independent of the expected returns is to concentrate only on portfolio risk, and in particular on the individual contributions of each asset in the portfolio to the overall risk. If $\Sigma \in S_+^n$ is the returns' covariance matrix, and $w \in \mathbb{R}^n$ is a portfolio, the overall risk, in terms of the overall return variance, is

[1] See R. C. Merton, "On Estimating the Expected Return on the Market: An Exploratory Investigation," *Journal of Financial Economics*, vol. 8(4), pp. 323–361, 1980.

$$\sigma^2(w) = w^\top \Sigma w = \sum_{i=1}^{n} w_i (\Sigma w)_i,$$

where $(\Sigma w)_i$ denotes the ith element of vector Σw. Observe that

$$\frac{\partial \sigma^2(w)}{\partial w_i} = 2(\Sigma w)_i$$

denotes the sensitivity of the total risk with respect to w_i. Therefore, the quantity $(\Sigma w)_i$ gives (up to a proportionality constant) the change in volatility of the portfolio induced by a small increase in the weight of ith component. We then define

$$\varsigma_i(w) \doteq w_i (\Sigma w)_i$$

as the *risk contribution* of the ith asset to the overall risk, so that

$$\sigma^2(w) = \sum_{i=1}^{n} \varsigma_i(w).$$

A *risk parity* portfolio is a portfolio in which each asset gives equal contribution to overall risk, that is, such that

$$\varsigma_i(w) = \varsigma_j(w), \quad i,j = 1, \ldots, n.$$

If we express each risk contribution in the form of a fraction (budget) θ_i of the overall risk, that is,

$$\varsigma_i(w) \doteq \theta_i \sigma^2(w), \quad \sum_{i=1}^{n} \theta_i = 1, \ \theta_i \geq 0, \ i = 1, \ldots, n,$$

then the risk parity portfolio is one that satisfies the above conditions with equal budgets $\theta_i = 1/n$, $i = 1, \ldots, n$. Enforcing risk parity corresponds to imposing n equality constraints on w:

$$w_i (\Sigma w)_i = \frac{1}{n} w^\top \Sigma w, \quad i = 1, \ldots, n,$$

and, similarly, enforcing generic given risk budgets $\theta_i \geq 0$, $\sum_i \theta_i = 1$, corresponds to imposing

$$w_i (\Sigma w)_i = \theta_i w^\top \Sigma w, \quad i = 1, \ldots, n.$$

These equations form a system of n quadratic equations, whose solution is not obvious in general.

In the sequel, we shall be interested specifically in the case of long-only portfolios, for which we require that $w \geq 0$. The risk budgeting conditions, in this case, are written as

$$w_i (\Sigma w)_i = \theta_i w^\top \Sigma w, \quad w_i \geq 0, \quad i = 1, \ldots, n. \tag{12.1}$$

It can now be proved that the unique optimal solution of the convex minimization problem

$$\min_w f(w) \doteq \tfrac{1}{2} w^\top \Sigma w - \lambda \sum_{i=1}^n \theta_i \log w_i, \qquad (12.2)$$

where $\lambda > 0$ is an arbitrary constant, satisfies conditions (12.1). Indeed, the objective function $f(w)$ is strictly convex on the domain $w > 0$,[2] and it achieves its minimum at the point where its gradient vanishes, that is,

$$\frac{\partial f(w)}{\partial w_i} = (\Sigma w)_i - \lambda \frac{\theta_i}{w_i} = 0, \quad i = 1, \ldots, n.$$

This means that for the optimal w_i^* it holds that

$$\lambda \theta_i = w_i^* (\Sigma w^*)_i, \quad i = 1, \ldots, n, \qquad (12.3)$$

and, summing both sides and considering that $\sum_i \theta_i = 1$,

$$\lambda = \sum_{i=1}^n w_i^* (\Sigma w^*)_i = w^{*\top} \Sigma w^*,$$

thus substituting this λ expression into (12.3) shows that the risk budget conditions (12.1) hold for the optimal solution of (12.2). Observe that the conditions (12.1) are homogeneous, in the sense that if w is any portfolio that satisfies them, then also αw satisfies the conditions for any scalar $\alpha > 0$. Indeed, fixing a λ value in (12.2) corresponds to fixing a specific normalization for the portfolio vector w; in particular, w^* gets normalized so that its total risk is equal to λ. Once we have $w^* > 0$ from the optimal solution of (12.2), we can also normalize it as

$$z^* = \frac{w^*}{\mathbf{1}^\top w^*},$$

to obtain a portfolio $z^* > 0$ which satisfies the risk budget equations (12.1) as well as the cash budget normalization $\mathbf{1}^\top z^* = 1$. Notice further that the conditions (12.1) fully define the portfolio composition, up to a scaling. Such binding conditions, therefore, do not allow for much flexibility in the design, since the introduction of an objective function (e.g., related to expected return, to be maximized under the risk budget constraints) will have no impact on the solution, while the addition of constraints would make the problem infeasible. We discuss in the next section how to overcome this issue.

[2] Observe that the Hessian matrix of f is

$$\nabla^2 f(w) = \Sigma + \lambda \mathrm{diag}\left(\frac{\theta_1}{w_1^2}, \ldots, \frac{\theta_n}{w_n^2}\right),$$

which is positive definite on the domain $w > 0$, for $\theta > 0$. This means that f is strictly convex, and hence its minimum is unique.

12.1.1 Approximate risk budgeting

We first note that the inequality conditions

$$w_i (\Sigma w)_i \geq \theta_i w^\top \Sigma w, \quad w_i \geq 0, \quad i = 1, \ldots, n, \qquad (12.4)$$

are equivalent to the equality conditions in (12.1). A proof of this fact is easily obtained by contradiction: if we assume that there exist a strict inequality $w_j(\Sigma w)_j > \theta_j w^\top \Sigma w$ in (12.4), then summing over i and recalling that $\sum_{i=1}^n \theta_i = 1$, we would get

$$w^\top \Sigma w = \sum_{i=1}^n w_i(\Sigma w)_i > w^\top \Sigma w \sum_{i=1}^n \theta_i = w^\top \Sigma w,$$

which would be a contradiction. Observe also that, since $\theta_i w^\top \Sigma w \geq 0$ and $w_i \geq 0$, conditions (12.4) imply that $(\Sigma w)_i \geq 0$ for $i = 1, \ldots, n$. Thus, letting $z_i \doteq (\Sigma w)_i \geq 0$ and $\Sigma = LL^\top$, we see that (12.4) have the form of hyperbolic constraints

$$w_i z_i \geq \|\sqrt{\theta_i} L^\top w\|_2^2, \ w_i \geq 0, z_i \geq 0, \quad i = 1, \ldots, n.$$

Using (10.12) these can be converted into the equivalent convex SOC constraints: for $i = 1, \ldots, n$,

$$\left\| \begin{bmatrix} 2\sqrt{\theta_i} L^\top w \\ w_i - (\Sigma w)_i \end{bmatrix} \right\|_2 \leq w_i + (\Sigma w)_i, \quad (\Sigma w)_i \geq 0, w_i \geq 0. \tag{12.5}$$

The above reasoning proves that (12.5) are equivalent to (12.4), which are in turn equivalent to the original risk budget conditions in (12.1), and all of these conditions are binding, that is, they fully define the portfolio, up to scaling. One way for obtaining more flexibility from the model is to allow *controlled violations* of the budget constraints. In particular, we may introduce violation variables $v_i \geq 0$ in (12.5),

$$\left\| \begin{bmatrix} 2\sqrt{\theta_i} L^\top w \\ w_i - (\Sigma w)_i \end{bmatrix} \right\|_2 \leq w_i + (\Sigma w)_i + v_i, \tag{12.6}$$

and then use the mean violation $V = \frac{1}{n} \sum_{i=1}^n v_i$ as a penalty term in a minimization objective. In this way, one may re-introduce in the picture the classical expected return and risk tradeoff, together with risk budget requirements. For instance, we may consider a problem in the form

$$\max_{w,v} \ \hat{r}^\top w - \gamma w^\top \Sigma w - \eta \frac{1}{n} \sum_{i=1}^n v_i, \tag{12.7}$$

$$\text{s.t.:} \ w \geq 0, \ \mathbf{1}^\top w = 1,$$
$$\left\| \begin{bmatrix} 2\sqrt{\theta_i} L^\top w \\ w_i - (\Sigma w)_i \end{bmatrix} \right\|_2 \leq w_i + (\Sigma w)_i + v_i, \quad i = 1, \ldots, n,$$
$$(\Sigma w)_i \geq 0, \ v_i \geq 0, \quad i = 1, \ldots, n,$$

where $\gamma \geq 0$ controls the tradeoff between return and risk, and $\eta \geq 0$ controls the amount of admissible violation of the given risk budgets.

Example 12.1 (*Risk parity portfolio with DJI components*) As a numerical example, we considered daily price data from $n = 30$ components of the Dow Jones Industrial Index, from Oct. 11, 2005 to Oct. 9, 2006, for a total of 251 data points. From this data we computed the daily simple returns and estimated their sample mean return vector and plain sample covariance matrix. Then, we solved problem (12.7) with $\theta_i = 1/n$, $\gamma = 10$, and $\eta = 70$. The resulting portfolio composition is shown in Figure 12.1(a); the individual risk budgets are shown in Figure 12.1(b). We observe that risk parity is achieved exactly for most of the contributions, and approximately for some of them.

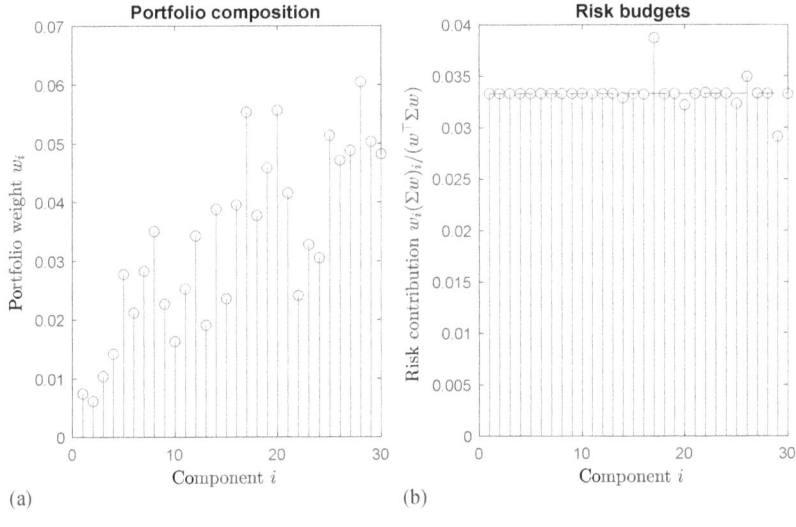

Figure 12.1 (a) Portfolio composition for the DJI risk parity example. (b) Risk contributions; see code file risk_parity_port_test.m in the online resources.

12.2 Index tracking

A financial index represents a quantitative proxy for the status of an economic sector (e.g., commodities, oil, technology, etc.), of groups of financial instruments (e.g., equities, mutual funds), of a geographic region, and so on. For instance, the Standard & Poor's 500 (S&P 500) is a stock market index that tracks the performance of 500 large companies listed on US stock exchanges. Often, professional investors aim at constructing portfolios that replicate the behavior of an index. Replicating the S&P 500 exactly, however, would require managing all the 500 component assets with the proper weights, which may be inconvenient. The *index tracking* problem refers to the problem of finding simplified portfolios that approximately replicate the behavior of a given index.

Suppose we are given T samples of the returns $y_k \in \mathbb{R}$ of the index of interest at time periods $k = 1, \ldots, T$, as well as the T samples of the returns $r(1), \ldots, r(T) \in \mathbb{R}^n$ of a universe of n assets, where $r_i(k)$ denotes the return of the ith asset over the kth period. If $w \in \mathbb{R}^n$ is a portfolio composition vector, then the portfolio return at k is given by $\rho(k) \doteq w^\top r(k)$. The tracking error at k between the portfolio return and the index return is $e(k) \doteq y_k - \rho(k) = y_k - w^\top r(k)$. Considering the total quadratic error, the tracking problem can be cast as

$$\min_{w \in \mathcal{W}} \|R^\top w - y\|_2^2, \qquad (12.8)$$

where $y = [y_1 \cdots y_T]^\top$ is the vector of the index returns, and

$$R \doteq [r(1) \cdots r(T)] \in \mathbb{R}^{n,T}$$

is the matrix containing by columns the assets' returns. The set $\mathcal{W}$ allows to specify the constraints on the portfolio composition, such as, for instance, $w \geq 0$, $\mathbf{1}^\top w = 1$. With such standard constraints, problem (12.8) is explicitly written as a convex QP

$$\min_{w \in \mathbb{R}^n} \quad w^\top R R^\top w - 2w^\top R y + \|y\|_2^2, \qquad (12.9)$$

$$\text{s.t.:} \quad w \geq 0, \quad \mathbf{1}^\top w = 1.$$

While the above approach is based directly on samples of the returns (i.e., it is a direct, data-driven, optimization approach), an alternative indirect approach is based on a stochastic model of the returns and index. If we assume that the assets return vector r and index return v are random with known mean and covariance

$$\mathrm{E}\left\{\begin{bmatrix} r \\ v \end{bmatrix}\right\} = \begin{bmatrix} \hat{r} \\ \hat{v} \end{bmatrix},$$

$$\mathrm{var}\left\{\begin{bmatrix} r \\ v \end{bmatrix}\right\} = \begin{bmatrix} \Sigma & c \\ c^\top & s^2 \end{bmatrix},$$

then the objective would be to minimize the expected value of the squared tracking error $e \doteq w^\top r - v$, that is,

$$\mathrm{E}\{e^2\} = \mathrm{E}\{(w^\top r - v)^2\} = \mathrm{E}\{(w^\top (r - \hat{r}) - (v - \hat{v}) + (w^\top \hat{r} - \hat{v}))^2\}$$
$$= w^\top \Sigma w - 2w^\top c + (w^\top \hat{r} - \hat{v})^2$$
$$= w^\top (\Sigma + \hat{r}\hat{r}^\top) w - 2w^\top (c + \hat{v}\hat{r}) + \hat{v}^2.$$

With the standard constraints $w \geq 0$, $\mathbf{1}^\top w = 1$, the problem is again a convex QP

$$\min_{w \in \mathbb{R}^n} \quad w^\top (\Sigma + \hat{r}\hat{r}^\top) w - 2w^\top (c + \hat{v}\hat{r}), \qquad (12.10)$$

$$\text{s.t.:} \quad w \geq 0, \quad \mathbf{1}^\top w = 1.$$

12.2.1 Sparse tracking

An interesting version of the index tracking problem arises when we impose sparsity on the portfolio vector w. Tracking an index via a sparse portfolio is indeed attractive since it only requires management of a small number of assets. The basic formulation of the problem is

$$\min_{w \in \mathbb{R}^n} \quad \|R^\top w - y\|_2^2, \qquad (12.11)$$
$$\text{s.t.:} \quad w \geq 0, \quad \mathbf{1}^\top w = 1,$$
$$\|w\|_0 \leq k,$$

where $k < n$ is a given upper bound on the cardinality of the portfolio. This problem formulation is nonconvex, due to the presence of the cardinality constraint. Further, we observe that the usual convex relaxation that entails plain substitution of the ℓ_1 norm in place of the cardinality function has no effect in the present case since, due to the constraints in the problem, we simply have $\|w\|_1 = 1$, that is, the ℓ_1 norm of w is constant. However, we observe that for all w it holds that

$$\|w\|_1 = \sum_{i=1}^n |w_i| = \sum_{i:\, w_i \neq 0} |w_i|$$
$$\leq \sum_{i:\, w_i \neq 0} \|w\|_\infty = \|w\|_0 \|w\|_\infty.$$

Therefore,

$$\|w\|_0 \leq k \quad \Rightarrow \quad \frac{\|w\|_1}{\|w\|_\infty} \leq k,$$

which means that the constraint $\|w\|_1 / \|w\|_\infty \leq k$ is a *relaxation* of the original constraint $\|w\|_0 \leq k$. The relaxed version of problem (12.11) is therefore

$$\min_{w \in \mathbb{R}^n} \quad \|R^\top w - y\|_2^2, \qquad (12.12)$$
$$\text{s.t.:} \quad w \geq 0, \quad \mathbf{1}^\top w = 1,$$
$$\|w\|_\infty \geq \frac{1}{k}.$$

Note that this latter problem formulation is still nonconvex, due to the constraint $\|w\|_\infty \geq 1/k$. However, under the condition $w \geq 0$, we have that

$$\|w\|_\infty \geq 1/k \quad \Leftrightarrow \quad \max_{i=1,\ldots,n} w_i \geq 1/k \quad \Leftrightarrow \quad w_i \geq 1/k \text{ for some } i.$$

Since we do not know which one of the w_is should be $\geq 1/k$, we try them all by solving a collection of n convex QP problems (P_i), $i = 1, \ldots, n$,

$$(P_i): \quad p_i^* = \min_{w \in \mathbb{R}^n} \ \|R^\top w - y\|_2^2, \tag{12.13}$$
$$\text{s.t.:} \quad w \geq 0, \quad \mathbf{1}^\top w = 1,$$
$$w_i \geq \frac{1}{k}.$$

Finally, we select the index i corresponding to the smallest p_i^*, and the solution to such (P_i) is the optimal solution of (12.12). In this case, the nonconvex problem (12.12) could be solved exactly and efficiently, at the price of solving the n convex problems (P_i), for $i = 1, \ldots, n$. We stress, however, that (12.12) is still only a relaxation of the original problem (12.11), which remains hard to solve exactly.

The above approach should promote sparse solutions. However, if the solution to (12.13) still happens to be non-sparse, we can resort to the following iterative greedy approach: after a first round of solutions to (12.13) we find the index i_1 corresponding to the best objective value $p_{i_1}^*$, and fix w_{i_1} to its optimal value $w_{i_1}^*$. With such $w_{i_1} = w_{i_1}^*$ fixed, we solve a second round of a suitably adjusted version of problem (12.13), optimizing over all variables except w_{i_1}, and we so find the second index i_2 and the second optimal variable $w_{i_2} = w_{i_2}^*$. We proceed this way up to k times, after which we have the nonzero values $w_{i_1}^*, \ldots, w_{i_k}^*$, and we set the values of all other indices to zero. The goal of the described procedure is to find the support (i.e., the indices of the nonzero values) of the solution, which is given by the indices $i_1, \ldots, i_k$. Finally, we can minimize $\|R^\top w - y\|_2^2$ only over the support variables, the rest being fixed to zero. We obtain in this way a sparse sub-optimal solution to problem (12.11).

Example 12.2 (*Sparse proxy of the DJI index*) For a numerical example, we considered as the reference index the Dow Jones Industrial Average (DJI) index, which reflects the course of 30 prominent companies exchanged at the New York Stock Exchange. The data consists of 251 daily returns of each of the 30 component assets and of the DJI index itself, in the period from Oct. 11, 2005 to Oct. 9, 2006, see Figure 12.2(a). By running the above heuristics,[3] we computed a sparse portfolio with $k = 6$ components which approximates the behavior of the index, see Figure 12.2(b). The six nonzero components of the proxy are AIG, BA, IBM, JPM, KO, and XOM, with weights 0.1614, 0.1322, 0.1698, 0.1590, 0.2660, and 0.1116, respectively.

[3] See `sparse_index_tracking_DJI.m` in the online resources.

12.3 Robust portfolio optimization

We next discuss mean/variance portfolio optimization in the presence of uncertainty on the problem data. As extensively discussed in Chapter 11, the basic mean/variance tradeoff problem

12.3 ROBUST PORTFOLIO OPTIMIZATION

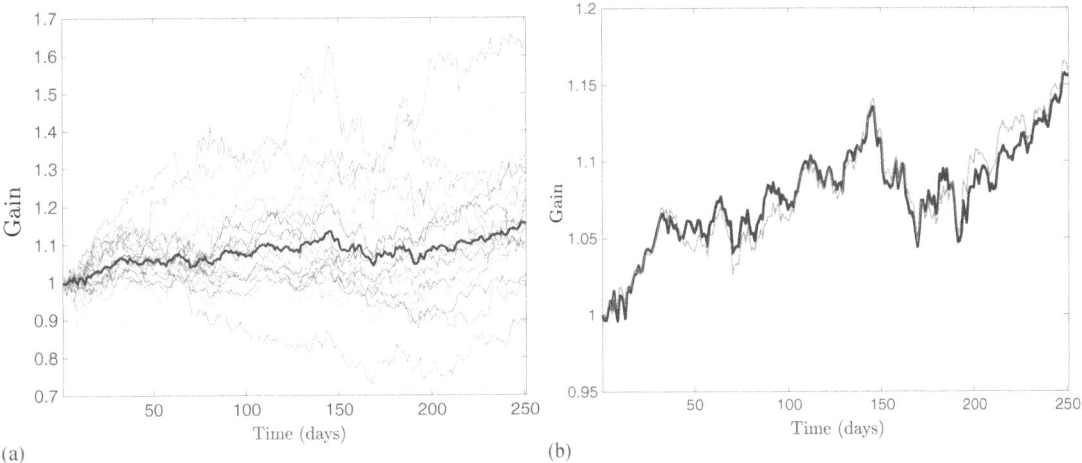

Figure 12.2 (a) Component assets and DJI index. (b) DJI index (thicker line) and its six-components sparse proxy.

$$\max_{w \in \mathcal{W}} \hat{r}^\top w - \gamma w^\top \Sigma w$$

hinges on the knowledge of the vector $\hat{r}$ of expected returns and of the covariance matrix Σ. In practice, these quantities are not known exactly, and need be estimated from data and/or expert knowledge. Since estimation is inevitably inexact, it is of crucial importance to take into account the uncertainty on the value of the parameters in the design of the optimal portfolio. In doing so, we shall take the approach of robust optimization discussed in Section 10.4 and assume that the actual expected return vector ranges over a given uncertainty set $\mathcal{U}_r$ around the nominal value $\hat{r}$, and the actual covariance matrix ranges over a given uncertainty set $\mathcal{U}_\Sigma$ around the nominal value Σ. Given the nominal values and the uncertainty sets, we may cast our robust portfolio optimization problem in the following min–max setting:

$$\max_{w \in \mathcal{W}} \min_{r \in \mathcal{U}_r, S \in \mathcal{U}_\Sigma} r^\top w - \gamma w^\top S w. \quad (12.14)$$

For simplicity of the discussion and notation, we next treat separately the cases of uncertainty on r and on S; the case of simultataneous uncertainty can be dealt with by simply summing the respective worst-case values of $r^\top w$ and of $-\gamma w^\top S w$.

12.3.1 Uncertainty on the expected return vector

We discuss the case of uncertainty on the expected return vector first. Different models are possible, depending on the description of the uncertainty set $\mathcal{U}_r$. We next overview the most common models.

Scenario uncertainty In the scenario uncertainty model, $\mathcal{U}_r = \{r^{(1)}, \ldots, r^{(K)}\}$ is simply the collection of possible realizations, or scenarios, of the expected return vector, see, for example, Figure 12.3.

The robust portfolio problem (12.14) becomes in this case

$$\max_{w \in \mathcal{W}, t} \; t - \gamma w^\top S w, \quad \text{s.t.:} \; r^{(i)\top} w \geq t, \; i = 1, \ldots, K, \tag{12.15}$$

which is still an easily solvable convex quadratic program.

Interval uncertainty In the interval uncertainty model we take $\mathcal{U}_r = \{r = \hat{r} + u, |u| \leq \rho\}$, where $\hat{r}$ is the nominal return vector and ρ is a vector of interval widths, see, for example, Figure 12.4.

The robust portfolio problem (12.14) becomes in this case (see Section 10.4.1)

$$\max_{w \in \mathcal{W}} \; \hat{r}^\top w - \rho^\top |x| - \gamma w^\top S w. \tag{12.16}$$

This problem can be cast as a convex QP, by introducing a slack vector variable ζ to replace $|x|$, and adding the linear inequality constraints $-\zeta \leq x \leq \zeta$.

Spherical uncertainty In the spherical uncertainty uncertainty model we take $\mathcal{U}_r = \{r = \hat{r} + \rho u, \|u\|_2 \leq \rho\}$, where $\hat{r}$ is the nominal return vector and $\rho \geq 0$ is the radius of the uncertainty sphere, see, for example, Figure 12.5.

The robust portfolio problem (12.14) becomes in this case (see Section 10.4.1)

$$\max_{w \in \mathcal{W}} \; \hat{r}^\top w - \rho \|x\|_2 - \gamma w^\top S w. \tag{12.17}$$

This problem can be cast as an SOCP by considering the square-root factorization $S = S^{1/2} S^{1/2}$ and writing the above as

$$\max_{w \in \mathcal{W}, t_1, t_2, \tau} \; \hat{r}^\top w - \rho t_1 - \gamma \tau,$$
$$\text{s.t.:} \quad \|w\|_2 \leq t_1,$$
$$\|S^{1/2} w\|_2 \leq t_2,$$
$$\left\| \begin{bmatrix} 2 t_2 \\ \tau - 1 \end{bmatrix} \right\|_2 \leq \tau + 1,$$

where the last constraint is the second-order cone representation of the constraint $t_2^2 \leq \tau$.

Mahalanobis uncertainty In the Mahalanobis uncertainty model we take $\mathcal{U}_r = \{r : (r - \hat{r})^\top C^{-1} (r - \hat{r}) \leq \rho\}$, where $\hat{r}$ is the nominal return vector, $\rho \geq 0$ modulates the size of the uncertainty set, and C is a

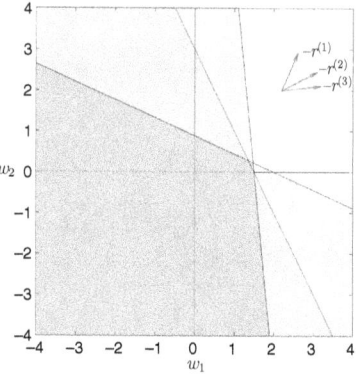

Figure 12.3 Example of the region where $r^{(i)\top} w \geq 1$, for scenarios $i = 1, 2, 3$.

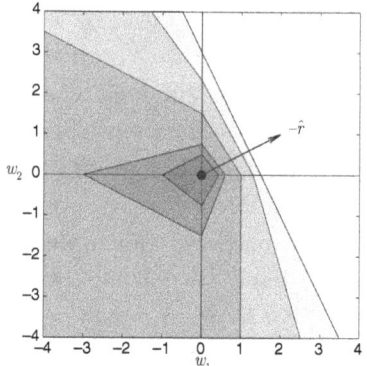

Figure 12.4 Example of the region where $r^\top w \geq 1$, for all r in the interval uncertainty set $\mathcal{U}_r$, with $\rho = \varrho \mathbf{1}$ and different levels of ϱ.

given positive semidefinite matrix. The set $\mathcal{U}_r$ is an ellipsoid, which can be represented equivalently as $\mathcal{U}_r = \{r = \hat{r} + Ru, \ \|u\|_2 \leq \rho\}$, where R is such that $C = RR^\top$. The robust portfolio problem (12.14) becomes in this case

$$\max_{w \in \mathcal{W}} \hat{r}^\top w - \rho \|R^\top x\|_2 - \gamma w^\top S w, \qquad (12.18)$$

which can be cast as an SOCP in a way analogous to (12.17).

12.3.2 Uncertainty on the covariance matrix

We next consider the case when the uncertainty acts on the covariance matrix of the returns. In this case, we rewrite problem (12.14) as

$$\max_{w \in \mathcal{W}} r^\top w - \gamma \sigma^2_{\text{wc}}(w), \qquad (12.19)$$

where $\sigma^2_{\text{wc}}(w)$ is the worst-case variance of portfolio w, defined as

$$\sigma^2_{\text{wc}}(w) \doteq \max_{S \in \mathcal{U}_\Sigma} w^\top S w. \qquad (12.20)$$

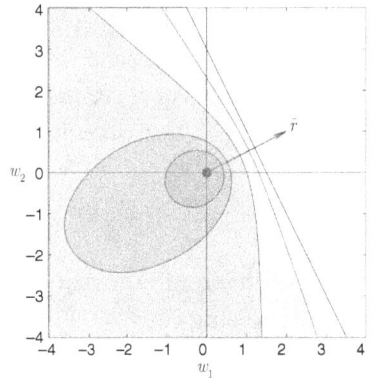

Figure 12.5 Example of the region where $r^\top w \geq 1$, for all r in the spherical uncertainty set $\mathcal{U}_r$, for different uncertainty radii ρ.

We observe that, in order for $w^\top S w$ to correctly represent a variance, we must impose that $S \succeq 0$ for all $S \in \mathcal{U}_\Sigma$. Thus, $\mathcal{U}_\Sigma$ is usually defined as the intersection of an uncertainty set (e.g., norm bounded) with the positive semidefinite cone. Notice also that, as for $S \succeq 0$ the quadratic form $w^\top S w$ is convex in w, then the function $\sigma^2_{\text{wc}}(w)$ is also convex, since it is the pointwise maximum of convex functions over $S \in \mathcal{U}_\Sigma$.

Scenario uncertainty In the scenario uncertainty model, $\mathcal{U}_\Sigma = \{\Sigma^{(1)}, \ldots, \Sigma^{(K)}\}$ is a collection of several possible instances of the covariance matrix. We have that $\sigma^2_{\text{wc}}(w) = \max_{i=1,\ldots,K} w^\top S^{(i)} w$, and the robust portfolio problem (12.14) can be formulated as

$$\max_{w \in \mathcal{W}, t} \hat{r}^\top w - \gamma t, \ \text{s.t.: } w^\top S^{(i)} w \leq t, \ i = 1, \ldots, K, \qquad (12.21)$$

which is a convex program with quadratic constraints. The problem can also be cast as an SOCP by using the factorizations $S^{(i)} = R^{(i)\top} R^{(i)}, \ i = 1, \ldots, K$:

$$\max_{w \in \mathcal{W}, t, \tau} \hat{r}^\top w - \gamma \tau,$$
$$\text{s.t.: } \|R^{(i)} w\|_2 \leq t, \ i = 1, \ldots, K,$$
$$\left\| \begin{bmatrix} 2\tau \\ t-1 \end{bmatrix} \right\|_2 \leq t + 1,$$

where the last constraint is equivalent to $\tau^2 \leq t$.

Norm-bounded uncertainty In the norm-bounded uncertainty model we consider $\mathcal{U}_\Sigma = \{S = \Sigma + \Delta, \|\Delta\| \leq \rho, S \succeq 0\}$, where $\Sigma \succeq 0$ is the nominal covariance matrix, and the matrix norm on Δ is either the Frobenius norm or the maximum singular value norm. Following the approach described in Section 10.4.2 we find that in this case

$$\sigma^2_{\text{wc}}(w) = \max_{S \in \mathcal{U}_\Sigma} w^\top S w = w^\top (\Sigma + \rho I) w.$$

Therefore, the robust portfolio problem (12.14) is formulated as the following convex QP:

$$\max_{w \in \mathcal{W}, t} \hat{r}^\top w - \gamma w^\top (\Sigma + \rho I) w. \tag{12.22}$$

Norm-bounded uncertainty on the precision matrix The precision matrix is defined as the inverse of the covariance matrix. Suppose that the nominal covariance matrix is $\Sigma \succ 0$, and let $P = \Sigma^{-1}$ be the corresponding nominal precision matrix. We consider an uncertainty set based on norm-bounded perturbations of the precision matrix:

$$\mathcal{U}_\Sigma = \{S \colon S^{-1} = P + \Delta, P + \Delta \succ 0, \|P^{-1/2} \Delta P^{-1/2}\|_2 \leq \rho < 1\},$$

where $\|\cdot\|_2$ denotes the maximum singular value (spectral) norm. Such a seemingly unnatural uncertainty description actually corresponds to the confidence region that arises when using a maximum likelihood procedure for estimating the covariance matrix. By defining $\tilde{\Delta} = P^{-1/2} \Delta P^{-1/2}$ we have that the above uncertainty set can also be expressed equivalently as

$$\mathcal{U}_\Sigma = \{S \colon S^{-1} = P^{1/2}(I + \tilde{\Delta}) P^{1/2}, I + \tilde{\Delta} \succ 0, \|\tilde{\Delta}\|_2 \leq \rho < 1\}.$$

Considering that $\tilde{\Delta}$ is symmetric and can be factored as $\tilde{\Delta} = U \Lambda U^\top$, where $\Lambda = \text{diag}(\lambda_1, \ldots, \lambda_n)$ and U is an orthogonal matrix, for each w we have that

$$\begin{aligned} \sigma^2(w) &= w^\top S w = w^\top P^{-1/2}(I + \tilde{\Delta})^{-1} P^{-1/2} w \\ &= w^\top P^{-1/2} U (I + \Lambda)^{-1} U^\top P^{-1/2} w \\ &= \sum_{i=1}^n \frac{z_i^2}{1 + \lambda_i(\tilde{\Delta})}, \end{aligned}$$

where $z \doteq U^\top P^{-1/2} w$. Observe that $\|\tilde{\Delta}\|_2 \leq \rho < 1$ implies that $|\lambda_i(\tilde{\Delta})| \leq \rho < 1$ for all i, and then that $\sigma^2(w)$ is maximized when all λ_i are at their minimal value, that is, for $\lambda_i(\tilde{\Delta}) = -\rho$, that is, for $\tilde{\Delta} = -\rho I$, for which we obtain

$$\sigma_{\text{wc}}^2(w) = \max_{S \in \mathcal{U}_\Sigma} w^\top S w = \sum_{i=1}^n \frac{z_i^2}{1-\rho}$$
$$= \frac{1}{1-\rho}\|z\|_2^2 = \frac{1}{1-\rho}\|U^\top P^{-1/2} w\|_2^2$$
$$= \frac{1}{1-\rho}\|P^{-1/2} w\|_2^2 = \frac{1}{1-\rho} w^\top \Sigma w.$$

The robust portfolio problem (12.14) is hence formulated as the following convex QP:

$$\max_{w \in \mathcal{W}} \hat{r}^\top w - \frac{\gamma}{1-\rho} w^\top \Sigma w. \qquad (12.23)$$

12.4 Multi-period decision problems

In this section we discuss the problem of taking decisions over a time horizon composed of many periods, as opposed to the single-period approach discussed so far. In the single-period approach an investor decides their allocation at $t = 0$ by looking at the expected performance of their decision at $t = 1$. At $t = 1$ they may repeat the operation looking at the performance at $t = 2$, and so on as already described in Section 11.6. Even if the process can be repeated indefinitely in a sliding-horizon fashion, each decision at t is made according to a *myopic* view that only takes into account what happens after one step. A multi-stage decision problem considers instead an horizon composed of $T > 1$ periods, and computes at time $t = 0$ the whole sequence of decisions for $t = 0, 1, \ldots, T-1$, including in the objective the final performance at T. A suggestive similitude can be made with the game of chess, whereby a single-period myopic player decides their move by maximizing their position on the checkerboard after one round, considering all possible counter-moves the opponent can make; a multi-stage player will instead solve the much more complex problem of maximizing their position considering $T > 1$ rounds ahead, by taking into account all successive moves and counter-moves.

While a multi-stage approach solved at $t = 0$ over an horizon $T > 1$ will likely provide a better decision at $t = 0$ with respect to a single-stage approach, one should refrain from applying in reality the computed decisions at later steps $t = 1, \ldots, T-1$. This is due to the fact that while at $t = 0$ the whole future is unknown, when we arrive at $t = 1$ (and we "played" our decision at $t = 0$) we actually saw what happened during the first period. Therefore, at $t = 1$ it would be quite inefficient to disregard this information and play the decision for $t = 1$ that we computed at $t = 0$! It is instead intuitively better to

324 12 BEYOND THE MEAN/VARIANCE MODEL

"reconsider" our decision on the basis of what just happened during the first period. In the chess example, a good player would carefully think about their sequence of moves and predicted counter-moves, and then play their first move in the sequence at $t = 0$; at $t = 1$ they will most likely *not* play the second move originally devised, but will instead reconsider their strategy on the basis of the *actual* move that the opponent played in the first round. In such dynamic approach, rather than optimizing over a fixed sequence of decisions, we should devise optimal strategies (or *policies*) that indicate how to replan in response to observed events (planning to re-plan).

Example 12.3 (*Nominal, robust and dynamic decisions.*) In this example we highlight the different approaches to multi-stage decisions under uncertainty. We consider the example of short-term financial planning described in Example 10.2. That example concerned a problem of decisions over $T = 6$ time periods (months) in a nominal setting, that is, under no uncertainty. The nominal problem was formulated as an LP in vector variables x, y, z:[4]

$$\max_{x,y,z}: z_6: A_x x + A_y y + A_z z \leq b, \ 0 \leq x \leq 100, \ x \geq 0, \ z \geq 0,$$

where $b \in \mathbb{R}^6$ represented the given nominal stream of expected payments and liabilities:

$$b^\top = \hat{b}^\top \doteq \begin{bmatrix} 150 & 100 & -200 & 200 & -50 & -300 \end{bmatrix}.$$

We next consider the case when the liability vector b is subject to uncertainty. Specifically, at each time t, we assume that the liability is

$$b_t = \hat{b}_t(1 + 0.06 u_t + 0.02 u_{t-1})$$

for some values u_t, u_{t-1} that are only known to be in $[-1, 1]$, and where $\hat{b}$ contains the nominal liability values. This models *relative* errors in b_t, which depends on noise at times t and $t - 1$. The liability vector, in function of the uncertainty, can now be expressed as

$$b(u) = \hat{b} + Bu, \ \|u\|_\infty \leq 1,$$

where $u \in \mathbb{R}^6$, and

$$B = \text{diag}\left(\hat{b}\right) \begin{bmatrix} 0.06 & 0 & 0 & 0 & 0 & 0 \\ 0.02 & 0.06 & 0 & 0 & 0 & 0 \\ 0 & 0.02 & 0.06 & 0 & 0 & 0 \\ 0 & 0 & 0.02 & 0.06 & 0 & 0 \\ 0 & 0 & 0 & 0.02 & 0.06 & 0 \\ 0 & 0 & 0 & 0 & 0.02 & 0.06 \end{bmatrix}.$$

We can then follow a robust approach and look for the sequence of decisions that are worst-case optimal. This means solving the robust optimization problem

[4] See short_term_financing.m in the online resources.

$$\max_{x,y,z} c^\top v,$$
$$\text{s.t.:} \quad Av \geq \hat{b} + Bu, \ \forall u\colon \|u\|_\infty \leq 1,$$
$$v = \begin{bmatrix} x \\ y \\ z \end{bmatrix} \geq 0, \ x \leq 100,$$

where $A = [A_x \ A_y \ A_z]$ and $c^\top = [0 \cdots 0 1]$. This problem is a robust LP with interval uncertainty, treated in Section 10.4.1. Each row in the inequality $Av \geq \hat{b} + Bu$ is given by $A_t^\top v \geq \hat{b}_t + B_t^\top u$, where $A_t^\top$, $B_t^\top$ denote the t-th row of A and B, respectively, for $t = 1,\ldots,6$. Each of these t inequalities is satisfied for all u such that $|u| \leq 1$ if

$$A_t^\top v \geq \hat{b}_t + \max_{|u|\leq 1} B_t^\top u = \hat{b}_t + \|B_t\|_1.$$

The robust version of the problem can therefore be written explicitly as

$$\max_{x,y,z} c^\top v,$$
$$\text{s.t.:} \quad A^\top v \geq \hat{b} + |B|\mathbf{1},$$
$$v = \begin{bmatrix} x \\ y \\ z \end{bmatrix} \geq 0, \ x \leq 100,$$

where $|B|$ denotes the matrix of absolute values of the elements of B. Solving the above LP we find the optimal robust solution[5]

[5] See short_term_financing_robust.m in the online resources.

month	x	y	z
1	0.00	159	0.00
2	62.90	45.10	0.00
3	0.00	256.58	377.05
4	0.00	—	0.00
5	0.00	—	0.00
6	—	—	14.29

which can be compared to the nominal solution in Example 10.2: the robust solution is substantially worse than the nominal one in terms of terminal wealth, which goes from 92.5 in the nominal case to 14.29 in the robust case. However, the terminal wealth delivered by the robust solution has to be compared to the one delivered by the nominal one *when uncertainty is present*. Indeed, with uncertainty, the nominal policy is a failure, since the cash-flow requirements are simply not met!

Observe further that the robust solution is based on a multi-stage problem formulation; however, it is a *static* solution, in the sense that the whole sequence of actions is computed at the initial time and is intended to be executed as is, without modifications. As already argued before, such an approach is potentially inefficient, since it neglects the fact that not all decisions need be made at the initial time (here-and-now variables). In the current example, one may apply the initial decisions x_1, y_1, z_1 and then wait-and-see what the actual outcome of the uncertainty is in the first step and let the future decisions depend on such outcome. In this approach rather than computing a solution we seek for a *policy* that prescribes how to replan the decisions in reaction to the uncertainty outcomes. Determining optimal policies, in general, is a formidable problem. However, we can find sub-optimal practical approaches by fixing in advance a simple functional structure for the policy.

In this example, for instance, we let the decisions be strictly causal and affine functions of the uncertainty, that is, we fix policies of the form

$$x(u) = x + Xu, \quad y(u) = y + Yu, \quad z(u) = z + Zu,$$

where X, Y, Z are strictly lower-triangular matrices, so that x_t, y_t, z_t depend only on $(u_1, \ldots, u_{t-1})$. Here, the variables are x, y, z, and X, Y, Z. The above structure is called an *affine recourse* policy. We can write $v = (x, y, z)$ as a function of u:

$$v(u) = v + Vu, \quad V = \begin{bmatrix} X \\ Y \\ Z \end{bmatrix},$$

where v is a vector, and V is a matrix, whose structure is constrained by the triangular structure of X, Y, Z. The robust problem, under the affine recourse policy, can now be written as

$$\begin{aligned}
\max_{x,y,z,X,Y,Z} \min_{u:\, \|u\|_\infty \leq 1} \quad & c^\top(v + Vu), \\
\text{s.t.:} \quad & X, Y, Z \text{ strictly lower-triangular}, \\
& A(v + Vu) \geq \hat{b} + Bu, \quad \forall u, \ \|u\|_\infty \leq 1, \\
& v + Vu \geq 0, \ x + Xu \leq 100.
\end{aligned}$$

This is again a robust LP with interval uncertainty, which we can recast explicitly as

$$\begin{aligned}
\max_{x,y,z,X,Y,Z} \quad & c^\top v - \|V^\top c\|_1, \\
\text{s.t.:} \quad & Av \geq \hat{b} + |B - AV|^\top \mathbf{1}, \\
& v \geq |V|\mathbf{1}, \ x + |X|\mathbf{1} \leq 100, \\
& X, Y, Z \text{ strictly lower-triangular}, \\
& v = \begin{bmatrix} x \\ y \\ z \end{bmatrix}, \ V = \begin{bmatrix} X \\ Y \\ Z \end{bmatrix},
\end{aligned}$$

which is an LP in the vector variables x, y, z and in the matrix variables X, Y, Z. Solving the problem numerically[6] with an LP solver we obtain the following solution:

[6] See short_term_financing_recourse.m in the online resources.

Month	x	y	z
1	0.00	159	0.00
2	41.86	69.12	4.72
3	9.91	217.85	377.24
4	23.79	–	15.45
5	48.14	–	14.05
6	–	–	21.21

and policy matrices:

$$X = \begin{bmatrix} 0 & 0 & 0 & 0 & 0 & 0 \\ -14.30 & 0 & 0 & 0 & 0 & 0 \\ -4.79 & 5.11 & 0 & 0 & 0 & 0 \\ -12.59 & 6.65 & -4.55 & 0 & 0 & 0 \\ -2.05 & 5.21 & -2.06 & 2.76 & 0 & 0 \end{bmatrix},$$

$$Y = \begin{bmatrix} 0 & 0 & 0 & 0 & 0 & 0 \\ 11.33 & 0 & 0 & 0 & 0 & 0 \\ -0.97 & -4.00 & 0 & 0 & 0 & 0 \end{bmatrix},$$

$$Z = \begin{bmatrix} 0 & 0 & 0 & 0 & 0 & 0 \\ -4.71 & 0 & 0 & 0 & 0 & 0 \\ 4.24 & 4.43 & 0 & 0 & 0 & 0 \\ -3.09 & 5.00 & -7.37 & 0 & 0 & 0 \\ -3.70 & 2.85 & -4.04 & 3.45 & 0 & 0 \\ -0.39 & 0.99 & -1.20 & 0.40 & 3.95 & 0 \end{bmatrix}.$$

We observe an optimal final wealth value which is in between the nominal solution and the static robust solution: the dynamic (affine recourse) approach improved the worst case objective with respect to the static robust solution while maintaining robustness against the uncertainty. Such improvement is obtained by exploiting the dynamic nature of the problem, in which the uncertainty is revealed as time flows, and such knowledge is used to suitably adjust the decision variables according to the policy structure.

It is to be mentioned in closing that in most practical application even the affine recourse approach is used with the only purpose of devising a good here-and-now decision at the initial time. Such decision is "played" and then at the next stage, when part of the uncertainty is revealed, instead of playing the second decision according to the policy we computed previously we rather solve the whole problem again on a reduced forward time horizon, and so on until the end of the planning horizon.

12.4.1 Multi-period portfolio optimization

We next discuss a possible approach to portfolio optimization in a multi-step setting. The notation and definitions of return vectors $r(k)$, gains $g(k) = \mathbf{1} + r(k)$, and portfolio vectors $w(k)$, $k = 1, \ldots$, are those set in Section 11.1. We consider a decision horizon composed of $T \geq 1$ periods, where the kth period starts at time $k-1$ and ends at time k, see Figure 12.6.

We let $w(0)$ denote the initial portfolio composition at $k = 0$, before we start market transactions. At $k = 0$, we can adjust the initial portfolio by increasing or decreasing the amount invested in each asset. Just after transactions, the adjusted portfolio is

$$w^+(0) = w(0) + u(0),$$

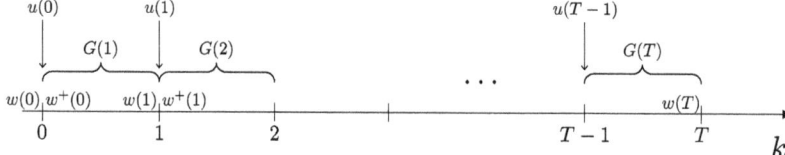

Figure 12.6 Horizon and periods for multi-period portfolio optimization.

where $u_i(0) > 0$ if we increase the position on the ith asset, $u_i(0) < 0$ if we decrease it, and $u_i(0) = 0$ if we leave it unchanged. The portfolio $w^+(0)$ is held fixed for the first period of time Δ. At the end of this first period, the portfolio composition is

$$w(1) = G(1)w^+(0) = G(1)w(0) + G(1)u(0),$$

where $G(1) = \mathrm{diag}\,(g_1(1),\ldots,g_n(1))$ is a diagonal matrix of the asset gains over the period from time 0 to time 1. At time $k = 1$, we perform again an adjustment $u(1)$ of the portfolio: $w^+(1) = w(1) + u(1)$, and then hold the updated portfolio for another period of duration Δ. At time $k = 2$ the portfolio composition is hence

$$w(2) = G(2)w^+(1) = G(2)w(1) + G(2)u(1).$$

Proceeding in this way for $k = 0, 1, 2, \ldots$, we determine the iterative dynamic equations of the portfolio composition at the end of period $(k+1)$:

$$w(k+1) = G(k+1)w(k) + G(k+1)u(k), \tag{12.24}$$

as well as the equations for portfolio composition just after the $(k+1)$th transaction

$$w^+(k) = w(k) + u(k). \tag{12.25}$$

From (12.24) we find that the portfolio composition at time $k = 1,\ldots,T$, is

$$w(k) = \Phi(1,k)w(0)$$
$$+ \begin{bmatrix} \Phi(1,k) & \cdots & \Phi(k-1,k) & \Phi(k,k) \end{bmatrix} \begin{bmatrix} u(0) \\ \vdots \\ u(k-2) \\ u(k-1) \end{bmatrix}$$
$$= \Phi(1,k)w(0) + \Omega_k \mathbf{u}, \tag{12.26}$$

where we defined $\Phi(v,k)$, $v \leq k$, as the *compounded gain* matrix from the beginning of period v to the end of period k:

$$\Phi(v,k) \doteq G(k)G(k-1)\cdots G(v), \quad \Phi(k,k) \doteq G(k),$$

and
$$\mathbf{u} \doteq \begin{bmatrix} u(0)^\top & \cdots & u(T-2)^\top & u(T-1)^\top \end{bmatrix}^\top,$$
$$\Omega_k \doteq \begin{bmatrix} \Phi(1,k) & \cdots & \Phi(k-1,k) & \Phi(k,k) \big| 0 & \cdots & 0 \end{bmatrix}.$$

Alongside the portfolio $w(k)$ we consider a cash account $c(k)$ from which we may draw cash for financing the portfolio, where $c(0) \geq 0$ denotes the initially available cash. The dynamics of the cash account are assumed to be

$$c(k+1) = c(k) - \sum_{i=1}^{n} u_i(k) - \alpha \sum_{i=1}^{n} |u_i(k)|, \quad (12.27)$$

where $\alpha \geq 0$ denotes the transaction cost rate (here the transaction costs are assumed to be linear, that is, proportional to the amount bought/sold, and equal for buy and selling). The overall wealth increase at each $k = 0, 1, \ldots, T-1$, is expressed as

$$\begin{aligned} \delta(k+1) &\doteq \sum_{i=1}^{n} \left(w_i(k+1) - w^+(k) \right) - u_i(k) - \alpha |u_i(k)| \\ &= r(k+1)^\top w^+(k) - \mathbf{1}^\top u(k) - \alpha \|u(k)\|_1 \\ &= r(k+1)^\top (w(k) + u(k)) - \mathbf{1}^\top u(k) - \alpha \|u(k)\|_1. \end{aligned}$$

Suppose first that the whole return sequence $r(1), \ldots, r(T)$ is precisely known at $t = 0$. Then, we can formulate our portfolio optimization problem as the maximization of the sum of wealth increases over the whole horizon:

$$\begin{aligned} \max_{\mathbf{u}} \quad & \sum_{t=1}^{T} \delta(t), \\ \text{s.t.:} \quad & w(k) + u(k) \in \mathcal{W}, \; k = 0, \ldots, T-1, \\ & w(k+1) = G(k+1)(w(k) + u(k)), \; k = 0, \ldots, T-1, \\ & c(0) - \sum_{k=0}^{t} \left(\mathbf{1}^\top u(k) + \alpha \|u(k)\|_1 \right) \geq 0, \; t = 0, \ldots, T-1 \end{aligned}$$

where $w(0)$ and $c(0)$ are given, $\mathcal{W}$ is a given polytope describing possible constraints on the portfolio composition, such as maximum exposure limits, and so on, and the last constraint imposes $c(t) \geq 0$ at all t. The problem above can be cast as an LP and readily solved.

Static solution under scenario uncertainty Clearly, in practice the sequence of returns is *not* known exactly. The approach we take next is a *scenario* one, in which we assume that N scenarios are available, in the form of possible realizations of the stream of returns $r(1), \ldots, r(T)$. We shall denote with $\{r^{(i)}(1), \ldots, r^{(i)}(T)\}$ the ith scenario, with $i = 1, \ldots, N$, and we define accordingly $g^{(i)}(t) = 1 + r^{(i)}(t)$

and $G^{(i)}(t) = \text{diag}(g^{(i)}(t))$, for $i = 1,\ldots,N$, and $t = 1,\ldots,T$. Analogously, we define for $k = 0,\ldots, T-1$,

$$\delta^{(i)}(k+1) \doteq r^{(i)}(k+1)^\top (w^{(i)}(k) + u(k)) - \mathbf{1}^\top u(k) - \alpha \|u(k)\|_1 \tag{12.28}$$

to be the wealth increases in the ith scenario, for $i = 1,\ldots,N$, and let

$$\bar{\delta}(t) \doteq \frac{1}{N} \sum_{i=1}^{N} \delta^{(i)}(t), \quad t = 1,\ldots,T$$

denote the average wealth increases. In order to quantify *risk* we introduce an empirical measure of deviation from the mean, namely

$$\text{MAD}(t) \doteq \frac{1}{N} \sum_{i=1}^{N} \left| \delta^{(i)}(t) - \bar{\delta}(t) \right|, \quad t = 1,\ldots,T. \tag{12.29}$$

We these ingredients we may cast a static multi-period portfolio problem in which we optimize a tradeoff objective that aims to maximize the average wealth increase and to minimize the mean absolute deviation:

$$\begin{aligned}
\max_{\mathbf{u}} \quad & \sum_{t=1}^{T} \left(\bar{\delta}(t) - \gamma \text{MAD}(t) \right), \tag{12.30}\\
\text{s.t.:} \quad & w^{(i)}(k) + u(k) \in \mathcal{W},\\
& w^{(i)}(k+1) = G^{(i)}(k+1) \left(w^{(i)}(k) + u(k) \right),\\
& c(0) - \sum_{k=0}^{t} \left(\mathbf{1}^\top u(k) + \alpha \|u(k)\|_1 \right) \geq 0,\\
& k = 0,\ldots, T-1; \ i = 1,\ldots,N, t = 0,\ldots,T-1.
\end{aligned}$$

where $\gamma \geq 0$ is a tradeoff parameter, and $w^{(i)}(0)$ are initialized to $w(0)$ for all $i = 1,\ldots,N$. The above problem can be cast as an LP; it can be a large-scale problem when n and N are large. The above static problem can be solved at time $t = 0$ for obtaining the whole optimal sequence $u(0),\ldots, u(T-1)$. Then, just the first element $u(0)$ is actually "played" in practice, and at $t = 1$ a version of the same problem on a shifted horizon is solved again, and so on. In the next paragraph we hint at how out strategy can be further improved by introducing a reactive policy in the problem formulation.

Solution with affine recourse We next consider the case in which the investment adjustments $u(k)$ may dynamically react to the random scenarios. In particular, we consider decisions prescribed by affine policies of the following form:

$$\begin{aligned}
u(0) &= \bar{u}(0),\\
u(k) &= \bar{u}(k) + \Theta(k) \left(g(k) - \bar{g}(k) \right), \quad k = 1,\ldots,T-1, \tag{12.31}
\end{aligned}$$

where $\bar{u}(k) \in \mathbb{R}^n$ are "nominal" allocation decision variables, $g(k)$ is the vector of gains (gross returns) over the kth period, $\bar{g}(k)$ is a given estimate of the expected value of $g(k)$, and $\Theta(k) \in \mathbb{R}^{n,n}$, $k = 1, \ldots, T-1$, are the policy "reaction matrices," whose role is to adjust the nominal allocation with a term proportional to the deviation of the gain $g(k)$ from its expected value (we fix henceforth $\Theta(0) \doteq 0$). Observe that the recourse rule is *causal*, since the gain $g(k)$ is observed, and hence becomes known, at time k, precisely when it is needed for deciding about the allocation $u(k)$.

Applying the recourse policy (12.31) to the portfolio dynamics equations (12.24), (12.25) we have (with $\Theta(0) \doteq 0$)

$$w^+(k) = w(k) + \bar{u}(k) + \Theta(k)\left(g(k) - \bar{g}(k)\right), \qquad (12.32)$$
$$w(k+1) = G(k+1)w^+(k), \quad k = 0, \ldots, T-1. \qquad (12.33)$$

From repeated application of this recursion, we obtain the expression for the portfolio composition at a generic instant $k = 1, \ldots, T$:

$$w(k) = \Phi(1,k)w(0) + \Omega_k \bar{\mathbf{u}} + \sum_{t=1}^{k} \Phi(t,k)\Theta(t-1)\tilde{g}(t-1), \qquad (12.34)$$

where

$$\bar{\mathbf{u}}^\top \doteq \begin{bmatrix} \bar{u}(0)^\top & \cdots & \bar{u}(T-2)^\top & \bar{u}(T-1)^\top \end{bmatrix},$$

and $\tilde{g}(k) \doteq g(k) - \bar{g}(k)$, for $k = 1, \ldots, T$. We observe that (12.34) is affine in the decision variables $\bar{u}(0), \ldots, \bar{u}(T-1)$ and $\Theta(1), \ldots, \Theta(T-1)$. Given N scenarios, we define $\delta^{(i)}(t)$ as in (12.28), in which we now substitute $u(k) \leftarrow \bar{u}(k) + \Theta(k)\left(g^{(i)}(k) - \bar{g}(k)\right)$, and we define $\bar{\delta}(t)$ and MAD(t) accordingly. We then cast the dynamic portfolio optimization problem as

$$\max_{\substack{\bar{u}(0),\ldots,\bar{u}(T-1),\\ \Theta(1),\ldots,\Theta(T-1)}} \quad \sum_{t=1}^{T}\left(\bar{\delta}(t) - \gamma \text{MAD}(t)\right), \qquad (12.35)$$

$$\begin{aligned}
\text{s.t.:} \quad & w^{(i)}(k) + u^{(i)}(k) \in \mathcal{W},\\
& w^{(i)}(k+1) = G^{(i)}(k+1)\left(w^{(i)}(k) + u^{(i)}(k)\right),\\
& c(0) - \sum_{k=0}^{t}\left(\mathbf{1}^\top u^{(i)}(k) + \alpha \|u^{(i)}(k)\|_1\right) \geq 0,\\
& u^{(i)}(k) = \bar{u}(k) + \Theta(k)\left(g^{(i)}(k) - \bar{g}(k)\right),\\
& k = 0,\ldots,T-1;\ i = 1,\ldots,N, t = 0,\ldots,T-1.
\end{aligned}$$

Problem 12.35 may be recast as an LP, and hence efficiently solvable numerically, although the problem may be of large scale. Again, in practical application of the method, problem 12.35 is solved at time

$k = 0$ mainly to obtain a good here-and-now decision $u(0)$. The future decisions or policies are rarely implemented in practice. Rather, at time $k = 1$ the decision maker collects the information coming from the realization of $G(1)$, which can be used to update the model used in the scenario generation, and then solves the whole problem again over a forward-shifted interval, to obtain $u(1)$, and the same process is iterated for all subsequent periods. At each decision time k, the optimization interval can either be held fixed (receding horizon), if the investment is to be iterated indefinitely in time, or shrunk in duration by one period (shrinking horizon), if the final investment time T is fixed.

12.5 Exercises

Exercise 12.1 *Robust portfolio optimization problem*
We consider a robust portfolio optimization problem, of the form

$$\max_{x \in \mathcal{X}} \min_{r \in \mathcal{U}} r^T w,$$

where $\mathcal{X} \subseteq \mathbb{R}^n$ is a polytope that encodes a set of affine inequality or equality constraints on the portfolio weight vector x. Here, $\mathcal{U}$ is a set that encodes the uncertainty on the return vector $r \in \mathbb{R}^n$.

For the following given choices of $\mathcal{U}$ formulate the above problem as a convex optimization problem; make sure to specify which of the acronyms (LP, QP, SOCP) best applies to your formulation.

1. ℓ_1 norm uncertainty: $\mathcal{U} = \{r = \hat{r} + \rho u : \|u\|_1 \leq 1\}$, with $\rho > 0$ given.

2. Diagonal uncertainty: $\mathcal{U} = \{r = \hat{r} + Du : \|u\|_2 \leq 1\}$, where D is a positive-definite diagonal matrix. What is the shape of the uncertainty set?

3. Factor-like model:
$$\mathcal{U} = \{r = \hat{r} + Fu + Dv : \|u\|_2 \leq 1, \ \|v\|_\infty \leq 1\},$$
where D is a positive-definite diagonal matrix, and $F \in \mathbb{R}^{n \times k}$, with $k \leq n$.

4. Convex hull:
$$\mathcal{U} = \left\{ r = \lambda_1 r^{(1)} + \cdots + \lambda_K r^{(K)} : \lambda \geq 0, \ \mathbf{1}^T \lambda = 1 \right\},$$
where $r^{(k)}$, $k = 1, \ldots, K$ are given scenarios.

Exercise 12.2 *Centrality and deviation measures in portfolio design*
Consider n assets whose returns over a fixed period of time are described by an n-dimensional random vector r. Assume that historical realizations of the returns from these assets are observed over m periods and collected in a return matrix

$$R = \begin{bmatrix} r^{(1)} & r^{(2)} & \cdots & r^{(m)} \end{bmatrix} \in \mathbb{R}^{n,m},$$

where $r^{(k)} \in \mathbb{R}^n$ is the vector of returns observed in the kth period. If an investor's position in these n assets were described by vector $w \in \mathbb{R}^n$, assuming that the total allocated budget is normalized to one, the historical returns of their portfolio would be given by the sequence $\rho_k(w) \doteq w^\top r^{(k)}$, $k = 1, \ldots, m$, which we collect in a vector

$$\rho(w) = [\rho_1(w), \ldots, \rho_m(w)].$$

334   12 BEYOND THE MEAN/VARIANCE MODEL

For given and fixed w, consider the following two problems:

$$\min_{\hat{\varrho}} \frac{1}{m} \sum_{k=1}^{m} |\rho_k(w) - \hat{\varrho}|, \quad (12.36)$$

$$\min_{\bar{\varrho}} \frac{1}{m} \sum_{k=1}^{m} (\rho_k(w) - \bar{\varrho})^2. \quad (12.37)$$

In the first problem above, the interpretation of the optimal $\hat{\varrho}$ is a "central" return value which minimizes the average absolute deviation from $\hat{\varrho}$ to the historical returns. In the second problem, the interpretation of the optimal $\bar{\varrho}$ is a central value that minimizes the average squared deviation from $\bar{\varrho}$ to the historical returns.

It turns out that that optimal solutions for both problem (12.36) and (12.37) can be found in "closed form." In particular, an optimal $\hat{\varrho}$ for problem (12.36) is the *median* of the returns values (i.e., a value that leaves the same number of return values on its left and on its right; you are not requested to prove this fact).

1. For the given historical returns in R and portfolio w, find explicit expressions of the optimal $\bar{\varrho}$ in problem (12.37), or explain an explicit way for computing this value.

2. Supposing that the portfolio composition $w \in \mathbb{R}^n$ is now variable, we can cast a portfolio design problem aimed at finding a portfolio w which minimizes either the average absolute deviation or the average squared deviation from the corresponding central values, while guaranteeing that the central value remains above a desired given threshold t. For the absolute deviations we consider specifically the problem

$$\min_{w,\hat{\varrho}} \frac{1}{m} \sum_{k=1}^{m} |\rho_k(w) - \hat{\varrho}|, \quad \text{s.t.: } \hat{\varrho} \geq t, \ w \geq 0, \mathbf{1}^\top w = 1. \quad (12.38)$$

Discuss if problem (12.38) is (or can be cast as) a known convex optimization model (e.g., LP, QP, SOCP, etc.) and, in the positive case, give an explicit formulation of the model in standard form.

3. Repeat the previous point for the squared deviations, that is, for the problem

$$\min_{w,\bar{\varrho}} \frac{1}{m} \sum_{k=1}^{m} (\rho_k(w) - \bar{\varrho})^2, \quad \text{s.t.: } \bar{\varrho} \geq t, \ w \geq 0, \mathbf{1}^\top w = 1. \quad (12.39)$$

Exercise 12.3 *Approximate risk parity*

In Section 12.1 we discussed risk budgeting and we said that a portfolio $x \in \mathbb{R}^n$ achieves given risk budgets $\theta_i \geq 0$, $i = 1, \ldots, n$, $\sum_i \theta_1 = 1$, if

$$x_i(Cx)_i = \theta_i(x^\top Cx), \quad i = 1,\ldots,n, \qquad (12.40)$$

where C is the covariance matrix of the returns. We also proved that these conditions are equivalent to n SOCP (hence convex) constraints on the portfolio, in the case when $x \geq 0$. In particular, a *risk parity* portfolio is one which satisfies the above conditions for $\theta_i = 1/n$, $i = 1,\ldots,n$.

It can be proved that in the "long-only" case (i.e., $x \geq 0$), the risk parity portfolio is uniquely determined by (12.40), which makes risk parity unsuitable for embedding into other portfolio design methodologies. For instance, we cannot impose risk parity *and* seek for a tradeoff between mean return and variance since the risk parity conditions alone fully determine the portfolio composition. Further, the convex characterization of the risk budgeting constraints only holds for the case $x \geq 0$.

Here, you are asked to generalize the risk budgeting approach to the case when x is not necessarily positive, and risk parity is allowed to be achieved only approximately. In particular, we shall consider the following optimization problem:

$$\min_{x \in \mathbb{R}^n, \eta} \sum_{i=1}^n (x_i(Cx)_i - \eta)^2 + \rho x^\top Cx, \qquad (12.41)$$

$$\text{s.t.:} \quad \ell_i \leq x_i \leq u_i, \quad i = 1,\ldots,n,$$
$$\sum_{i=1}^n x_i = 1,$$
$$\hat{r}^\top x \geq \varrho,$$

for given value $\rho \geq 0$ of the tradeoff parameter, given minimal desired return ϱ, and given upper and lower bounds on the portfolio entries, u_i and ℓ_i. We write for short $x \in \mathcal{X}$ to denote the portfolio constraints in problem (12.41), which we then express more compactly as

$$\min_{x \in \mathcal{X}, \eta} \sum_{i=1}^n (x_i(Cx)_i - \eta)^2 + \rho x^\top Cx. \qquad (12.42)$$

1. Prove that, for any x that is feasible for the above problem, there exists a unique optimal value of η, which is given by

$$\eta^* = \frac{1}{n} \sum_i x_i(Cx)_i.$$

2. Let $\bar{C} = 1/n\, C$. Show that problem (12.42) can be written as

$$\min_{x \in \mathcal{X}, y} \sum_{i=1}^n \left(x^\top (M_i - \bar{C})y\right)^2 + \rho x^\top Cx, \quad \text{s.t.:} y = x, \qquad (12.43)$$

for appropriate matrices M_i, $i = 1,\ldots,n$, that you should determine. Discuss about the convexity of such problem.

Exercise 12.4 *Penalized risk parity*

We consider a relaxation of the problem (12.43), in which we treat the equality constraint $y = x$ approximately through a penalization in the objective:

$$\min_{x,y \in \mathcal{X}} \sum_{i=1}^{n} \left(x^\top (M_i - \bar{C})y\right)^2 + \rho x^\top C x + \mu \|y - x\|_2^2.$$

Then, we may use an alternate minimization heuristic to try to solve the problem in the penalized form. Suppose you are given an initial portfolio $x^{(0)}$ (e.g., a minimum-variance portfolio), and a parameter $\mu^{(k)} > 0$. At each iteration $k = 0, 1, \ldots$ of the method, we do the following:

(a) Set $y^{(0)} = x^{(0)}$, and compute

$$x^{(k+1)} = \arg\min_{x \in \mathcal{X}} \sum_{i=1}^{n} \left(x^\top (M_i - \bar{C})y^{(k)}\right)^2$$
$$+ \rho x^\top C x + \frac{1}{\mu^{(k)}} \|y^{(k)} - x\|_2^2,$$

$$y^{(k+1)} = \arg\min_{y \in \mathcal{X}} \sum_{i=1}^{n} \left(x^{(k+1)\top} (M_i - \bar{C})y\right)^2$$
$$+ \rho y^\top C y + \frac{1}{\mu^{(k)}} \|y - x^{(k+1)}\|_2^2.$$

(b) Choose a new penalty parameter $\mu^{(k+1)} \in (0, \mu^{(k)})$.

(c) Let $k \leftarrow k + 1$, and repeat until some exit condition is met.

1. Discuss the convexity of the problems you need to solve in step (a), and write a code implementing this method.

2. Test your code using

$$C = \begin{bmatrix} 94.868 & 33.75 & 12.325 & 1.178 & 8.778 \\ 33.75 & 445.642 & 98.955 & 7.901 & 84.954 \\ 12.325 & 98.955 & 117.265 & 0.503 & 45.184 \\ 1.178 & 7.901 & 0.503 & 5.46 & 1.057 \\ 8.778 & 84.954 & 45.184 & 1.057 & 34.126 \end{bmatrix}, \quad \hat{r} = 0.$$

Experiment with various settings of bounds and parameters. Comment on your findings. Is the method working efficiently, or at all?

3. Test for comparison of some direct solution techniques on the non-convex problem (12.43) (e.g., using MATLAB®'s fmincon). Discuss and illustrate your results.

13
Financial networks

NETWORK STRUCTURES arise naturally in many financial and economic contexts. For instance, if one considers a group of banks and observes payments sent from one bank to another during a certain interval of time, then a graph structure emerges in which banks are the nodes of the graph and payment transactions form the links between nodes. Further, both the nodes and the links may possess certain properties (e.g., the node property may be the equity value of each bank, and the link property may be the amount of payment transacted over each link), and a graph with properties or operations performed on the nodes and/or links will be referred to as a *network*. Various types of networks are of interest in finance beyond payment networks, and some of these are discussed in Section 13.2. Once certain aspects of a financial system are modeled as a network, specific phenomena can be analysed which are related to the interconnection structure of the system and which may help one understand complex behaviors and feedback effects caused by the mutual interactions of the participating nodes. A typical example of such ensemble behavior is the avalanche effect that may cause a financial system to collapse in the presence of *defaults* at one or few nodes; this risk due to the interconnectedness (or *systemic risk*) is further discussed in Section 13.3. Graph theoretic and computational approaches may also be employed, for instance, for simplifying the network and visualizing its essential structure, for understanding the relative importance, or criticality, of the nodes (see the discussion on *node centrality* in Section 13.4), or for detecting clusters and communities with different roles in the network (see Section 13.5). Before delving into such themes, we introduce in the next section the basic terminology and formalism on graphs; such formal section is dense with definitions which are quite standard in graph theory, hence it can be safely skipped if the reader is already acquainted with the topic. For a thorough introduction to graph theory we direct the reader to classical references, such as D. B West, *Introduction to Graph Theory*, Pearson Education, 2001.

13.1 Graph definitions and terminology

A *graph* $G = (V, E)$ consists of a set of vertices or nodes $V(G)$ and a set of edges $E(G)$. Each edge has two associated vertices which constitute its endpoints. We shall here consider mainly *simple graphs*, that is graphs with no self-loops (edges for which endpoints coincide) and such that to each distinct pair of vertices there correspond at most one edge. An edge is said to *join* its endpoints, with which it is *incident*; two vertices are said to be *adjacent*, or *neighbors* if they are joined by an edge. If a graph G has adjacent vertices named u, v, then the edge e joining u and v is represented as $e = (u, v)$; note that the order of the nodes is irrelevant, hence $e(u, v) = e(v, u)$. The situation is different if we consider directed graphs, or *digraphs*, for which to each edge there is an associated direction: a *directed edge* (also called a *link* or *arc*) $e(u, v)$ is directed from its tail node u to its head node v, and a digraph is a graph in which each edge is directed. We consider mainly simple digraphs, that is, digraphs such that each ordered tail–head pair corresponds to at most one arc. An *orientation* $\widehat{G}$ of a simple graph G is a digraph obtained from G by assigning directions to all its edges.

The *degree* of a vertex v of a (simple) graph G is the number of neighbors of v in G, and it is denoted by $\deg(v)$. The degree matrix D_G of G is a diagonal matrix containing on the diagonal the degree of each vertex. In case of a digraph G, we define the *in-degree* of a vertex v as the number of arcs directed towards v, and the *out-degree* as the number of arcs departing from v. A node with degree zero is said to be *isolated*. If all nodes in a graph G have the same degree r, then G is said to be a *r-regular graph*.

If a simple graph G has n nodes $v_1, \ldots, v_n$, the associated *adjacency matrix* A_G is the $n \times n$ symmetric matrix whose (i,j)th element is defined as

$$[A_G]_{i,j} = \begin{cases} 1 & \text{if } v_i, v_j \text{ are adjacent,} \\ 0 & \text{otherwise.} \end{cases}$$

The *Laplacian matrix* L_G of G is a symmetric matrix defined as

$$L_G = D_G - A_G.$$

For a digraph G we define similarly the directed adjacency matrix A_G as the $n \times n$ matrix whose (i,j)th element is defined as

$$[A_G]_{i,j} = \begin{cases} 1 & \text{if there is an arc from } v_i \text{ to } v_j, \\ 0 & \text{otherwise.} \end{cases}$$

The in-degree of vertex v_i corresponds to the sum of the entries in the ith column of the (directed) adjacency matrix, and its out-degree corresponds to the sum of the entries in the ith row.

Clearly, the structure of a graph or a digraph is completely characterized by its adjacency matrix,[1] which thus constitutes an encoding of the graph. An alternative encoding is given by the *incidence matrix*: for a graph G with n ordered vertices $v_1, \ldots, v_n$, and m ordered edges $e_1, \ldots, e_m$, the incidence matrix B_G is an $n \times m$ matrix whose (i,j)th element is defined as

$$[B_G]_{i,j} = \begin{cases} 1 & \text{if } v_i \text{ is incident on } e_j, \\ 0 & \text{otherwise.} \end{cases}$$

For digraphs, the directed incidence matrix is defined as

$$[B_G]_{i,j} = \begin{cases} -1 & \text{if } v_i \text{ is the tail node of } e_j, \\ 1 & \text{if } v_i \text{ is the head node of } e_j, \\ 0 & \text{otherwise.} \end{cases}$$

[1] If edges of a graph have an associated value or weight, then we can define a *weighted* adjacency matrix, such that $A_G(i,j) = w_{ij}$, where w_{ij} is the weight associated with the edge incident on v_i, v_j, or its is zero if there is no such edge.

A weighted version of the incidence matrix is defined by replacing 1 (resp. -1) with a weight $w_{ij} \geq 0$ (resp. $-w_{ij}$) in the above definition. Similarly, a weighted adjacency matrix can be defined, whose element in position (i,j) is w_{ij}^2 if there is an arc from v_i to v_j, and it is zero otherwise.

There exists an algebraic relation between the Laplacian matrix of a simple graph G and the directed incidence matrix of any orientation $\overset{\frown}{G}$ of G, namely it holds that, independent of the chosen orientation,

$$L_G = B_{\overset{\frown}{G}} B_{\overset{\frown}{G}}^\top,$$

which also implies that the Laplacian is positive semidefinite. As an illustration, consider the graph G depicted in Figure 13.1. The degree matrix for this graph is $D = \text{diag}(1,3,3,2,3)$ and the adjacency matrix is

$$A = \begin{bmatrix} 0 & 0 & 0 & 0 & 1 \\ 0 & 0 & 1 & 1 & 1 \\ 0 & 1 & 0 & 1 & 1 \\ 0 & 1 & 1 & 0 & 0 \\ 1 & 1 & 1 & 0 & 0 \end{bmatrix}.$$

13 FINANCIAL NETWORKS

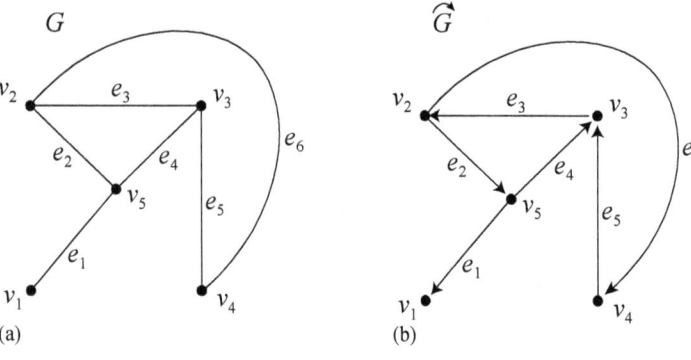

Figure 13.1 A graph with $n = 5$ vertices and $m = 6$ edges (a), and one of its possible orientations (b).

The incidence matrix of the orientation $\overset{\frown}{G}$ of the graph is

$$B = \begin{bmatrix} 1 & 0 & 0 & 0 & 0 & 0 \\ 0 & -1 & 1 & 0 & 0 & -1 \\ 0 & 0 & -1 & 1 & 1 & 0 \\ 0 & 0 & 0 & 0 & -1 & 1 \\ -1 & 1 & 0 & -1 & 0 & 0 \end{bmatrix},$$

and it holds that

$$L = D - A \equiv BB^\top = \begin{bmatrix} 1 & 0 & 0 & 0 & -1 \\ 0 & 3 & -1 & -1 & -1 \\ 0 & -1 & 3 & -1 & -1 \\ 0 & -1 & -1 & 2 & 0 \\ -1 & -1 & -1 & 0 & 3 \end{bmatrix}.$$

13.1.1 Potentials and flows

The incidence matrix $B \in \mathbb{R}^{|V|,|E|}$ of a digraph $G(V,E)$, with $|V| = n$, $|E| = m$, is useful also to express mappings from the edge space to the node space, or vice-versa. Let $q^\top = [q_1, \ldots, q_m]$ represent values attached to the edges of the graph, for instance q_j may represent the directed *flow* (of traffic, of liquid, of money, of current, etc.) across the jth edge of the graph. Then, it can be seen that

$$p = Bq$$

represents an n-vector of node values, where

$$p_i = \sum_j B_{ij} q_j$$

represents the net (possibly weighted) flow at node i, that is, the weighted sum of all incoming flows at i, minus the weighted sum of all outgoing flows from i. Vice versa, if $p^\top = [p_1, \ldots, p_n]$ represent values attached to the nodes of the graph (for instance, p_i may represent the *potential*, e.g., voltage, pressure, etc., value at node i), then it can be seen that

$$q = B^\top p$$

represents an m-vector of edge values, where

$$q_j = \sum_i B_{ij} p_i$$

represents the (possibly weighted) potential drop across the jth edge, that is, the potential at the head node minus the potential at the tail node of that edge, multiplied by the edge weight. For example, if p represents voltages, and edge weights represent resistances, then $q = B^\top p$ gives the currents across the graph edges. The squared Euclidean norm $\|q\|_2^2$ of the flows can hence be expressed as a quadratic form of the potentials as

$$\|q\|_2^2 = q^\top q = p^\top B B^\top p = p^\top L p,$$

where $L \succeq 0$ is the Laplacian of the undirected version of the digraph. Since it also holds that $L = D - A$, where D and $A \geq 0$ are the degree matrix and the (symmetric) adjacency matrix of the undirected version of the digraph, respectively, we have that

$$p^\top L p = p^\top (D - A) p = \frac{1}{2} \sum_{i,j=1}^n a_{ij}(p_i - p_j)^2.$$

13.1.2 Walks, paths, trails, and trees

A *walk* in a simple graph is an ordered sequence of adjacent vertices and edges connecting them, from an initial vertex v_{ini} to a terminal vertex v_{end}; the length of the walk is the number of edges crossed along the walk, counting repetitions. The walk is *closed* if the terminal vertex coincides with the initial one, and it is *open* otherwise. A *trail* is a walk in which each edge appears at most once. A *path* is a trail in which all visited vertices are distinct. A closed trail with of length ≥ 3 is called a *cycle*. A graph which contains no cycles is said to be *acyclic*.

Two vertices a, b of a graph G are said to be *connected* if there exists a walk in G that starts in a and ends in b; if all pairs of nodes in G are connected then G is said to be a connected graph. All vertices are considered to be connected to themselves via trivial walks of length zero. The *distance* $d(a, b)$ between two connected vertices a, b is the length of the shortest walk from a to b. Connectedness is an equivalence relation on the vertex set of G, and the equivalence classes of this relation are called the *connected components* of G, see, for example, Figure 13.2. Clearly, if G is connected it has a unique connected component, which is G itself.

A connected and acyclic graph is called a *tree*. A tree on n vertices has $n-1$ edges. Actually, for a graph G on n vertices any two of the following conditions (a), (b), (c) implies the third, where (a) G is connected, (b) G is acyclic, and (c) G has $n-1$ edges. A vertex of degree one in a tree is called a *leaf*. A *forest* is a graph whose connected components are trees, see Figure 13.3.

A *subgraph* of a graph G is a graph G′ such that $V(G') \subseteq V(G)$ and $E(G') \subseteq E(G)$. Given a vertex subset $V' \subset V(G)$, the *induced subgraph* G′ has vertex set V′ and as edge set the edges of G whose endpoints are in V′. A *spanning subgraph* of G is a subgraph of G having the same vertex set as G. A *component* of a graph G is a connected subgraph G′ of G such that no other subgraph of G that properly contains G′ is connected. Every graph with n vertices and m edges has at least $n-m$ components.

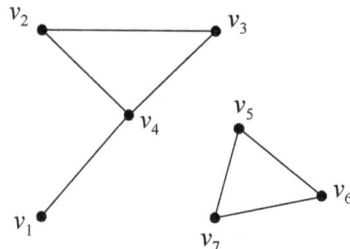

Figure 13.2 A graph with two connected components: $\{v_1, v_2, v_3, v_4\}$ and $\{v_5, v_6, v_7\}$.

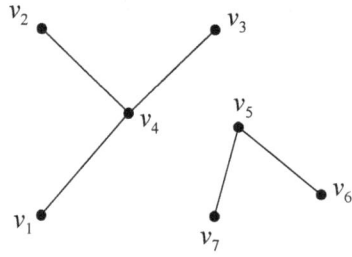

Figure 13.3 A forest with two trees.

13.2 Some types of financial networks

We next exemplify how various types of network models can be constructed from actual financial data. The first case we analyse is the construction of network models based on Pearson's correlation, which are hence known as correlation networks.

13.2.1 Correlation networks

A common situation when analysing financial data is that we have available the price or return data (e.g., daily) of n assets, and we can estimate the covariance matrix Σ of these assets using, for example, one of the methods described in Chapter 3. From the estimated covariance matrix, we obtain the Pearson's correlation coefficients as

$$q_{ij} = \frac{\Sigma_{ij}}{\sqrt{\Sigma_{ii}\Sigma_{jj}}} \in [-1, 1], \quad i, j = 1, \ldots, n.$$

A correlation network is created by considering n nodes representing the assets, and assigning an edge in the graph between node i and j whenever $|q_{ij}| > 0$. If Σ is estimated via the sample covariance matrix of the returns, we usually have that all the q_{ij} are nonzero and hence the resulting correlation network has a *complete graph* (i.e., each node is adjacent to each other node). In such situation, the correlation network does not help much in revealing and interpreting the correlation structure among the assets. One simple approach for filtering out the noise that creates possibly spurious correlations is to threshold the correlations, thus creating an edge between node i and j only when $|q_{ij}| > \gamma$, where $\gamma < 1$ is a threshold value to be appropriately selected. Clearly, as γ is increased the resulting correlation network becomes increasingly sparse, progressively revealing only the most important correlations among nodes. The downside of this approach is that it may be difficult to select an appropriate level of γ, and that possibly meaningful but small correlations are cut out by the thresholding. More sound methods than simple thresholding have been proposed in the literature for obtaining sparse estimates of the covariance matrix, and hence of the correlation structure. One approach, already discussed in Section 3.7.1, consists in minimizing

$$g_\lambda(\Sigma) \doteq \ln \det \Sigma + \text{Tr}(S\Sigma^{-1}) + \lambda \|\Sigma\|_1 \qquad (13.1)$$

with respect to $\Sigma \succ 0$, for suitable values of the sparsity-inducing parameter $\lambda \geq 0$. This problem, known as the "covariance graphical lasso," is nonconvex but there exist efficient methods for solving it approximately.[2]

[2] This approach was proposed and developed in J. Bien and R. Tibshirani, "Sparse Estimation of a Covariance Matrix," *Biometrika*, vol. 98(4), pp. 807–820, 2011. An alternative solution approach based on block coordinate minimization was later proposed in H. Wang, "Coordinate Descent Algorithm for Covariance Graphical Lasso," *Statistics and Computing*, vol. 24, pp. 521–529, 2014.

Minimum spanning tree (MST) Each edge in the original (typically complete) correlation graph obtained from Σ can be associated with an edge value measuring a pseudo-distance between the two adjacent nodes, for example, by defining $w_{ij} = 1 - \rho_{ij}^2$, so that i,j are close when w_{ij} is small (i.e., when i,j are highly positively or negatively correlated), and are far away when w_{ij} is near one (i.e., when i,j are loosely correlated). A further method for effectively sparsifying the network can then be obtained by extracting from it a spanning tree (i.e., a subset of the edges that connects all the vertices together, without any cycles) such that the sum of edge weights is as small as possible. Since small weights are associated with strong correlations, the tree structure will reveal the backbone of essential correlations in the network. A MST always exists for a connected network, and its extraction can be performed efficiently via standard algorithms such as Dijkstra's or Prim's algorithms, which are available in most common numerical computing software packages.

Validating correlation networks Once a sparse network model is constructed, it is important in practice to assess its reliability. This is a nontrivial task, since in the typical endeavor in which correlation networks are used there is no ground truth against which to compare the model, and also additional data may not be available or usable.

For instance, in the case of financial returns, one is typically interested in a correlation model based on a given data frame in a specific time period, and hence return data outside that period, even if available, may not be usable for validating the model. In such situation a *bootstrap* technique may be successfully used. Denoting with

$$R \doteq [r(1) \cdots r(T)] \in \mathbb{R}^{n,T}$$

the matrix containing by columns the assets' return vectors in the considered time frame from 1 to T, one can generate N bootstrap versions $R^{(i)}$ of R, for $i = 1, \ldots, N$, where each $R^{(i)}$ is obtained by sampling T times the columns of the original matrix R, uniformly at random with replacement.[3] Notice that when sampling T times with replacement from a set of T elements, there is a probability $(1 - T^{-1})^T \simeq e^{-1}$ of never picking a specific element, a probability $(1 - T^{-1})^{T-1}$ of picking a specific element exactly once, a probability $(T-1)/(2T)(1 - T^{-1})^{T-2}$ of picking it twice, and so on according to a Binomial distribution with success probability $1/T$.

Using bootstrap, N different sparse correlation networks can be derived from the N bootstrapped return matrices $R^{(i)}$. Then, to each edge in the original sparse network (the one constructed on the basis of the original R via thresholding or other methods) we can associate a bootstrap validation value equal to the proportion of bootstrap networks in which that edge appears, with higher values indicating more reliable and hence validated edges.

Example 13.1 (*Network structure of the DJI components*) We analyse the network structure of the correlation among the $n = 30$ assets that compose the Dow Jones Industrial Index (DJI).[4] The analysis is based on a price data frame of 251 days, from Oct. 11, 2005 to Oct. 9, 2006. Figure 13.4 shows the correlation graphs obtained at different threshold levels, the plot on the top-left corner corresponding to the complete graph obtained for $\gamma = 0$, see code file correlation_network_DJI_1.m in the online resources. Figure 13.5 shows instead the correlation graphs obtained by solving (13.1) at various λ levels. In Figure 13.6 we can see a comparison of the node degrees of the sparsest graphs obtained from thresholding (solid bars) and from the sparse covariance estimation approach (dashed bars), see code file correlation_network_DJI_2.m in the online resources. The node degrees are similar in the two approaches and show, for instance, that the four most influential nodes (in the sense of degree level) in the network are KO, JPM, GE, AIG and AXP. Interestingly, these assets are representative of market sectors (there are 11 stock market sectors, according to the most commonly used classification system: the Global Industry Classification Standard (GICS)), namely consumer (KO), industrials (GE), and financials (JPM, AIG, and AXP), with the financial sector being the most represented, as could be expected since it

[3] Observe that the total number of distinct k samples from an T-element set such that repetition is allowed and ordering does not matter is $\binom{T+k-1}{k}$. Therefore, there are $\binom{2T-1}{T}$ different (up to reordering) possible bootstrap versions of R. This number is so large even for moderate T that it can be considered unlimited in practice; for example, for $T = 50$, it is equal to 2.5×10^{28}.

[4] The components of the DJI are listed in the following table.

Symbol	Name
AA	ALCOA
AIG	AMER INTL GROUP
AXP	AMER EXPRESS
BA	BOEING
C	CITIGROUP
CAT	CATERPILLAR
DD	DU PONT DE NEM
DIS	WALT DISNEY
GE	GEN ELECTRIC
GM	GEN MOTORS
HD	HOME DEPOT
HON	HONEYWELL INTL
HPQ	HEWLETT PACKARD
IBM	INTL BUSINESS MACH
INTC	INTEL
JNJ	JOHNSON AND JOHNS
JPM	JP MORGAN CHASE
KO	COCA COLA
MCD	MCDONALDS
MMM	3M COMPANY
MO	ALTRIA GROUP
MRK	MERCK
MSFT	MICROSOFT
PFE	PFIZER
PG	PROCTER GAMBLE
T	AT&T
UTX	UNITED TECH
VZ	VERIZON
WMT	WAL MART STORES
XOM	EXXON MOBIL

has connections with most of the other market entities. Isolated vertices (e.g., AA, GM, MSFT, XOM), which are uncorrelated with all other vertices and among themselves, constitute good candidates for a diversified portfolio.

Figure 13.7 shows the minimum spanning tree of the original correlation network, based on the edge weights $w_{ij} = 1 - \rho_{ij}^2$, see code file `correlation_network_DJI_3.m` in the online resources. We observe that the banking sector (JPM) appears at the center, as might be expected. Here again, leaf nodes of the tree may be good candidates for entering a diversified portfolio.

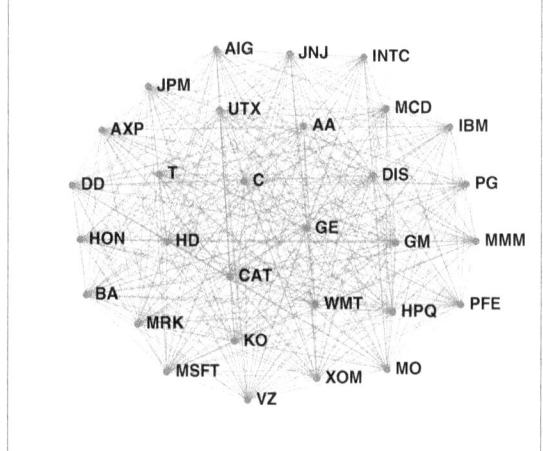
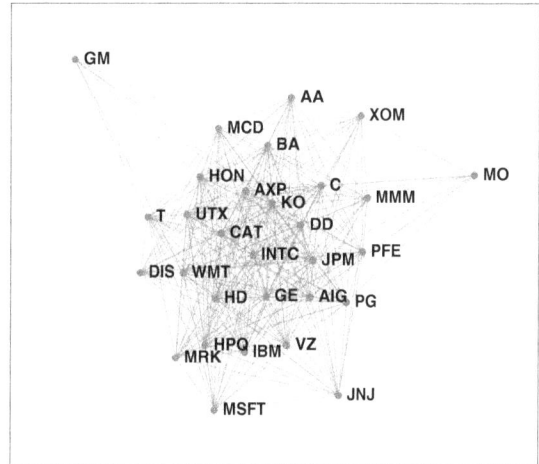
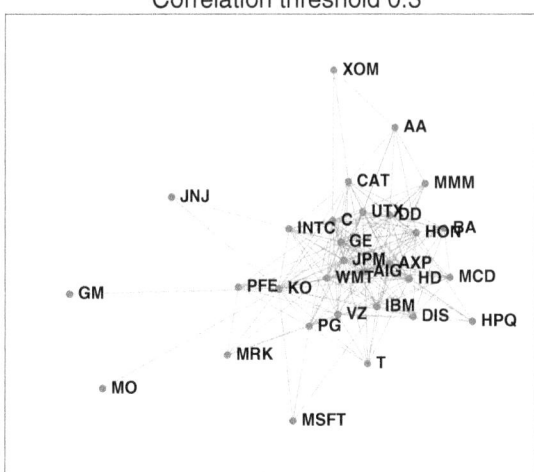
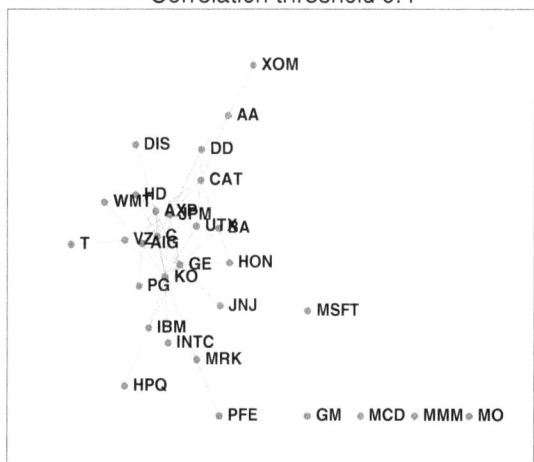

Figure 13.4 Correlation graphs for the DJI components obtained by thresholding the sample correlation matrix at different threshold levels.

346 13 FINANCIAL NETWORKS

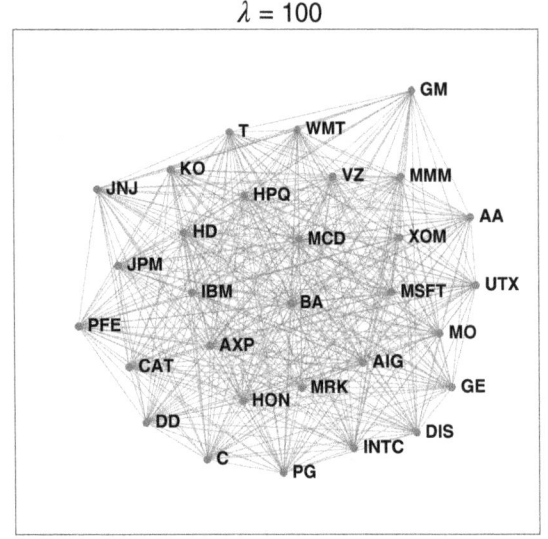
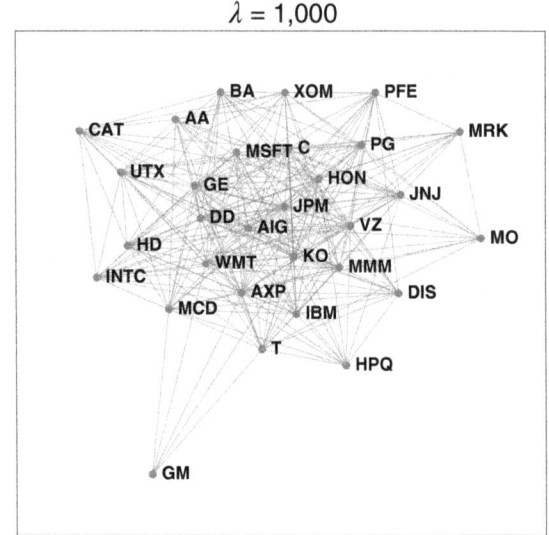
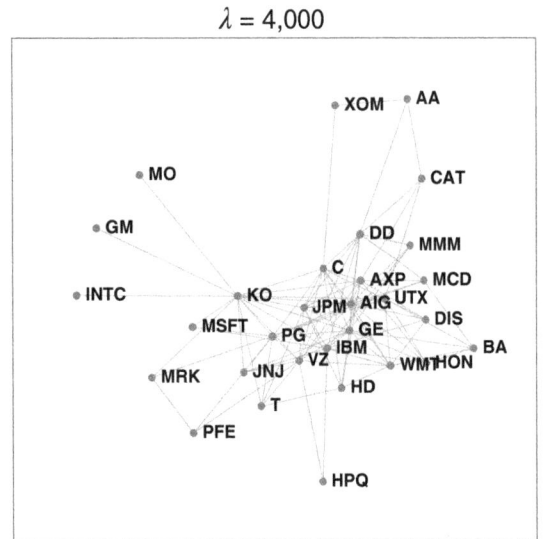
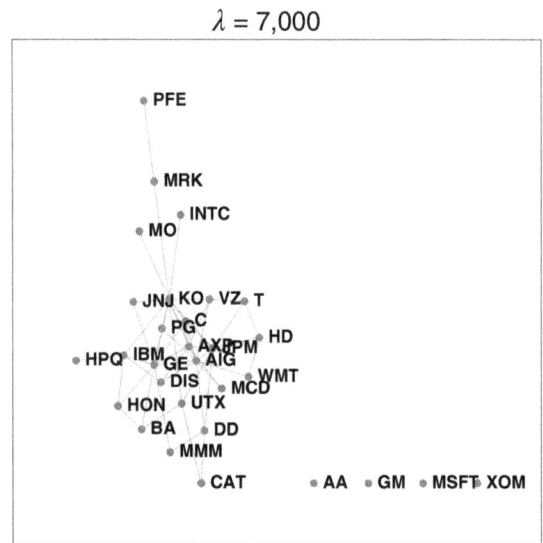

Figure 13.5 Correlation graphs for the DJI components obtained by solving the covariance graphical lasso at different threshold levels.

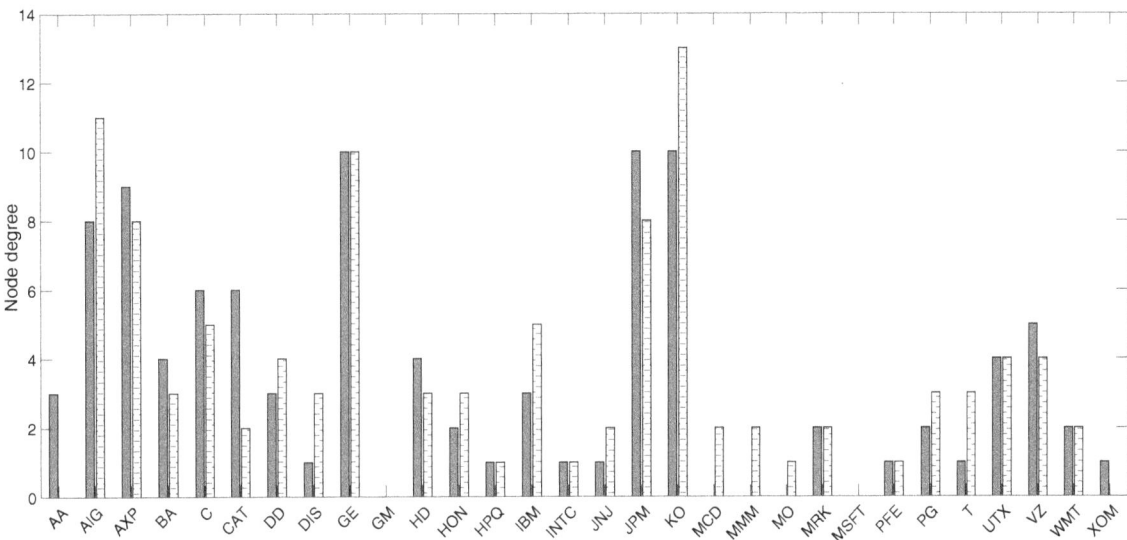

Figure 13.6 Node degrees of the thresholded graph with $\gamma = 0.4$ (solid) and of the graph obtained from the graphical lasso approach with $\lambda = 7{,}000$ (dashed).

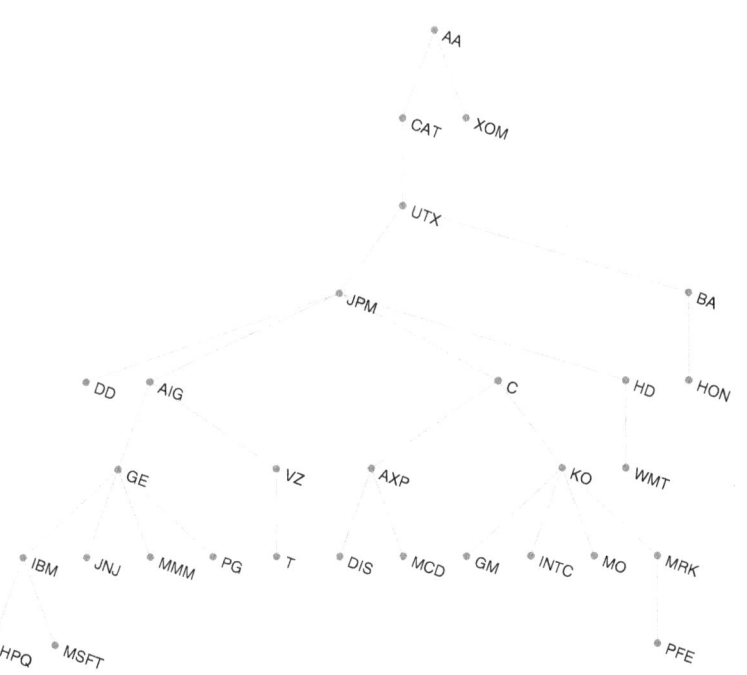

Figure 13.7 A minimum spanning tree for the DJI correlation network.

13.2.2 Payment networks

Payment networks describe the mutual payments executed among a set of financial entities (typically, banks) in a given period of time. A payment network is a directed graph in which the banks represent the nodes and a directed link exist from node i to node j if a payment is executed from bank i to bank j. The link weight w_{ij} represents the total number of payments from i to j in the considered time period, or the total value of such payments. A payment network can be easily created from tabular data containing the transactions in the period of interest. In practice, however, such data listing all transactions among a given groups of banks is rarely available, due to confidentiality issues. For this reason payment networks are typically studied by describing them via standard network structures such as a core–periphery structures,[5] which can be approximated by means of scale-free networks[6] with parameters tuned to match some of the known characteristics of the actual network that one wishes to describe.[7]

13.2.3 Investment and ownership networks

Investment networks are bipartite networks[8] in which the two vertex partitions represent investors (e.g., venture capital) and companies, respectively. Similarly, ownership networks are bipartite networks in which one vertex set represents owners (shareholders) and the other set represent companies. Constructing investment or ownership networks requires access to investment/ownership data, which is rarely publicly available for free. An analysis of a venture capital investment network using CrunchBase data is proposed, for instance, in Chapter 5 of *Network Theory and Financial Risk*.[9] A large ownership network of millions of firms and their shareholders at worldwide level is instead considered by Engel et al. (2021).[10]

From any bipartite network with disjoint node sets U, W, we can derive two *projection networks*, one involving the nodes in U (say, investors) and one involving the nodes in W (say, companies). The projection network on U is obtained by considering all pairs of nodes in U and, for each node pair, creating an edge if both nodes have at least a common neighbor in W in the original bipartite network; to the created edge one can also associate a weight value, equal to the number of common neighbors, or to the relative sum of money invested in the common companies. Therefore, in the case of an investment network, two investors are linked in the projection network if they have one or more investments in common. Similarly, a projection network can

---

[5] In a perfect core–periphery network all core nodes form a connected component and are also linked to at least one periphery node, while periphery nodes are only linked to core nodes and not to other periphery nodes. In practice, real-world networks may have an approximate rather than perfect core–periphery structure.

[6] A scale-free network is one in which the degree distribution follows a power law, that is, the fraction $p(k)$ of the n nodes in the network having k connections to other nodes is such that $p(k) \propto k^{-\alpha}$, with $\alpha > 0$ and $k = 1, \ldots, n$. The power law implies that the vast majority of nodes have very few connections, while a few important nodes (called *hubs*) have a very large number of connections. Random scale-free networks can be generated via the method of growth and preferential attachment described in A.-L. Barabási and R. Albert, "Emergence of Scaling in Random Networks," *Science*, vol. 286(5439), pp. 509–512, 1999.

[7] An interesting application of this approach to the world-wide SWIFT (Society for Worldwide Interbank Financial Telecommunication) payments is discussed in S. Cook and K. Soramäki, "The global Network of Payment Flows," SWIFT Institute working paper n. 2012-006, 2014.

[8] Bipartite networks are characterized by graphs in which the vertex set V is divided into two disjoint sets U, W such that $U \cap W = 0, U \cup W = V$, and edges exist only between elements of U and W, that is, no edge exists among elements of U or among elements of W.

[9] S. Cook and K. Soramäki, *Network Theory and Financial Risk*, London: RiskBooks, 2022.

[10] J. Engel, M. Nardo, and M. Rancan, "Network Analysis for Economics and Finance: An Application to Firm Ownership," in S. Consoli et al. (eds.) *Data Science for Economics and Finance*, New York: Cham, Switzerland: Springer, 2021.

be constructed on the W vertex set according to the same criterion, that is, there is an edge among two nodes in W if they have one or more common neighbors in U. Therefore, two companies are linked in the projection network on W if they have one or more investors in common. Figure 13.8 shows an example of a bipartite network and two of its associated projection networks.

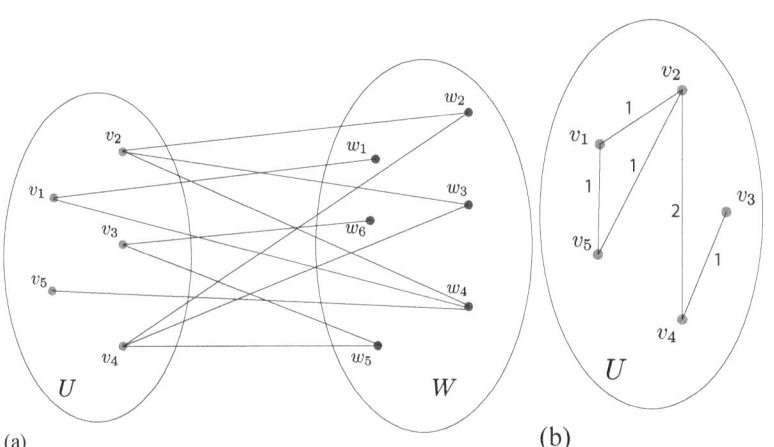
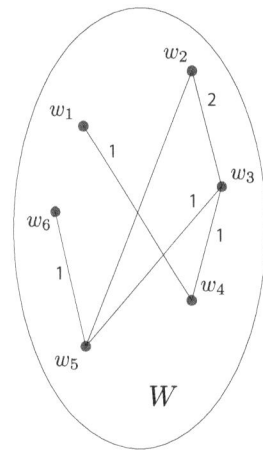

Figure 13.8 (a) A bipartite network and (b) its two projection networks.

13.3 Liability networks and default contagion

Liability networks are commonly used to study the systemic risk of default in financial interconnected systems. A liability network is described by means of n financial institutions (banks), which represent the nodes, and by a weighted directed adjacency matrix $\bar{P} = (\bar{p}_{ij})$, which represents the mutual liabilities among the financial institutions. Namely, $\bar{p}_{ij} \geq 0$ means that node i has an obligation to pay $\bar{p}_{ij}$ currency units to node j at the end of some fixed time period, and a directed link $e_{ij} = (i,j)$ from node i to node j exists if and only if $\bar{p}_{ij} > 0$. By definition, $\bar{p}_{ii} = 0$, $\forall i$, so the digraph contains no self-arcs. Along with mutual liabilities, the banks have outside assets. The outside asset $\bar{c}_i \geq 0$ represents the total payment due from non-financial entities (the external sector) to node i; these numbers constitute an external payments vector $\bar{c} = (\bar{c}_i)$. Similarly, one can consider the liability of node i towards the external sector as $\bar{b}_i \geq 0$. Often, the external liabilities are formally replaced by liabilities to an additional fictitious node representing the external sector. Adding this virtual node to the node set we shall henceforth assume without loss of generality that $\bar{b} = 0$.

For each node i we can write the classic two sides of the balance sheet, that is, the *asset* side, corresponding to the nominal cash inflow, and the *liability* side, corresponding to the nominal outflow:

$$\bar{\phi}_i^{\text{in}} \doteq \bar{c}_i + \sum_{k \neq i} \bar{p}_{ki}, \quad \bar{p}_i \doteq \bar{\phi}_i^{\text{out}} \doteq \sum_{k \neq i} \bar{p}_{ik}. \tag{13.2}$$

The nodes with $\bar{p}_i = 0$ have no outgoing arcs and, according to the graph-theoretical terminology, they are called *sinks*. As mentioned before, one such node can be fictitiously defined for the purpose of collecting the debts to the external sector. In general, however, other sinks may exist.

In regular operations it will hold that $\bar{\phi}_i^{\text{in}} \geq \bar{\phi}_i^{\text{out}}$, meaning that each bank is able to pay its liabilities at the end of the period. The main concern of systemic risk theory arises in the situation when a financial shock hits some nodes, meaning that the outside assets drop to smaller-than-expected values $c_i \in [0, \bar{c}_i)$. In this situation, it may happen that

$$c_i + \sum_{k \neq i} \bar{p}_{ki} < \bar{p}_i.$$

In such case, node i becomes unable to fully meet its payment obligations, and it then *defaults*. When in default, a node pays out according to its reduced capacity, thus reducing the amounts paid to the adjacent nodes, which in turn may also default and reduce their payments to other nodes, and so on in a cascaded fashion. As a result of default, the *actual* payments $p_{ij} \in [0, \bar{p}_{ij}]$ may be less than the nominal due payments $\bar{p}_{ij}$. The actual vectors of inflows and outflows are denoted by

$$\phi^{\text{in}} \doteq c + P^\top \mathbf{1}, \quad p \doteq \phi^{\text{out}} \doteq P\mathbf{1}. \tag{13.3}$$

The conditions to which the payments matrix $P = P(c, \bar{P})$ is subject to are as follows:[11]

(i) *limited liability*: the total payment of node i does not exceed the inflow, that is, $\phi^{\text{in}} \geq \phi^{\text{out}}$;

(ii) *absolute priority of debt claims*: either node i pays its obligations in full ($p_i = \bar{p}_i$), or it pays all its value to the creditors ($p_i = \phi_i^{\text{in}}$).

Recalling that $P \leq \bar{P}$ and $p = P\mathbf{1} \leq \bar{p}$, conditions (i) and (ii) may be reformulated compactly as

$$P\mathbf{1} = \min(\bar{p}, c + P^\top \mathbf{1}). \tag{13.4}$$

[11] For further details see the seminal reference by L. Eisenberg and T. H. Noe, "Systemic Risk in Financial Systems," *Management Science*, vol. 47(2), pp. 236–249, 2001.

A matrix P is called a *clearing matrix* (or matrix of clearing payments) corresponding to the vector of outside assets c, if $0 \leq P \leq \bar{P}$, and (13.4) holds.

Notice that (13.4) is a system of n nonlinear equations in the n^2 variables p_{ij}. Hence, one cannot expect to find a unique solution in general. To obtain uniqueness of the solution (in the generic situation), a third requirement is typically introduced, known as the *proportionality* or *pro-rata* rule, which expresses the requirement that all debts have equal priority and must be paid in proportion to the initial claims. The imposition of this rule reduces the number of variables to n. It is known that under the pro-rata rule a clearing vector always exists; also, one such vector can be found by solving a convex optimization problem with n variables or applying a standard iterative method known as the *fictitious default algorithm*.[12]

The pro-rata rule reflects an underlying criterion of local fairness among neighboring nodes, and it is a convention enforced in many contracts. Under the pro-rata rule the payments p_{ij}, of node i to the claimants, should be proportional to the nominal liabilities $\bar{p}_{ij}$. Define the matrix of relative liabilities

$$A = (a_{ij}), \quad a_{ij} = \begin{cases} \frac{\bar{p}_{ij}}{\bar{p}_i}, & \text{if } \bar{p}_i > 0, \\ 1, & \text{if } \bar{p}_i = 0 \text{ and } i = j, \\ 0, & \text{otherwise}. \end{cases} \quad (13.5)$$

[12] See the previously mentioned paper by Eisenberg and Noe, and the more recent review by P. Glasserman and H. P. Young, "Contagion in Financial Networks," *Journal of Economic Literature*, vol. 54(3), pp. 779–831, 2016.

By definition, matrix A is *stochastic*, that is, $a_{ij} \geq 0$ and $\sum_j a_{ij} = 1$ for all i or, equivalently, $A\mathbf{1} = \mathbf{1}$. The pro-rata rule can then be formulated as $P = \text{diag}(P\mathbf{1})A$ or, equivalently,

$$p_{ij} = p_i a_{ij}, \quad \forall i, j = 1, \ldots, n. \quad (13.6)$$

Under (13.6) it holds that

$$(P^\top \mathbf{1})_i = \sum_{j=1}^n (P^\top)_{ij} = \sum_{j=1}^n p_{ji} = \sum_{j=1}^n p_j a_{ji} = (A^\top p)_i, \quad \forall i,$$

which allows us to rewrite (13.4) in the equivalent vector form

$$p = \min(\bar{p}, c + A^\top p). \quad (13.7)$$

A vector $p \geq 0$ is said to be a *clearing* vector if it satisfies (13.7). A clearing vector can be found by solving the following linear program:

$$\min_{p \in \mathbb{R}^n} \sum_{i=1}^n (\bar{p}_i - p_i), \quad (13.8)$$
$$\text{s.t.: } \bar{p} \geq p \geq 0,$$
$$c + A^\top p \geq p.$$

This types of clearing vector is also unique under mild conditions, such as, for instance, that there exist a path from any node to the set $C^+ \doteq \{i : c_i > 0\}$. Observe that whenever defaults arise in the network, we have $p \leq \bar{p}$. In such case, the total loss in values due to reduced payments to both the financial and non-financial sectors is measured by $L(c) \doteq \sum_{i=1}^{n}(\bar{p}_i - p_i)$, which is the objective of the LP (13.8). The clearing vector therefore also minimizes the total loss, under the stated pro-rata, limited liability and absolute priority constraints.

The total loss $L(c)$ of a network corresponding to a given configuration of external payments c can thus be easily computed, in principle, via linear programming. In practice, however, the problem is that reliable data on the banks mutual liabilities, and hence on the $\bar{P}$ matrix, is rarely available. Also in this case, therefore, researchers typically work on proxy models of reality, constructed on the basis of certain structural characteristics that are empirically found in the available use cases, such as core–periphery structures and power-law distributions for the node degrees.[13]

[13] See, for example, Section 9 in the cited survey paper by Glasserman and Young (see note 10).

Besides the evaluation of the total loss for a collection of scenarios of the external payments, another key issue is related to the identification of the nodes that are most fragile or, contrary, those that are most resilient to the default contagion. A precise analysis would of course depend on the actual liability structure and on the value of the output payment vector. Researches have observed empirically that relations exist between a node's probability of default and its eigenvector *centrality*,[14] so that higher centrality of a node is likely to imply a lower probability of default for that node. However, a downside effect has also been observed, whereby higher centrality of a node also corresponds to higher contagiousness of that node's default to the rest of the network; we direct the reader to Section 9 of the cited paper by Glasserman and Young for further details on Network measures and their influence on systemic risk.

[14] See Section 13.4 for a discussion on centrality measures

Example 13.2 (*A simple liability network*) We consider a simple network composed of $n = 5$ nodes (including the fictitious node representing the external sector), with nominal liability matrix

$$\bar{P} = \begin{bmatrix} 0 & 180 & 0 & 0 & 180 \\ 0 & 0 & 100 & 0 & 100 \\ 90 & 0 & 0 & 100 & 50 \\ 150 & 0 & 0 & 0 & 150 \\ 0 & 0 & 0 & 0 & 0 \end{bmatrix},$$

where the last row refers to the fictitious node. Suppose there is a nominal scenario where external cash inflows are given as

$$c = c_{\text{nom}} \doteq [121,\ 21,\ 150,\ 204,\ 0]^\top.$$

It can be readily verified that in this nominal scenario all the nodes in the network remain solvent (no defaults), and the clearing payments coincide with the nominal liabilities. Consider next a situation in which a "shock" happens on the inflow at node 3, so that this inflow reduces from 150 to 120, that is,

$$c = c_{shock} \doteq [121, 21, 130, 204, 0]^\top.$$

Under the pro-rata rule, the (unique) clearing payments resulting from the solution of (13.8)[15] are shown in smaller font below the nominal liabilities in Figure 13.9: all nodes in the network default in a cascade fashion due to initial default of node 3. The total defaulted amount, defined as the sum of all the unpaid liabilities is in this case $L(c_{shock}) = 13.98$.

[15] See file ex_liabilitynetw.m in the online resources.

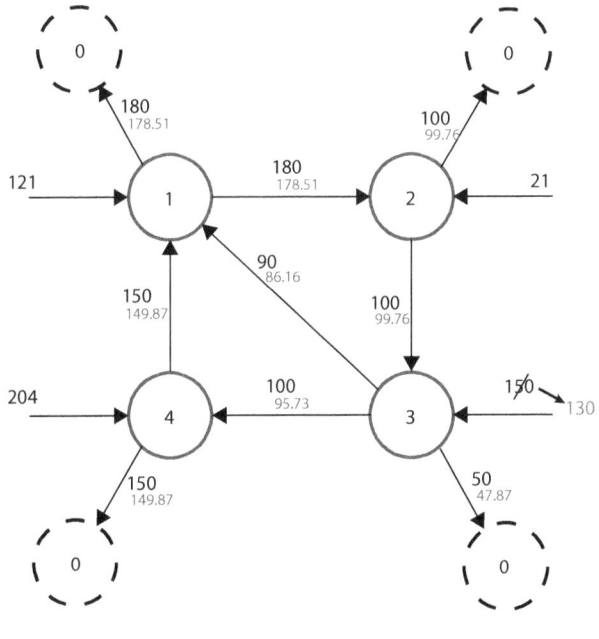

Figure 13.9 A five nodes network with a shock on the cash inflow at node 3. In larger font the nominal liabilities, in smaller font the clearing payments.

13.4 Centrality measures

Centrality measures seek to quantify the importance of a node in a network. Many definitions of centrality have been proposed in the literature, depending on the nature and purpose of the network model. Perhaps the simplest centrality measure in a network is just the node degree, that is, the number of edges connected to it. In social networks, for instance, a node's degree measures the connections that an individual has with other individuals, and it may be reasonable to think that individuals with many connections have more influence on the network's dynamics (e.g., of opinions) than individuals with fewer connections. Similarly, in the context of payment networks, banks with higher in-degree represent hubs that are critical to the functioning of the whole network, and in liability networks nodes with high out-degree are banks likely to have high debts towards other institutions, and thus potentially prone to default in case of shocks. In cases when a positive weight value is attached to the graph edges (representing, for instance, liability values), a more reliable measure of centrality is the weighted degree, or *strength*, defined for each node as the sum of the weight values of the incident edges. For digraphs, we define the weighted in-degree and the weighted out-degree as the sum of the weights of the incoming and outgoing links, respectively.[16] We shall henceforth assume that the weighted adjacency matrix A has all nonnegative entries. Degree and strength are local metrics, since they count only the immediate neighbors of each node, ignoring the rest of the network's structure.

[16] The weighted in-degree of vertex v_i corresponds to the sum of the entries in the ith column of the directed weighted adjacency matrix, and its weighted out-degree corresponds to the sum of the entries in the ith row.

13.4.1 Eigenvector centrality

A more sophisticated concept of centrality is given by *eigenvector centrality*, which is based on the principal eigenvector of the (possibly weighted) adjacency matrix $A \in \mathbb{S}^n$ of a (undirected) graph. The intuitive concept behind eigenvector centrality is that the importance value x_i of node i, rather than being simply given by the degree $\sum_j a_{ij}$, is given by a linear combination of the values of the neighboring nodes, that is,

$$x_i = \kappa \sum_j a_{ij} x_j, \quad i = 1, \ldots, n.$$

This reflects the intuition that a node in a network is important not only if it has many connections, but also and prominently if it is connected to other nodes that are also important: a node's importance is the weighted combination of its neighbors' importances. In vector notation the previous expression can be written as

$$Ax = \lambda x, \qquad (13.9)$$

where A is the adjacency matrix of the graph, $\lambda = \kappa^{-1}$, and x is the n-vector of nodes' importances. Equation (13.9) says that the vector of importances x is an eigenvector of A associated to its eigenvalue λ. We next recall a fundamental result on *nonnegative*[17] matrices, known as the Perron–Frobenius theorem.

[17] A real matrix A is *nonnegative* if all its entries are nonnegative numbers. The matrix is said to be *positive* if all its elements are positive numbers.

Theorem 13.1 (Perron–Frobenius) *If $A \in \mathbb{R}^{n,n}$ is nonnegative and irreducible,[18] then:*

[18] For an undirected graph, the adjacency matrix A is irreducible whenever the graph is connected. For a directed graph, A is irreducible whenever the digraph is *strongly connected*, meaning that there exist a directed path from any node to any other node.

1. *A has a positive eigenvalue $\bar{\lambda} > 0$ (the so-called Perron–Frobenius eigenvalue), and it holds that $|\lambda_i| \leq \bar{\lambda}$ for all other eigenvalues λ_i of A. Moreover, $\bar{\lambda}$ is simple, that is, it has unit multiplicity.*

2. *The left and right[19] eigenvectors u, v of A associated to $\bar{\lambda}$ have all positive components, and only the eigenvectors associated to $\bar{\lambda}$ have this property (u and v are unique, up to normalization).*

3. *It holds that*

$$\bar{\lambda} \geq \max(\min_i \text{outdeg}(i), \min_j \text{indeg}(j)),$$
$$\bar{\lambda} \leq \min(\max_i \text{outdeg}(i), \max_j \text{indeg}(j)),$$

where $\text{outdeg}(i), \text{indeg}(i)$ *denote the (possibly weighted) out-degree and in-degree of node i, respectively.*

[19] A left eigenvector u of A is such that $u^\top A = \lambda u^\top$; a right eigenvector v is such that $Av = \lambda v$.

A strengthened version of Theorem 13.1 holds for *primitive*[20] matrices.

Corollary 13.1 *If $A \in \mathbb{R}^{n,n}$ is primitive, then all statements of Theorem 13.1 hold and, moreover, $|\lambda_i| < \bar{\lambda}$ for all other eigenvalues λ_i of A, and*

$$\lim_{k \to \infty} \frac{A^k}{\bar{\lambda}^k} = vu^\top,$$

where u, v are the left and right eigenvectors of A, respectively, normalized so that $u^\top v = 1$.

[20] A nonnegative matrix A is *primitive* if there exists an integer k such that A^k has all positive elements. Positive matrices are of course primitive. Primitive matrices are irreducible. If A is irreducible and has at least one nonzero diagonal element, then A is primitive. A graph's adjacency matrix is primitive if and only if the underlying graph is connected (resp., strongly connected for digraphs) and *acyclic*, meaning that the greatest common divisor of the lengths of all loops in the graph is equal to one.

In the case of undirected graphs the adjacency matrix is symmetric and hence left and right eigenvectors coincide. Further, since nodes importances need be positive, the λ value of interest in equation (13.9) is the Perron–Frobenius eigenvalue $\lambda = \bar{\lambda}$, and the importances x are found as the corresponding eigenvector of A. Nodes' importances are typically normalized so to sum to 1.

In the case of directed graphs the situation is slightly different, since there are two notions of centrality, one related to the incoming links at a node, which we shall call in-bound centrality, or *hub centrality*, and one related to outgoing links, which we shall call out-bound centrality, or *authority centrality*. The most commonly used notion is that of in-bound centrality. Given the adjacency matrix A (non necessarily symmetric, for digraphs), we define the in-bound centrality of node j as

$$x_j^{\text{in}} = \kappa \sum_i x_i^{\text{in}} a_{ij},$$

or equivalently, in vector notation and letting $\lambda = \kappa^{-1}$,

$$A^\top x^{\text{in}} = \lambda x^{\text{in}},$$

which tells us that x^{in} is a left eigenvector of A. If A is irreducible, we conclude from Theorem 13.1 that the desired vector x^{in} of in-bound centrality values is given by the left eigenvector of A associated to the Perron–Frobenius eigenvalue $\lambda = \bar\lambda$. Similarly, the out-bound centrality of nodes is given by the equation

$$Ax^{\text{out}} = \lambda x^{\text{out}},$$

and, if A is irreducible, we conclude from Theorem 13.1 that the desired vector x^{out} of out-bound centrality values is given by the right eigenvector of A associated to the Perron–Frobenius eigenvalue $\lambda = \bar\lambda$.

In-bound and out-bound centralities have specific meanings depending on the context and type of network under consideration. In the case of liability networks, for instance, nodes with high in-bound centrality (funding centrality) are lending money to – and hence have claims on – nodes with high in-bound centrality. On the other hand, nodes with high out-bound centrality (borrowing centrality) are borrowing money from – and hence have debts with – nodes with high out-bound centrality. A default of a node with high funding centrality may cause a liquidity shock in the network, due to withdrawal of funds. On the other hand, a default of a node with high borrowing centrality may cause failure to pay back a large amount of debts, and hence cause a default cascade. In the context of social networks and web analytics the most used metric is in-bound centrality, since the importance of an individual in a social network is dictated mostly by the number of nodes that point at that individual (e.g., followers), rather than by the number of nodes that are pointed by that individual. Similarly, in web analytics the importance of a web page depends mostly on the pages that point to that page rather than on the pages that are pointed; in-bound eigenvector centrality is indeed the concept applied in the celebrated Google PageRank algorithm for ranking web pages.

Markov centrality Let $A \in \mathbb{R}^{n,n}$, $A \geq 0$, be the adjacency matrix of a strongly connected digraph, and let

$$p_{ij} \doteq \frac{a_{ij}}{d_i}, \quad d_i \doteq \sum_k a_{ik}, \quad i,j = 1,\ldots,n.$$

Matrix $P = (p_{ij})$ is a normalized version of A, in which each row i is divided by the sum of its entries (representing the weighted out-degree of node i); letting $D \doteq \text{diag}(d_1,\ldots,d_n)$, it can be expressed as $P = D^{-1}A$. In particular, P can be interpreted as the transition probability matrix of a Markov chain on the nodes of the graph, where p_{ij} represents the conditional probability of the chain to transition from node i to node j in the next step, given it was in i. It holds by construction that $P\mathbf{1} = \mathbf{1}$, which means that P is a (row) stochastic matrix and has $\bar{\lambda} = 1$ as its Perron–Frobenius eigenvalue. If we use this normalized version P of the adjacency matrix for defining the in-bound centrality, we have that the centrality vector x is given by the left eigenvector of P associated to the unit eigenvalue, that is, $x^\top P = x^\top$. Here x, normalized so that $\sum_i x_i = 1$, corresponds to the steady-state probability of the Markov chain, and is referred to as the *Markov centrality* vector. An interpretation, for instance in a liability network context, can be given in terms of the average dwell time of a dollar circulating in the network: if such dollar jumps from node to node according to the Markov transition probabilities in P, then in the long run it will pass $x_i\%$ of its time visiting node i.

Example 13.3 We consider as a numerical example the adjacency matrix

$$A = \begin{bmatrix} 0 & 0 & 0 & 1 & 0 \\ 0 & 0 & 0 & 1 & 1 \\ 0 & 1 & 0 & 0 & 0 \\ 0 & 0 & 1 & 0 & 0 \\ 1 & 0 & 1 & 0 & 0 \end{bmatrix},$$

which corresponds to a modification of the digraph shown on the right in Figure 13.1, with a link added from v_1 to v_4. This modified digraph is strongly connected, hence A is irreducible. The diagonal matrices of in-degrees and out-degrees are, respectively $D_{\text{in}} = \text{diag}(1,1,2,2,1)$ and $D_{\text{out}} = \text{diag}(1,2,1,1,2)$. The Perron–Frobenius eigenvalue of A results to be $\bar{\lambda} = 1.3640$, and the left and right eigenvectors are, respectively,

$$u = \begin{bmatrix} 0.1129 \\ 0.2100 \\ 0.2864 \\ 0.2367 \\ 0.1540 \end{bmatrix}, \quad v = \begin{bmatrix} 0.1129 \\ 0.2864 \\ 0.2100 \\ 0.1540 \\ 0.2367 \end{bmatrix}.$$

The left eigenvector u represents in-bound relative centralitites, and node 3 is the most central, according to this metric. The right eigenvector v represents out-bound relative centralitites, and node 2 is the most central in this case.

We next compute Markov centralities: the transition probability matrix $P = D_{\text{out}}^{-1} A$ is

$$P = \begin{bmatrix} 0 & 0 & 0 & 1 & 0 \\ 0 & 0 & 0 & \frac{1}{2} & \frac{1}{2} \\ 0 & 1 & 0 & 0 & 0 \\ 0 & 0 & 1 & 0 & 0 \\ \frac{1}{2} & 0 & \frac{1}{2} & 0 & 0 \end{bmatrix},$$

and the left eigenvector if P associated with the unit Perron–Frobenius eigenvalue is

$$u_{\text{mkv}} = \begin{bmatrix} 0.0714 \\ 0.2857 \\ 0.2857 \\ 0.2143 \\ 0.1429 \end{bmatrix}.$$

Observe that Markov centrality gives a result which is different from both the plain left and right eigenvectors of A, and it assigns equal maximum importance to nodes 2 and 3.

As a further example, we consider the liability network of Example 13.2, and use the nominal liabilities matrix $\bar{P}$ restricted to the subgraph composed of the financial sector nodes (i.e., excluding the fictitious 0 node) as the weighted adjacency matrix of the graph, which results in

$$A = \begin{bmatrix} 0 & 180 & 0 & 0 \\ 0 & 0 & 100 & 0 \\ 90 & 0 & 0 & 100 \\ 150 & 0 & 0 & 0 \end{bmatrix}.$$

In this case the left and right Perron–Frobenius eigenvectors are given respectively by

$$u = \begin{bmatrix} 0.2767 \\ 0.3308 \\ 0.2197 \\ 0.1459 \end{bmatrix}, \quad v = \begin{bmatrix} 0.2444 \\ 0.2044 \\ 0.3078 \\ 0.2435 \end{bmatrix}.$$

Node 3 has the largest borrowing centrality, and indeed we saw in Example 13.2 that a default in this node provoked, in the given scenario, the cascaded defaults of all nodes in the network.

Google PageRank The network structure may not always be strongly connected. For instance, there could exist *sink* nodes that have no outgoing links, which would produce a row of zeros in the adjacency matrix, or more generally the network graph may have more than one connected component. In all these cases the adjacency matrix is not primitive, nor irreducible, in general. Let A be the original adjacency matrix of the network, and let P represent the corresponding Markov transition matrix, obtained as $P = D^{-1}A$, where $D = \text{diag}(d_1, \ldots, d_n)$ being $d_i = \text{outdeg}(i)$ if $\text{outdeg}(i) \neq 0$, and $d_i = 1$ if $\text{outdeg}(i) = 0$. In the context of web analytics, matrix $P = (p_{ij})$ describes the probability with which a generic user visiting page i will jump to page j. In reality, however, such probability is not simply dictated by the local network structure (outgoing links available at page i), since the user may decide to randomly jump to any other page, even if not directly adjacent to the current page. A network in which a user jumps uniformly at random to any page would be described by a transition matrix of the form $E \doteq \frac{1}{n}\mathbf{1}\mathbf{1}^\top$. The idea behind Google's PageRank is thus to blend the transition probability derived from the network structure and summarized by P with a purely uniform random transition model. This is achieved by considering a transition probability matrix of the form

$$P(\alpha) \doteq (1-\alpha)P + \alpha E,$$

where $\alpha \in [0,1]$ is the blending parameter that controls how much randomness to add to the original link-based transition matrix P; a value of $\alpha = 0.15$ has seemingly been used by Google in some of the original implementations of the PageRank algorithm. We further observe that, for any $\alpha \in (0,1]$, matrix $P(\alpha)$ is (row) stochastic and positive, and therefore it is also primitive, and the results of Corollary 13.1 apply. The idea of adding to the normalized adjacency matrix a small multiple of a matrix of the form $\frac{1}{n}\mathbf{1}\mathbf{1}^\top$ constitutes in practice a common trick used to insure that the blended matrix is primitive, which not only guarantees the existence and uniqueness of a positive Perron–Frobenius eigenvector, but also insures that a standard power iteration algorithm for computing such eigenvector converges exponentially fast to the desired solution, as is further discussed in the following paragraph.

The Power Iteration algorithm Let $A \in \mathbb{R}^{n,n}$ be primitive, and let $v(0)$ be any given positive vector in $\mathbb{R}^n$, normalized so that $\|v(0)\|_2 = 1$. Consider the recursion

$$v(k+1) = \frac{Av(k)}{\|Av(k)\|_2}, \quad k = 0,1,\ldots. \quad (13.10)$$

At generic $k > 0$, we have that

$$v(k) = \frac{A^k v(0)}{\|A^k v(0)\|_2}, \quad k = 1,2,\ldots.$$

By applying the second statement of Corollary 13.1 we have that $A^k \to \bar{\lambda}^k v u^\top$ for $k \to \infty$, where $\bar{\lambda} > 0$ is the Perron–Frobenius eigenvalue of A, and u, v are the corresponding left and right eigenvectors. Therefore,

$$v(k) = \frac{A^k v(0)}{\|A^k v(0)\|_2} \to \frac{\bar{\lambda}^k v u^\top v(0)}{\|\bar{\lambda}^k v u^\top v(0)\|_2} = \frac{v}{\|v\|_2}, \text{ for } k \to \infty,$$

where the last passage holds since all quantities above are positive, and provided that $u^\top v(0) \neq 0$.[21] The recursion in (13.10) is a version of the Power Iteration algorithm already discussed in Section 4.2.2.

[21] Since the initial vector $v(0)$ of the recursion is typically chosen at random, there is zero probability that $u^\top v(0)$ is exactly zero.

Under the stated assumptions on A this algorithm produces a sequence of normalized vectors $v(k)$ that converges to the (normalized) right Perron–Frobenius eigenvector of A. Further, by Corollary 13.1 the strict inequality holds $\bar{\lambda} > |\lambda_i|$ for all other eigenvalues of A, hence we can define the *spectral ratio* $\delta \doteq |\lambda_2|/\bar{\lambda} < 1$, where λ_2 denotes an eigenvalue of A having the second-largest modulus, and it can be proved that the convergence of the power iterations happens at a geometric rate δ^k, hence the smaller δ the faster will be the convergence.

An analogous recursion can be constructed for the left eigenvector: let $u(0)$ be any given positive vector in $\mathbb{R}^n$, normalized so that $\|u(0)\|_2 = 1$, and consider the recursion

$$u(k+1) = \frac{A^\top u(k)}{\|A^\top u(k)\|_2}, \quad k = 0, 1, \ldots. \qquad (13.11)$$

This recursion produces a sequence of normalized vectors $u(k)$ that converge geometrically to $u/\|u\|_2$, that is, to the normalized left Perron–Frobenius eigenvector.

13.4.2 Closeness and betweenness centrality

The *closeness* centrality C_i of a node i measures the centrality of i in terms of the inverse of its mean distance from the other vertices, where the distance d_{ij} from node i to node j is the (possibly weighted) length of the shortest path joining the two vertices:

$$C_i \doteq \left(\frac{\sum_j d_{ij}}{n} \right)^{-1}.$$

This measure of centrality suffers from some drawbacks. First, it is zero for all nodes whenever the network has more than one connected component, since the distance from a node in one component to a node in a different component is infinite, hence the inverse is zero. Another problem is that the range of C_i values is typically small, in the sense that high centrality values are numerically close to low centrality values, and this yields instability in the nodes' centrality ranking as soon as the network is perturbed. Closeness centrality is nevertheless a popular centrality measure in social networks and opinion dynamics, where it is used to determine which individuals are placed to influence the entire network most quickly.

Another different concept of centrality is given by *betweenness centrality*, which essentially counts how many shortest paths pass through a given vertex. Letting n_{od}^i denote the number of shortest paths from origin o to destination d that pass though node i, and letting q_{od} the total number of shortest paths from o to d, we define the betweenness of i as

$$x_i = \sum_{o,d} \frac{n^i_{od}}{q_{od}},$$

where the summation is extended to all possible origin-destination pairs, with the convention that $n^i_{od}/q_{od} = 0$ when both factors are zero. Nodes with high betweenness represent intermediary nodes through which a large part of the network traffic passes. If we think, for instance, of a payments network and we let payments in the network flow over time, nodes with high betweenness are those through which a large part of the money passes. In communication or traffic networks, these nodes may represent bottlenecks or bridges that are important for controlling the traffic through the network.

Example 13.4 (*Global economies and direct investments*) A direct investment arises when an investor resident in one economy (country) makes an investment that gives control or a significant degree of influence over the management of an enterprise that is resident in another economy. We used as an example global data from the International Monetary Fund (IMF) that results from a Coordinated Direct Investment Survey (CDIS). The CDIS database[22] contains breakdowns of inward direct investment position data for a total of $n = 246$ (after data cleaning) countries, as of end-2021. Data preprocessing involved removing countries with no in-bound and no out-bound investments, and setting to zero negative values and values marked as confidential. We hence created a directed network with adjacency matrix $A = (a_{ij})$, where $a_{ij} \geq 0$ is the direct investment (in million US dollars) that country i makes in country j. The resulting graph has three nodes with zero in-degree, that is, countries that receive no direct investment, and 135 nodes with zero out-degree, that is, countries that have reported no direct investment abroad. The graph is therefore not strongly connected, it has 130 singleton components, and one largest strongly connected component composed of 108 countries (see Figure 13.10).

The distributions of the in-degree and of the out-degree are shown in Figure 13.11, with the out-degree following approximately a power law.

The five countries with largest weighted in-degree centrality[23] are, in descending order: United States, the Netherlands, China, Lithuania, and United Kingdom. The five countries with largest weighted out-degree centrality are, in descending order: United States, the Netherlands, United Kingdom, Luxembourg, and China.

Focussing on the strong component of the network, we have that the five countries with the largest in-bound (hubs) centrality are, in descending order: Honduras, Bosnia-Herzegovina, Bulgaria, Comoros, and Ghana, while the five countries with the largest out-bound centrality are, in descending order: the Falklands Islands, Cambodia, Iran, Indonesia, and Côte d'Ivoire. It can be observed that some of these central countries correspond to more or lesser known tax havens. This aspect is confirmed if

[22] See https://data.imf.org/, see also the data file DI_inward.xls in this book's online resources.

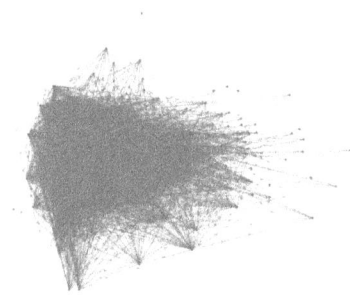

Figure 13.10 The subgraph corresponding to the 108 nodes in the strong component of the original network.

[23] The weighted in-degree of a node here corresponds to the total investment received by that node; the weighted out-degree of a node corresponds to the total investment made by that node.

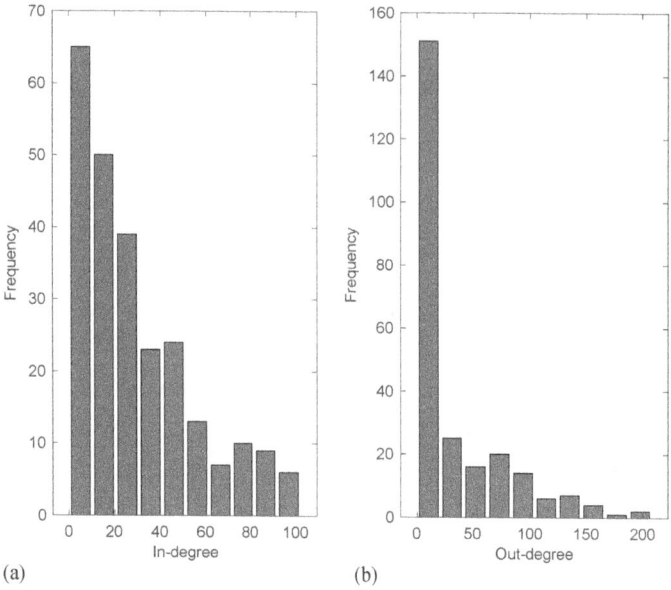

Figure 13.11 (a) In-degree and (b) out-degreee distribution of the investment network; see file Foreign_investments_network.m in the online resources.

we compute the betweenness centrality of the nodes in the strong component, which highlights which countries effectively act as bridges for direct investment flows; the five countries with the largest betweenness centrality are, in descending order: Bosnia-Herzegovina, the Falklands Islands Indonesia, Cambodia, and Honduras. To obtain a simplified view of the network, we extracted the subgraph containing the 17 nodes with highest weighted in-degreee, which account for about 80% of the total investments; the resulting sub-network is shown in Figure 13.12.

Figure 13.12 The subnetwork formed by the 17 countries of highest weighted in-degree.

13.5 Clustering and community detection

Communities in a network are dense groups of vertices which are tightly coupled to each other inside the group and loosely coupled to the rest of the vertices in the network. An example of a community is given by a maximum *clique*, which is the largest complete[24] subgraph in a graph, or by an *r*-clique, which is the largest subgraph such that the distance between each pair of its nodes is at most *r*. Community detection plays a key role in understanding the structure of complex networks, and it is an extensively studied topic. It is out of the scope of this monograph to give a complete account of community detection approaches, so we shall here concentrate only on the mainstream approach based on the notion of *modularity*, and refer the reader to the existing literature and surveys[25] for further information on this topic.

Modularity measures the quality of a network subdivision into communities (subnetworks) by evaluating the fraction of the edges in the network that connect vertices in the same community (within-community edges), minus the expected value of the same quantity in a network with the same community divisions but random connections between the vertices. If the number of within-community edges is no better than random, the modularity will be zero; values of the modularity approaching one indicate networks with strong community structure. For a simple undirected graph, given a community subdivision C, the modularity $Q \in [-1, 1]$ is defined as

$$Q(C) = \frac{1}{w_{\text{tot}}} \sum_{ij} \left(a_{ij} - \frac{w_i w_j}{w_{\text{tot}}} \right) C(i,j), \qquad (13.12)$$

where $A = (a_{ij})$ is the weighted adjacency matrix of the graph, $w_i \doteq \sum_j a_{ij}$ is the total weight of the edges incident on i, $w_{\text{tot}} = \mathbf{1}^\top A \mathbf{1}$ is the sum of all edges weights, and $C(i,j)$ is the community indicator function, which is one if i and j belong to the same community and it is zero otherwise. Community detection can therefore be cast as an optimization problem where one seeks to maximize the modularity with respect to the network subdivision C. This problem, however, is an NP-hard integer programming problem, which is difficult to solve in practice to global optimality. Nevertheless, efficient heuristic algorithms exist for finding possibly good solutions. Newman and Girvan (2004) propose an approach based on the concept of edge betweenness, which is similar to the node betweenness discussed in Section 13.4.2. For instance, shortest-path betweenness for an edge is computed by finding the shortest paths between all pairs of vertices and counting what fraction runs along that edge. Their

[24] A graph is *complete* when all nodes are connected to all other nodes.

[25] See, for instance, M. E. J. Newman and M. Girvan, "Finding and Evaluating Community Structure in Networks," *Physical Review E*, vol. 69, 026113, 2004, and the survey by S. Fortunato and D. Hric, "Community Detection in Networks: A User Guide," *Physics Reports*, vol. 659, pp. 1–44, 2016.

community structure finding algorithm is a divisive hierarchical clustering method which proceeds iteratively starting from the full (connected) network and executing the following steps:[26]

1. Compute the betweenness scores for all edges in the network.

2. Find the edge with the highest score and remove it from the network.

3. Recompute betweenness for all remaining edges.

4. Repeat from step 2.

[26] See Newman and Girvan (2004) for implementation details (see note 23).

Step 2 finds the edge that is, in some sense, responsible for connecting many pairs of vertices; by removing repeatedly such type of edge the network gets divided into smaller and smaller components. The process can be stopped at any stage and the components at that stage constitute the network communities. The process can be represented as a dendrogram which depicts the successive splits of the network into smaller and smaller groups. At each split we can evaluate the modularity of the subdivision, and stop the procedure at a satisfactory local maximum of the modularity.

An alternative heuristic that efficiently finds high-modularity partitions of large networks and that unfolds a complete hierarchical community structure for the network is known as the Louvain method.[27] The Louvain method is an iterative agglomerative method that starts by assigning a different community to each node and then iterates two phases: a first phase where modularity is optimized by allowing only local changes of communities, and a second phase where the found communities are aggregated in order to build a new network of communities. The passes are repeated iteratively until no increase of modularity is possible.

[27] See V. D. Blondel, J.-L. Guillaume, R. Lambiotte, and E. Lefebvre, "Fast Unfolding of Communities in Large Networks," *Journal of Statistical Mechanics*, 2008.

Example 13.5 Consider the global direct investment network discussed in Example 13.4. The Louvain algorithm applied to the strong component of this network detects $k = 5$ communities, with a weak modularity of $Q = 0.25$. The subgraphs corresponding to these five communities are shown in Figure 13.13.

We further run the Louvain algorithm on the correlation network introduced in Example 13.1; in particular, we used the correlation network obtained via covariance graphical lasso with $\lambda = 2,000$. The algorithm finds in this case $k = 4$ communities of size $10, 8, 8$, and 4, with a final modularity of $Q = 0.27$, which denotes rather weak community structure, see Figure 13.14. Code for both examples can be found in file Community_example.m in the online resources.

13.5 CLUSTERING AND COMMUNITY DETECTION 365

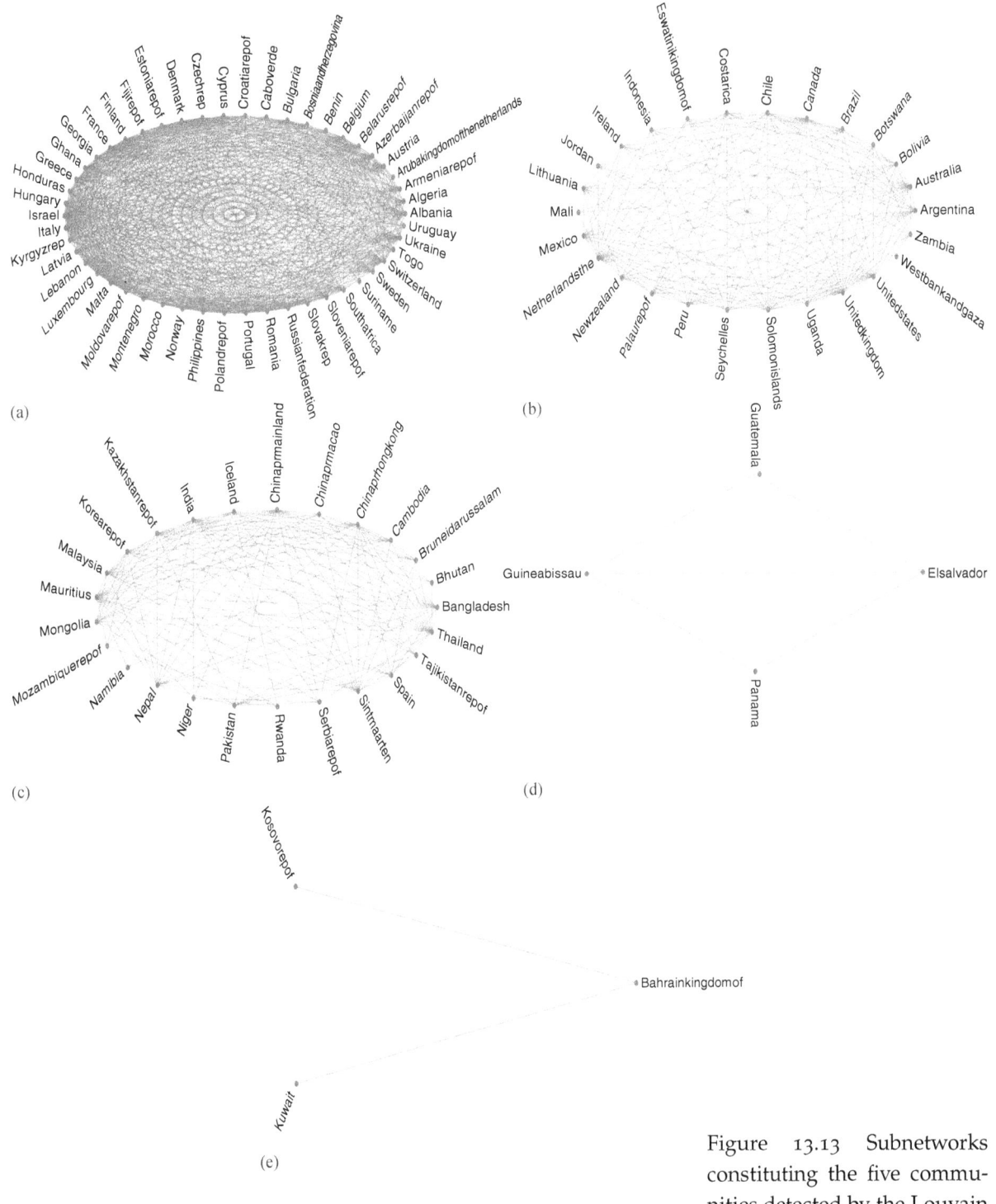

Figure 13.13 Subnetworks constituting the five communities detected by the Louvain algorithm on the global direct investment network. The sizes of the communities are (a) 50, (b) 26, (c) 25, (d) 4, (e) 3.

13 FINANCIAL NETWORKS

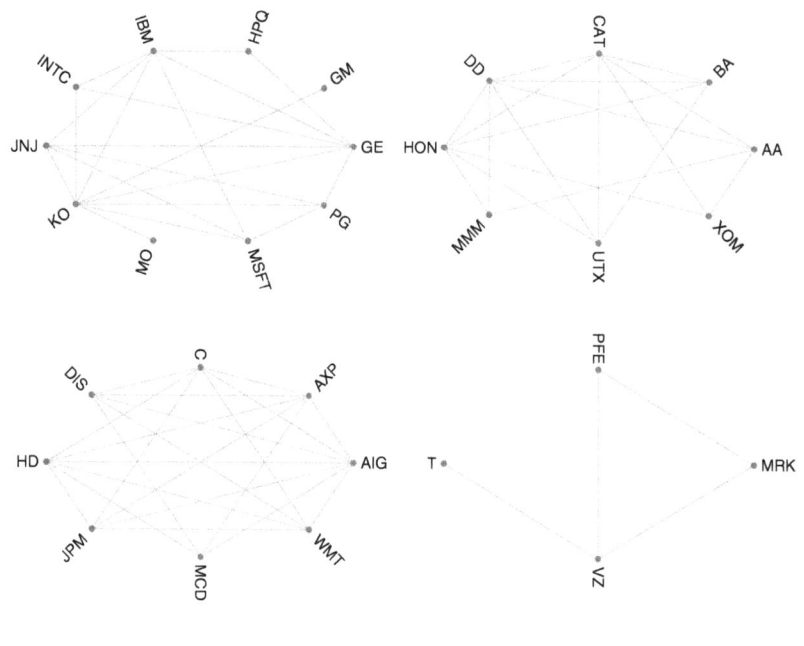

Figure 13.14 Subnetworks constituting the four communities detected by the Louvain algorithm on the DJI correlation network. The sizes of the communities are (a) 10, (b) 8, (c) 8, and (d) 4.

13.6 Exercises

Exercise 13.1 *Graphs' edges*
Prove that a connected graph on n vertices has at least $n-1$ edges. Prove that an acyclic graph on n vertices has at most $n-1$ edges.

Exercise 13.2 *A money routing problem*
Consider a network composed of n financial entities (e.g., banks) connected via a set of m directed edges representing the admissible transaction channels, that is, bank i can send money to bank j only if there is a directed edge from i to j. The network structure is described by means of the directed incidence matrix $B \in \mathbb{R}^{n,m}$, where $B_{ij} = 1$ if edge j enters node i, $B_{ij} = -1$ if edge j exits node i, and $B_{ij} = 0$ if edge j is not incident on i. Node 1 is the source node, generating an amount $B > 0$ of money that needs to be sent to its destination node n through the network. All other nodes $i = 2, \ldots, n-1$, are relaying nodes, who receive a money flow and pass it on towards the destination. Each edge $j = 1, \ldots, m$, has an associated cost $c_j \geq 0$, which represents the cost of transmitting one unit of money though that edge. Each edge $j = 1, \ldots, m$, also has a maximum capacity $u_j > 0$, which represents the maximum amount that can be transmitted over that edge. The problem is to find the optimal routing of the funds B from origin to destination, so to minimize the overall transaction cost. Explain how this problem can be formalized and solved efficiently by means of linear programming.

Exercise 13.3 *Organizations' values with cross-holdings*
Consider n organizations. The value of each organization depends on the owned shares of m primitive assets and of other organizations. The present value (or market price) of asset k is denoted by p_k, $k = 1, \ldots, m$, and we let $D_{ik} \geq 0$ be the share of the value of asset k held by (i.e., flowing directly into) organization i, with $\sum_i D_{ik} = 1$ for all k. An organization can also hold shares of other organizations: for any i, j the number $C_{ij} \geq 0$ is the fraction of organization j owned by organization i, where $C_{ii} = 0$ for each $i = 1, \ldots, n$. The matrix C can be thought of as a network in which there is a directed link from i to j if i owns a positive share of j. After all these cross-holding shares are accounted for, there remains a share $\hat{C}_{ii} \doteq 1 - \sum_{j=1}^{n} C_{ji}$ of organization i not owned by any organization in the system – a share assumed to be positive. This is the part that is owned by outside shareholders of i, external to the system of cross-holdings; the off-diagonal entries of the matrix $\hat{C}$ are defined to be 0.

The *equity* or *book value* V_i of an organization i is the total value of its shares – those held by other organizations as well as those held by outside shareholders. This is equal to the value of organization i's primitive assets plus the value of its claims on other organizations:

$$V_i = \sum_k D_{ik} p_k + \sum_j C_{ij} V_j, \quad i = 1,\ldots,n.$$

The *market value* v_i of an organization i is well captured by the equity value of that organization held by its outside investors, that is, $v_i = \hat{C}_{ii} V_i$.

1. Show that under the stated assumption that $\hat{C}_{ii} > 0$ for all i, all organizations' book values V_i and the market values v_i are uniquely defined (find an explicit expression for these values) for all $i = 1,\ldots,n$.

2. For given cross-holdings C and direct holdings D, assume that the asset prices p are such that $v_i > 0$ for all $i = 1,\ldots,n$. Explain how to find the smallest $\gamma \in [0,1]$ such that for asset prices $p' = \gamma p$ it holds that $v_i \geq 0$ for all $i = 1,\ldots,n$. The optimal γ^* measures by how much the prices can be reduced while maintaining all organizations solvent. Also explain how to determine the critical organizations, that is, those who are about to fail when prices fall to $\gamma^* p$.

Exercise 13.4 *Node centrality*

Let us consider the liability network represented in Figure 13.15. Find the node with the highest in-bound centrality and the one with the highest out-bound centrality.

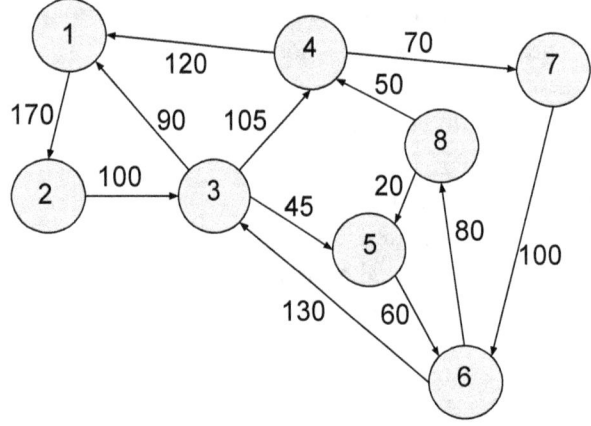

Figure 13.15 An eight nodes liability network.

Exercise 13.5 *Liability network and default propagation*

Let us consider again the liability network represented in Figure 13.15. We also suppose that the nodes have the following external assets:

$$c_{\text{nom}} = [5, 0, 15, 140, 25, 130, 40, 35].$$

We now inject shocks at each node (one at a time) by applying a reduction of $d\%$ in the node external asset. We consider increasing the shock amplitude, with $d = 25, 50, 75, 100$. For each value of d and node of the network, compute numerically the overall loss and the number of defaulted nodes. For each node, plot a curve of these values as a function of d.

14
Text analytics

TEXT ANALYTICS have substantially affected financial industries and have seen a significant increase in demand over the past few years. Coupled with the increasing amount of large, structured and unstructured textual data, and the advances of deep neural networks, the popularity of text analytics in the financial sector has grown rapidly. Text analytics is a process through which the user derives high-quality information from a given piece of text. Finance is one major sector that can benefit from these techniques; the analysis of large volumes of textual data is both a need and an advantage for companies, government, and the general public. This section discusses some important and widely used techniques in the analysis of textual data in the context of finance.

14.1 Representing words

The finance industry generates and consumes vast amounts of information in news articles, financial reports, earnings call transcripts, and other types of textual data. Natural language processing (NLP) and text analytics can help financial institutions and professionals to more efficiently and accurately process and analyze this information, enabling them to make more informed decisions, identify trends and patterns, and automate various tasks. Some examples of how text data is used in finance include sentiment analysis of financial news articles, extraction of information from financial reports and earnings call transcripts, aiding information for fraud detection and credit underwriting from client information, and automation of the generation of trading signals and risk alerts.

In order to work with any form of text data, the first question is how to represent words as numerical data to input to any model.

Once we have a representation of words, the next natural question is how to measure the similarity and difference between words. With word vectors, we can quite easily encode this ability in the vectors themselves and use distance measures such as Cosine or Euclidean, and so on. The simplest word vectors, called *one-hot vectors*, represent every word as an $\mathbb{R}^{|\mathcal{V}| \times 1}$ vector with all 0s and one 1. Here $|\mathcal{V}|$ is the size of the total vocabulary in the corpus.

Example 14.1 (*One-hot vector for financial news*) A one-hot vector for words in a financial news article is shown below. Each represents a word with all 0s and one 1 at the index of that particular word:

$$w^{\text{The}} = \begin{bmatrix} 1 \\ 0 \\ 0 \\ 0 \\ \vdots \\ 0 \end{bmatrix}, w^{\text{S\&P}} = \begin{bmatrix} 0 \\ 1 \\ 0 \\ 0 \\ \vdots \\ 0 \end{bmatrix}, w^{500} = \begin{bmatrix} 0 \\ 0 \\ 1 \\ 0 \\ \vdots \\ 0 \end{bmatrix}, w^{\text{rose}} = \begin{bmatrix} 0 \\ 0 \\ 0 \\ 1 \\ \vdots \\ 0 \end{bmatrix}.$$

When we represent words as one-hot vectors, each word is a completely independent entity. The one-hot vector representation does not directly give us any notion of similarity as the inner product of any pair of word vectors will always be 0. For instance, $\langle w^{\text{S\&P}}, w^{500} \rangle = \langle w^{\text{S\&P}}, w^{\text{cat}} \rangle = 0$.

14.1.1 Co-occurrence matrices

The co-occurrence matrix represents a word by the distribution of other words that appear nearby. The process of constructing a co-occurrence matrix is as follows:

- We first determine a vocabulary $\mathcal{V}$. Typically, $\mathcal{V}$ is all the words in a corpus.[1]

- Construct a matrix X of size $|\mathcal{V}| \times |\mathcal{V}|$.

- For each document, for each word w in the document, add all the counts of the other words w' in the document to the row corresponding to w at the column corresponding to w'.

- Optionally, we can normalize the rows by the sum.

The matrix X is a document-level co-occurrence matrix. Each row i of X, $X_i \in \mathbb{R}^{|\mathcal{V}|}$, is the word embedding for the ith word in the vocabulary $\mathcal{V}$. In the above example, we count a word w' whenever it appears in the same document, called a document-level window.

[1] A text corpus is the dataset that we work with. It can be a collection of documents, paragraphs, sentences, or other entities.

We can also adjust the *context window*. Instead of counting a word w' whenever it appears in the same document as w, we could say that a word w' is only co-occurring with w if w' appears within a window of size n. Supposing our fixed window size is n, then this is the n preceding words, $(w_{i-n}, \ldots, w_{i-1})$, and n subsequent words, $(w_{i+1}, \ldots, w_{i+n})$, in that document. Intuitively, for a large window size such as a document-level window, we are representing words by what kinds of document types they appear in (sports, law, medicine, etc.). For example, the word "investment" is represented by the other words that co-occur in financial documents while the word "vaccine" is represented by words that co-occur in medical documents. The co-occurrence matrix X is a symmetric matrix in which X_{ij} is the number of times w_j appears inside w_i's window across all documents.

Example 14.2 (*Sentence co-occurrence matrix for financial news*) Consider the following financial news document with three sentences:

- The S&P 500 rose.
- Tesla was the strongest force lifting the S&P 500 upward.
- The company's sales rose.

The sentence-level co-occurrence matrix for the document is as follows:

-	S&P	500	rise	Tesla	strong	force	lift	upward	company	sales
S&P	2	2	1	1	1	1	1	1	0	0
500	2	2	1	1	1	1	1	1	0	0
rise	1	1	0	0	0	0	0	0	1	1
Tesla	1	1	0	1	1	1	1	1	0	0
strong	1	1	1	1	1	1	1	1	0	0
force	1	1	1	1	1	1	1	1	0	0
lift	1	1	1	1	1	1	1	1	0	0
upward	1	1	1	1	1	1	1	1	0	0
company	0	0	1	0	0	0	0	0	1	1
sales	0	0	1	0	0	0	0	0	1	1

14.1.2 Apply SVD to co-occurrence matrix

The word-by-word co-occurrence matrix X is generally very large as the vocabulary size $|\mathcal{V}|$ is typically enormous. We can further perform singular value decomposition on the co-occurrence matrix X to get a $USV^\top$ decomposition. We then use the rows of U as the word vectors for all words in our vocabulary:

$$|\mathcal{V}| \begin{bmatrix} & & \\ & X & \\ & & \end{bmatrix} \overset{|\mathcal{V}|}{=} \begin{bmatrix} | & | & \\ u_1 & u_2 & \cdots \\ | & | & \end{bmatrix} \begin{bmatrix} \sigma_1 & 0 & \cdots \\ 0 & \sigma_2 & \cdots \\ \vdots & \vdots & \ddots \end{bmatrix} \begin{bmatrix} \text{---} & v_1 & \text{---} \\ \text{---} & v_2 & \text{---} \\ & \vdots & \end{bmatrix}.$$

14 TEXT ANALYTICS

We can reduce the dimensionality by selecting the first k singular vectors based on the desired percentage variance captured:

$$\frac{\sum_{i=1}^{k} \sigma_i}{\sum_{i=1}^{|\mathcal{V}|} \sigma_i}.$$

Then each row of the submatrix $U_{1:|\mathcal{V}|, 1:k}$ is a k-dimensional representation of a word in the vocabulary:

$$|\mathcal{V}| \begin{bmatrix} | & & | \\ u_1 & \cdots & u_k \\ | & & | \end{bmatrix} k \begin{bmatrix} \sigma_1 & 0 & \cdots \\ 0 & \ddots & \cdots \\ \vdots & \vdots & \sigma_k \end{bmatrix} k \begin{bmatrix} - & v_1 & - \\ & \vdots & \\ - & v_k & - \end{bmatrix}.$$

However, SVD-based methods do not scale well for big matrices and it is hard to incorporate new words or documents. Instead of computing and storing global information about all the datasets (which might be billions of sentences), we will learn how to encode the probability of a word given its context in Section 14.2

14.1.3 The bag-of-words

So far, we have discussed how to represent words. However, in some tasks, we may need to represent the entire document as opposed to individual words. The bag-of-words model is a representation that describes a document by the occurrence of words within a document. We first construct a vocabulary of known words from the documents. We then count the number of occurrences of each word in the documents. It is called a *bag* of words because it is an unordered set of words with their position ignored (see Figure 14.1). Any information about the order of words in the document is not included. The model is only concerned with whether known words occur in the document, not where in the document. The intuition is that documents are similar if they have similar content. For instance, "banks," "investment," "stocks," "bonds," and so on are more likely to appear together than "banks," "cucumber," "starfish," "yoga." Further, from the content alone we can learn something about the meaning of the document.

Example 14.3 (*Bag-of-words for financial news*) The following models a text document using a bag-of-words. Here is a simple text document:

- The S&P 500 rose upward. Tesla was the strongest force lifting the S&P 500 upward.

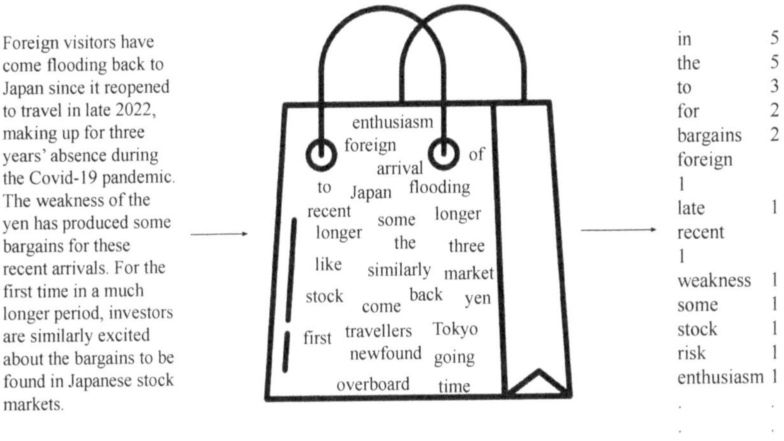

Foreign visitors have come flooding back to Japan since it reopened to travel in late 2022, making up for three years' absence during the Covid-19 pandemic. The weakness of the yen has produced some bargains for these recent arrivals. For the first time in a much longer period, investors are similarly excited about the bargains to be found in Japanese stock markets.

enthusiasm foreign of arrival to Japan flooding recent some longer longer the three like similarly market stock come back yen first travellers Tokyo newfound going overboard time

in 5
the 5
to 3
for 2
bargains 2
foreign 1
late 1
recent 1
weakness 1
some 1
stock 1
risk 1
enthusiasm 1

Figure 14.1 The *bag-of-words* representation of a document.

We first construct a vocabulary $\mathcal{V}$ from the document. Intuitively, words such as "the" or "was" are present in almost every document and are insignificant in helping us obtain the information. These common words are called *stop words* and are usually filtered out before constructing the vocabulary. Once the stop words are filtered, we then have the following vocabulary:[2] {S&P, 500, rise, Tesla, strong, force, lift, upward}. The bag-of-word representation then counts the frequency of each word in the document to form the representation:

$$x = \begin{bmatrix} 2 & 2 & 1 & 1 & 1 & 1 & 2 \end{bmatrix}.$$

[2] Here we lemmatize the document to simplify the vocabulary. Lemmatization is the process of grouping different forms of a word into one single form; for instance, reducing "rise," "rising," or "rose" to the same lemma "rise."

14.1.4 Text classification and naive Bayes

Given a text document, the task of *text classification* learns to predict a discrete label $y \in \mathcal{Y}$, where $\mathcal{Y}$ is the set of possible labels. Text classification has many applications, from spam filtering to the analysis of news and financial records. In text classification, we wish to learn a *classification function* given a training set of labeled documents. Suppose we have a dataset of N labeled documents, $\{(x^{(i)}, y^{(i)})\}_{i=1}^{N}$, assuming that they are i.i.d.. The joint probability of a document x and its ground truth label y is written $P_{X,Y}(x, y)$.[3] The first classification function we introduce is the *multinomial naive Bayes* (multinomial NB). Using the chain rule, the joint probability $P_{X,Y}(x, y)$ can be factored into

$$P_{X,Y}(x, y) = P_{X|Y}(x|y) \cdot P_Y(y), \tag{14.1}$$

where $P_{X|Y}(x|y)$ is the conditional probability of a document x being in class y and $P_Y(y)$ is the prior probability of a document being in

[3] The notation indicates the joint probability of random variables X and Y take the specific values x and y respectively.

class y. The idea is to use Bayes' rule[4] to transform (14.1) into

[4] Bayes' rule: $P(x|y) = \frac{P(y|x)P(x)}{P(y)}$

$$P_{X,Y}(x,y) = \frac{P_{Y|X}(y|x)P_X(x)}{P_Y(y)}P_Y(y) \quad (14.2)$$
$$= P_{Y|X}(y|x)P_X(x)P_Y(y). \quad (14.3)$$

We can interpret (14.3) as a document *generative process*: first, we sample a class from $P_Y(y)$, and then the words are generated by sampling from $P_{X|Y}(x|y)$. For classification, we compute the most probable class $\hat{y}$ given some document x by choosing the class that gives the highest product of the *likelihood* of the document $P(x|y)$[5] and the *prior probability* of the class $P(y)$:

[5] The subscript will often be omitted when it is clear from the context.

$$\hat{y} = \arg\max_y P(x|y)P(y). \quad (14.4)$$

From section 14.1.3, we can represent a document using bag-of-words. Given a bag-of-words document x with word features $w_1, w_2, w_3, \ldots, w_k$, we make the *naive Bayes assumption* (or conditional independence assumption) that the probabilities of a word w_i given the class y, $P(w_i|y)$ are independent given the class y and therefore can be "naively" multiplied together:

$$P(x|y) = P(w_1, w_2, \ldots, x_k|y) = P(w_1|y) \cdot P(w_2|y) \cdots P(w_k|y). \quad (14.5)$$

The final Naive Bayes classifier is thus:

$$\hat{y} = \arg\max_y P(y) \prod_{i=1}^{k} P(w_i|y). \quad (14.6)$$

To learn the probability of $P(y)$ and $P(w_i|y)$, we simply use their frequencies in the data. Let N_y be the number of documents in our labeled training data with class y and N be the total number of documents, then

$$\hat{P}(y) = \frac{N_y}{N}. \quad (14.7)$$

To learn $P(w_i|y)$, we count how many times the word w_i appears among all words in all documents of class y:

$$\hat{P}(w_i|y) = \frac{\text{count}(w_i, y)}{\sum_{w \in V} \text{count}(w, y)}, \quad (14.8)$$

where the vocabulary V consists of the union of all the words in all classes, not just the words in one class. A common problem in naive Bayes is that some words, say w_j, can show up in one class, say y_1, but not the others, for example y_2. Since naive Bayes multiplies all the feature likelihoods together, a zero likelihood $P(w_j|y_2) = 0$ will cause the entire probability to be zero. A simple solution called *add-one (Laplace) smoothing* is commonly used in naive Bayes where 1 is added to each likelihood to avoid having 0 probability:

$$\hat{P}(w_i|y) = \frac{\text{count}(w_i, y) + 1}{\sum_{w \in V}(\text{count}(w, y) + 1)} = \frac{\text{count}(w_i, y) + 1}{\left(\sum_{w \in V} \text{count}(w, y)\right) + |V|}. \tag{14.9}$$

Example 14.4 (*Financial sentiment analysis on stock market headlines*) Financial news is a rich source of text data. Sentiment analysis is a common task that assigns a label of positive, negative, or neutral to each document.[6] Consider the following headlines and the analyst rating:

[6] A document is a general term. It can be a sentence, a paragraph, or an entire article depending on the task at hand. In this example, a document is a news headline.

	Headlines	Rating
Training	Earnings increase from estimates	Positive
	Earnings expected to increase	Positive
	Earnings miss estimates	Negative
Test	Earnings estimates increase	?

To make a prediction for this example, the parameters we need to estimate are the priors $\hat{P}(\text{positive}) = 2/3$ and $\hat{P}(\text{negative}) = 1/3$ and the following conditional probabilities:

$$\hat{P}(\text{earnings}|\text{positive}) = (2+1)/(8+7) = 1/5,$$
$$\hat{P}(\text{estimates}|\text{positive}) = (1+1)/(8+7) = 2/15,$$
$$\hat{P}(\text{increase}|\text{positive}) = (2+1)/(8+7) = 1/5,$$
$$\hat{P}(\text{earnings}|\text{negative}) = (1+1)/(3+7) = 1/5,$$
$$\hat{P}(\text{estimates}|\text{negative}) = (1+1)/(3+7) = 1/5,$$
$$\hat{P}(\text{increase}|\text{negative}) = (0+1)/(3+7) = 1/10.$$

The denominators are $(8+7)$ and $(3+7)$ because the total number of words of the positive and negative classes is 8 and 3, respectively, and the total vocabulary $|V|$ consists of 7 terms. The final prediction is then:

$$\hat{P}(\text{positive}|x) \approx \left(\frac{1}{5}\right)^2 \cdot \frac{2}{15} \approx 0.0053,$$
$$\hat{P}(\text{negative}|x) \approx \left(\frac{1}{5}\right)^2 \cdot \frac{1}{10} \approx 0.004.$$

Thus, the classifier assigns the test document to $y = \text{positive}$.

14.2 Distributed word representation

As we have seen so far, the role of context is important, words that occur in *similar contexts* or *similar position* tend to have *similar meaning*. This connection between the similarity in how words are distributed and the similarity in what they mean is called the *distributional hypothesis*,[7] in which "a word is characterized by the company it keeps." At a high level, the meaning of a word is defined by the distribution

[7] J. R. Firth, "A Synopsis of Linguistic Theory 1930-1955," in *Studies in Linguistic Analysis*, pp. 1–32, Oxford: Philological Society, 1957.

of words that show up around it. Since "profit" shows up around "sales," "economy," "market," and so on, words that are similar to "market," such as "revenue," will have similar distributions of surrounding words.

14.2.1 Term frequency-inverse document frequency

Vectors or distributed representations are generally based on a co-occurrence matrix, a way of representing how often words co-occur. Term frequency-inverse document frequency (TF-IDF) can be broken down into two parts: TF (term frequency) and IDF (inverse document frequency). Term frequency[8] is the frequency (raw count) of the word t in the document d:

$$\text{tf}(t,d) = \frac{f_{t,d}}{\sum_{t' \in d} f_{t',d}},$$

where $f_{t,d}$ is the raw count of a term in a document and the denominator is simply the total number of terms in document d, counting each occurrence of the same term separately. Intuitively, words that are limited to a few documents are useful for discriminating those documents from the rest of the collection while words that occur frequently across the entire collection aren't as helpful.

The document frequency $\text{df}(t, D)$ of a term t is the number of documents it occurs in. The inverse document frequency (IDF) is defined using the fraction $\frac{N}{\text{df}(t,D)}$, where N is the total number of documents in the collection, and $\text{df}(t, D)$ is the number of documents in which t occurs. The fewer documents in which t occurs, the higher the IDF. It measures how much information the word provides by looking at how common (or uncommon) a word is amongst the entire collection of documents D:

$$\text{idf}(t, D) = \log \frac{N}{\text{df}(t, D)} = \log \frac{N}{|d \in D: t \in d|}.$$

The key intuition motivating TF-IDF is the importance of a term is inversely related to its frequency across all documents. TF gives us information on how often a term appears in a document and IDF gives us information about the relative rarity of a term in the collection of documents. By multiplying these values together we can get our final TF-IDF value:

$$\text{tf-idf}(t, D) = \text{tf}(t, D) \cdot \text{idf}(t, D).$$

[8] H. P. Luhn, "A Statistical Approach to Mechanized Encoding and Searching of Literary Information," *IBM Journal of Research and Development*, vol. 1(4), pp. 309–317, 1957.

Example 14.5 (*TF-IDF for document similarity*) TF-IDF model is commonly used to represent documents numerically and measure their similarity. Below is a TF-IDF weighted term-document matrix for three documents—The Wall Street Journal, Harry Potter, and The Economist—where each value represents the TF-IDF score of a specific word in a given document. Using these TF-IDF values, we can compute the similarity between documents using cosine similarity (see Section 2.4).

	Wall Street Journal	Harry Potter	The Economist
S&P	0.07	0	0.08
revenue	0.06	0.002	0.07
wand	0.001	0.07	0.002
muggle	0	0.09	0

Given two documents d_1 and d_2, we can then estimate their similarity by $\cos(d_1, d_2)$. The similarity between *Wall Street Journal* d_W and *The Economist* d_E is much higher than that between Wall Street Journal d_W and *Harry Potter* d_H:

$$\cos(d_W, d_E) = \frac{\langle d_W, d_E \rangle}{\|d_W\| \cdot \|d_E\|}$$
$$= \frac{\langle (0.07, 0.06, 0.001, 0), (0.08, 0.07, 0.002, 0) \rangle}{0.092 \cdot 0.11} = 0.97,$$
$$\cos(d_W, d_H) = \frac{\langle (0.07, 0.06, 0.001, 0), (0, 0.002, 0.07, 0.09) \rangle}{0.092 \cdot 0.11} = 0.019.$$

14.2.2 The n-gram model (language model)

The bag-of-words model is an order-less document representation that only considers the counts of words. In the above example, the bag-of-words representation will not reveal that the number "500" always follows "S&P" in these documents. As an alternative, instead of treating each word independently, we can try to predict upcoming words given some sequence of words, preserving the spatial information. We begin with the task of calculating $P(w_n|w_{1:\,n-1})$, the probability of a word w_n given a sequence of preceding words $(w_1, w_2, \ldots, w_{n-1})$. In practice, computing the exact probability of a word given a long sequence of preceding words for any possible combination is infeasible. Instead of computing the probability of a word given its entire history, we approximate the history by just the last few $n-1$ words, hence the name n-gram model. For example, the *bigram* model approximates the probability $P(w_n|w_{1:\,n-1})$ by considering only the conditional probability of the previous word $P(w_n|w_{n-1})$:

$$P(w_n|w_{1:\,n-1}) \approx P(w_n|w_{n-1}).$$

To generalize, a bigram ($n = 2$) model looks at one word into the past, a trigram ($n = 3$) model looks at two words into the past and an n-gram model looks at $n-1$ words into the past.

Example 14.6 (*n-gram for financial news*) Applying to the example:
- Doc 1: The S&P 500 rose.
- Doc 2: Tesla was the strongest force lifting the S&P 500 upward.

A bi-gram model will parse the text into the following units:

- "The S&P", "S&P 500", '500 rose", "rose 5.21", "5.21 or", "or 0.1".
- "Tesla was", "was the", "the strongest", "strongest forces", "force lifting", "lifting the", "the S&P", "S&P 500", "500 upward".

Conceptually, we can view the bag-of-word model as a special case of the *n*-gram model, with $n = 1$. The idea can be traced back to Claude Shannon's work[9] in information theory. Given a sequence of letters, Shannon measured the likelihood of the next letter. For example, given the sequence *stoc*, what is the probability of the next letter being *k*? Similarly, an *n*-gram model predicts word w_i based on previous $n - 1$ words (one word in a bigram model, two words in a trigram model, etc.), that is:

$$P(w_i \mid w_{i-1}, \ldots, w_{i-(n-1)}).$$

[9] C. E. Shannon, "Prediction and Entropy of Printed English," *Bell Systems Technical Journal*, vol. 30, pp. 51–64, 1951.

Here, we make a critical Markov assumption, which approximates the true underlying language by looking only at the last $n - 1$ words, that is, $P(w_i \mid w_{i-1}, \ldots, w_1) \approx P(w_i \mid w_{i-1}, \ldots, w_{i-(n-1)})$. This assumption is important because it massively simplifies the problem of estimating the language model without looking too far back into the past. To compute the probability of the entire word sequence or sentence $P(w_1, w_2, \ldots, w_N) = P(w_{1:N})$, we apply the chain rule of probability,

$$P(w_{1:N}) = \prod_{i=1}^{N} P(w_i \mid w_{i-1}, \ldots, w_{i-(n-1)}).$$

Hence, a unigram language model will be

$$P(w_1, w_2, \ldots, w_N) = \prod_{i=1}^{N} P(w_i),$$

and a bigram language model will be

$$P(w_1, w_2, \ldots, w_N) = \prod_{i=1}^{N} P(w_i \mid w_{i-1}).$$

To estimate the n-gram probability, we use the relative frequency estimate by counting the occurrence of a word in a corpus and then normalizing the counts. For example, the bigram probability of $(w_n \mid w_{n-1})$ can be computed by counting the bigram $C(w_{n-1}, w_n)$ and then normalizing it by the sum of all the bigrams that share the same first word w_{n-1}, which equals to the unigram count for w_{n-1}:

$$P(w_n \mid w_{n-1}) = \frac{C(w_{n-1}, w_n)}{\sum_w C(w_{n-1}, w)} = \frac{C(w_{n-1}, w_n)}{C(w_{n-1})}.$$

Example 14.7 (n-*gram language model for financial news*) To compute the probability of an entire sentence, it is common to pad the beginning and end of the sentence with special tokens such as <bos> and <eos>.[10] Then the bigram (n = 2) approximation to the probability of sentence *"The S&P 500 rose"* is

[10] Indicating "beginning of sentence" or "end of sentence."

$$P(\text{The S\&P 500 rose}) = P(\text{The}|\text{<bos>}) \times P(\text{S\&P}|\text{The})$$
$$\times P(500|\text{S\&P}) \times P(\text{rose}|500) \times P(\text{<eos>}|\text{rose}).$$

Suppose that the bigram probabilities after normalization is

$$P(\text{The}|\text{<bos>}) = 0.25, \quad P(\text{S\&P}|\text{The}) = 0.0011, \quad P(500|\text{S\&P}) = 0.68,$$
$$P(\text{rose}|500) = 0.002, \quad P(\text{<eos>}|\text{rose}) = 0.03.$$

Then we can compute the probability of sentences by multiplying the bigram probabilities together, as follows:

$$P(\text{<bos> The S\&P 500 rose <eos>}) = 0.25 \times 0.0011 \times 0.68 \times 0.002 \times 0.03.$$

In practice, it is common to compute language model probabilities in log format as log probabilities instead of raw probabilities to avoid numerical underflow. We only need to convert the log probabilities back when we need to report them:

$$P(w_1, w_2, \ldots, w_N) = \prod_{i=1}^{N} P(w_i|w_{i-1}) = \exp\left(\sum_{i=1}^{N} P(w_i|w_{i-1})\right).$$

14.2.3 Skip-gram word2vec

The skip-gram word2vec model[11] represents each word as a low-dimensional vector that is usually much smaller than the vocabulary size. The model learns the vector for each word by predicting the surrounding words in a short context. Consider the following example: *"Compared to the Dow Jones, the S&P 500 index rise."* Given the center word "S&P," the model will be able to predict or generate the surrounding context words "Dow," "Jones," "500," "index," and so on. Consider a finite vocabulary $\mathcal{V}$, skipgram predicts the probability of context (surrounding) words (e.g., "dow," "jones") from a center word (e.g., "s&p"). First, we define two matrices $\mathcal{C} \in \mathbb{R}^{d \times |\mathcal{V}|}$ and $\mathcal{O} \in \mathbb{R}^{d \times |\mathcal{V}|}$ where column i of $\mathcal{C}$ corresponds to the word vector for word w_i being the center word while column j of $\mathcal{O}$ corresponds to word w_j being the context word. Let (u, v) be a random pair of *center* and *context* words, the skipgram word2vec model estimates the probability of the context word v given the center word u as follows:

[11] T. Mikolov et al. "Efficient Estimation of Word Representations in Vector Space," *International Conference on Learning Representations* (2013).

$$P(v \mid u) = \frac{\exp(O_v^\top C_u)}{\sum_{w \in \mathcal{V}} \exp(O_w^\top C_u)}, \qquad (14.10)$$

where O_w refers to the column of O corresponding to word $w \in |\mathcal{V}|$. The inner product $O_v^\top C_u$ represents the compatibility between center word u and context word v while the *softmax* function[12] produces a probability distribution where larger-scored words get higher probability. The vector of probabilities over all words in the vocabulary $\mathcal{V}$ given a center word u,

[12] The softmax function: $\frac{\exp x_i}{\sum_{j=1}^n \exp x_j}$

$$\begin{bmatrix} P(w_1|u) & P(w_2|u) & \ldots & P(w_n|u) \end{bmatrix} \in \mathbb{R}^{|\mathcal{V}|},$$

is like a row of a co-occurrence matrix.

Let $\mathcal{D}$ be a collection of documents d. Each document d consists of a sequence of words $w_1^{(d)}, w_2^{(d)}, \ldots, w_n^{(d)}$ where $w \in \mathcal{V}$. Let $k \leq n$ be a positive-integer window size. For each document d, we loop over all words $w_i^{(d)}$ in the document as the center word. Then we loop over all words $w_{i-k}^{(d)}, \ldots, w_{i-1}^{(d)}, w_{i+1}^{(d)}, \ldots, w_{i+k}^{(d)}$ occurring in the window of the center word $w_i^{(d)}$. The task is to maximize the probability of all context words given the center word in $\mathcal{D}$:

$$\sum_{d \in \mathcal{D}} \sum_{i=1}^n \sum_{j=1}^k \left(\log P\left(w_{i-j}^{(d)} \mid w_i^{(d)}\right) + \log P\left(w_{i+j}^{(d)} \mid w_i^{(d)}\right) \right). \qquad (14.11)$$

To estimate the values of the parameters C and O, we minimize the *cross-entropy loss* in predicting the context words. The cross-entropy loss for skip-gram is determined by the probability the model assigns to the correct context word since only the correct word has a true probability of 1 while all the other words have a probability of 0. The cross-entropy loss of skip-gram is thus equivalent to maximizing the likelihood of (14.11):

$$\mathcal{L}(C, O) = \sum_{d \in \mathcal{D}} \sum_{i=1}^n \sum_{j=1}^k -\left(\log \frac{\exp(O_v^\top C_u)}{\sum_{w \in \mathcal{V}} \exp(O_w^\top C_u)} \right) \quad v = w_{i \pm j}^{(d)},\ u = w_i^{(d)}$$

$$= \sum_{d \in \mathcal{D}} \sum_{i=1}^n \sum_{j=1}^k -\left(\log P\left(w_{i-j}^{(d)} \mid w_i^{(d)}\right) + \log P\left(w_{i+j}^{(d)} \mid w_i^{(d)}\right) \right).$$

The intuition of word2vec is that instead of counting how often word w_j occurs near word w_i, we instead train a binary classifier to predict whether word w_j is likely to show up near w_i. We don't actually care about this prediction task but we take the learned classifier weights as the word embeddings. The skip-gram model learns two separate embeddings for each word w_i, the center word embedding C_{w_i} and the context word embedding O_{w_i}, stored in two matrices C and O. To represent the word w_i, it is common to add them together $C_{w_i} + O_{w_i}$ or throw away the C matrix and just represent each word by the vector O_{w_i}. The context window size k is a hyper-parameter that is often tuned on a development set. The word2vec embeddings are *static embeddings*, meaning that the method learns one fixed embedding for each word in the vocabulary.

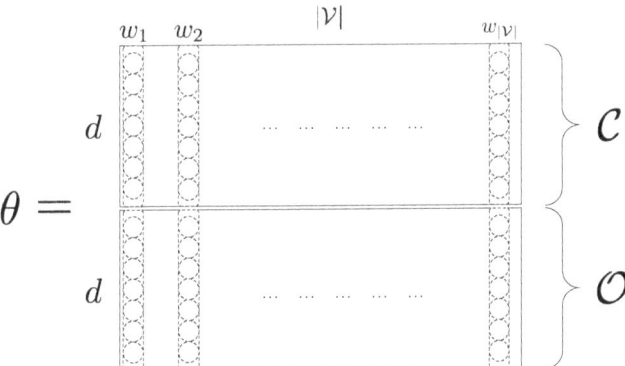

Figure 14.2 The embeddings learned by the skip-gram model. The model stores two embeddings for each word, the center word embedding C_{w_i}, and the context word embedding O_{w_i}.

14.2.4 Pre-trained language models

The idea of distributed hypothesis, as seen in the skip-gram word2vec model, is that word meanings can be learned solely based on the content of the texts and the association of the words they co-occur with. The idea of self-supervisedly learning rich representations from the data unlocks the recent advances of pre-trained language models. Self-supervised learning uses parts of the data itself (e.g., a word in a sentence) and attempts to learn other parts of the data (e.g., other words). This allows us to train gigantic models (usually in millions or billions, and sometimes even trillions) to learn the representation of words by processing a large amount of text. This idea of training on some large amount of text is called *pre-training* and the resulting large models are called *pre-trained language model* or recently *large language models* (LLMs). Various deep learning architectures such as RNNs and Transformers or even feed-forward networks can be used to train a (large) language model. In the next section, we'll discuss some common deep learning architectures for training language models.

14.3 Deep neural networks for text analytics

14.3.1 RNNs as language models

As seen in the n-gram language model, we are interested in predicting the next word in a sequence given some preceding context. The n-gram language models of Section 14.2.2 compute the probability of a word by the counts of its occurrence given the $n-1$ prior words. This gives us the ability to assign probabilities to entire sequences by combining these conditional probabilities with the chain rule:

$$P(w_{1:n}) = \prod_{i=1}^{n} P(w_i|w_{<i}). \tag{14.12}$$

Let $X = (x_1, x_2, \ldots, x_n)$ be the input sequence consisting of a series of words that are represented as a one-hot vector of size $|V| \times 1$. At each time step t, the model retrieves the embedding e_t for the current word from the embedding matrix E by multiplying it with the one-hot vector x_t. Then it combines the current word embedding e_t with the hidden layer h_{t-1} from the previous step to compute a new hidden layer h_t for the current step. This hidden layer is then used to generate an output layer o_t which is passed through a softmax function to generate a probability distribution over the entire vocabulary V. In summary, at time t, we have

$$e_t = Ex_t, \tag{14.13}$$
$$h_t = g(Uh_{t-1} + We_t), \tag{14.14}$$
$$o_t = Vh_t, \tag{14.15}$$
$$\hat{y}_t = \mathrm{softmax}(o_t). \tag{14.16}$$

The output layer resulting from Vh can be thought of as a set of scores over the entire vocabulary given the context provided in h. The softmax function then normalizes these scores into a probability distribution over V. The probability that a particular word i is the next word is represented by $y_t[i]$, the ith component of y_t:

$$P(w_{t+1} = i|w_1, \ldots, w_t) = y_t[i].$$

From (14.12), the probability of an entire sequence is then the product of the probabilities of each word in the sequence:

$$P(w_{1:n}) = \prod_{i=1}^{n} P(w_i|w_{1:i-1}). \tag{14.17}$$

The RNN language models[13] process the input sequence one word at a time, attempting to predict the next word from the current word

[13] T. Mikolov et al. "Recurrent neural network based language model," Interspeech (2010).

14.3 DEEP NEURAL NETWORKS FOR TEXT ANALYTICS

and the previous hidden state. The hidden state in theory represents information about all of the preceding words all the way back to the beginning of the sequence. Recurrent neural networks thus don't have the limited context window size that n-gram models have as the context is recurrently updated while moving through the sequence. Although at each step, the models depend only on the previous hidden state h_{t-1}, this vector is in turn influenced by all previous words through the recurrence operation, allowing RNN language models to handle long-range dependencies. This is an important distinction from n-gram language models, where any information outside the context window is ignored.

To train an RNN as a language model, we minimize the cross-entropy loss in predicting the next word. This is determined by the probability the model assigns to the correct next word. So at time t the cross-entropy loss is the negative log probability the model assigns to the next word in the training sequence:

$$L_{CE}(\hat{y}_t, y_t) = -\log \hat{y}_t[w_{t+1}]. \tag{14.18}$$

We then average the losses over the entire training corpus and train the weights via gradient descent.

$$\mathcal{L} = \frac{1}{T} \sum_{t=1}^{T} -\log \hat{y}_t[w_{t+1}]. \tag{14.19}$$

The training process for RNN language models is schematized in Figure 14.3.

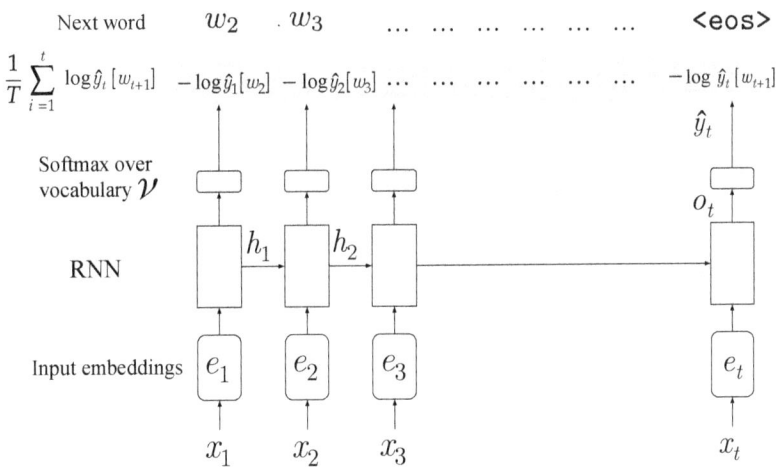

Figure 14.3 Training process for RNN language models.

14.3.2 RNNs for text classification

We can apply RNNs for text classification by passing the text through the RNN one word at a time. At each step t, we generate a new hidden layer h_t and pass it to the next step $t + 1$. This process is continued until we reach the last word. We then take the hidden layer for the last word of the text, h_n, and pass it as an input to a (one-layer) feedforward network (FFNN) that chooses a class via a softmax over the possible classes as illustrated in Figure 14.4:

$$e_t = Ex_t, \tag{14.20}$$
$$h_t = g(Uh_{t-1} + We_t), \tag{14.21}$$
$$\hat{y} = \text{softmax}(Mh_n). \tag{14.22}$$

In principle, the last hidden layer h_n constitutes a compressed representation of the entire sequence, which is then used to perform classification. In (14.20), there are no intermediate outputs for each word in the sequence preceding the last element. Therefore, there are no intermediate loss terms over the sequence. Instead, the loss function is based solely on the final classification task. The weights are trained via backpropagation where the error signal from the classification is backpropagated all the way through the weights in the feedforward classifier and then through to the sets of weights in the RNN.

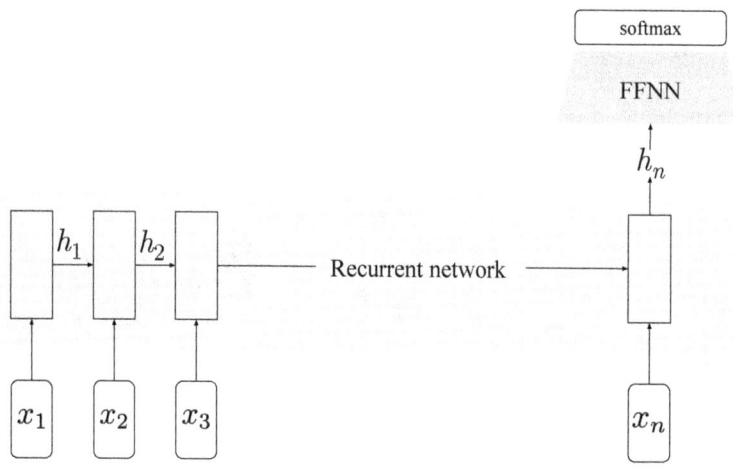

Figure 14.4 Text classification using an RNN combined with a feedforward network.

14.3.3 Tranformers as language models

Similar to training RNN language models, we use the same self-supervised task of predicting the next word given its preceding words in a sequence, using cross-entropy loss. Figure 14.5 illustrates the general training approach with Transformers architecture. At each step t, given all the preceding words, the final transformer layer produces an output distribution over the entire vocabulary $\mathcal{V}$. As with

RNNs, the training loss is the average cross-entropy loss over the entire corpus. The key difference between Transformers and RNN language models is parallelization. In RNN, the calculation of the outputs and the losses at each step depends on the previous hidden states, thus making it inherently serial. However, with transformers, the training loss at each step can be processed in parallel since the output for each step in the sequence can be computed separately.

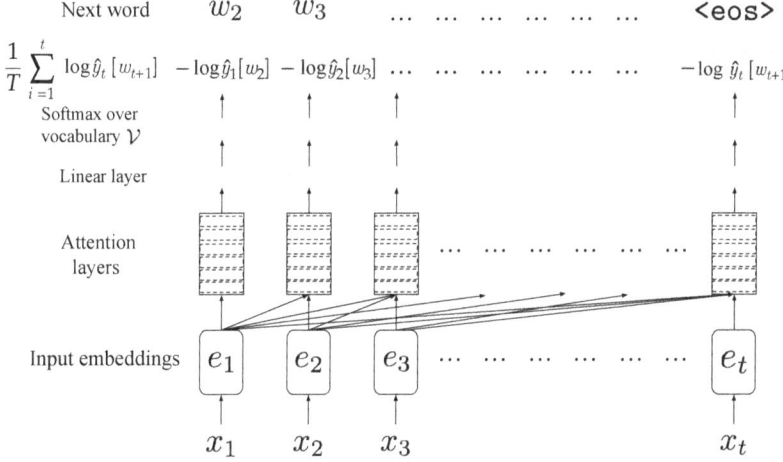

Figure 14.5 Large language model based on the Transformers architecture.

14.3.4 Text generation with language models

Text generation has many applications such as question-answering, text summarization, machine translation, conversational dialogue, story generation, and so on. In principle, any task that produces text conditioned on some other text can be viewed as text generation. The use of language models for text generation is one of the most impactful areas in natural language processing. To generate text using language models, we first give a sequence of words as input. The input sequence can be short or long depending on the tasks at hand. For question-answering, the input sequence is the question while for summarization the input sequence can be a document or even as long as multiple documents. Given the input sequence, we first pass the input through a neural network such as RNN or Transformers to encode the input and create a compressed representation of the input, often called context, similar to the idea of text classification with RNN in Section 14.3.2. We then pass this contextualized representation to another neural network to decode the representation and generate a task-specific output sequence. The model generates the output sequence by sampling words conditioned on the previous words until it reaches a pre-determined length or an end-of-sequence (e.g., <eos>) token is generated. This framework is often called the *encoder–decoder* model as illustrated in Figure 14.6.

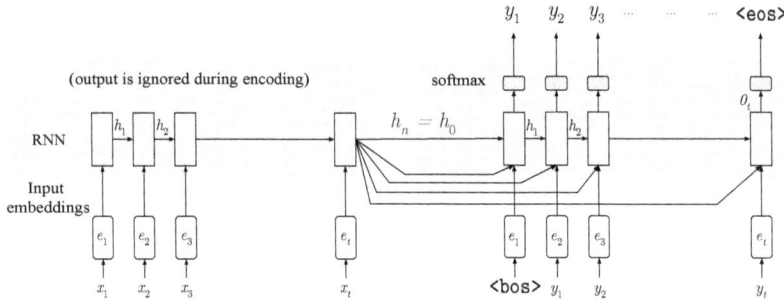

Figure 14.6 The basic RNN version of the encoder–decoder approach. Source and target sentences are concatenated with a separator token in between, and the decoder uses context information from the encoder's last hidden state.

14.4 Evaluating text analytics tasks

As with any machine learning application, it is critical to have some form of metric that measures the performance of the model. In this section, we discuss some commonly used metrics that are used for evaluating text analytics tasks. For text classification tasks, the simplest is *accuracy*:

$$\frac{1}{N}\sum_{i=1}^{N}(y_i = \hat{y}_i).$$

The accuracy can go wrong when there is a class imbalance. Suppose we are building a classifier to predict the loan default using the client's past financial reports, which appears in only 1% of all loan borrowers. A classifier that predicts negative for all applicants would achieve 99% accuracy but this classifier would be practically useless. As a result, we need other metrics that are capable of detecting the classifier's ability to discriminate between classes, even when the distribution is skewed.

14.4.1 Evaluating text classification

In Section 7.1.2 we discuss detailed performance metrics for classification. Here we highlight some common metrics used in text classification. Suppose we have $\mathcal{K}$ classes, $c_1, c_2, \ldots, c_\mathcal{K}$, then for any label c_k, there are two possible errors:

- false positive (FP): the model incorrectly predicts $\hat{y} = c_k$ when ground truth $y \neq c_k$;
- false negative (FN): the model incorrectly predicts $\hat{y} \neq c_k$ when ground truth $y = c_k$.

Similarly, for any label c_k, there are two possible correct outcomes:

- true positive (TP): the model correctly predicts $\hat{y} = c_k$ when ground truth $y = c_k$;
- true negative (TN): the model correctly predicts $\hat{y} \neq c_k$ when ground truth $y \neq c_k$.

Next, we can define two sources of error:

$$\text{recall}(y, \hat{y}, k) = \frac{\text{TP}}{\text{TP} + \text{FN}} = \frac{\text{TP}}{\text{count}(y = c_k)},$$
$$\text{precision}(y, \hat{y}, k) = \frac{\text{TP}}{\text{TP} + \text{FP}} = \frac{\text{TP}}{\text{count}(\hat{y} = c_k)}.$$

Recall evaluates the percent of true positives identified given the class $y = c_k$. Classifiers that make a lot of false negatives result in low recall, failing to predict the class c_k even when they are. On the other hand, *precision* evaluates the percent of true positives identified given all predicted values of $\hat{y} = c_k$. Classifiers that make a lot of false positives have low precision, predicting the class c_k when they are not. Recall and precision are complementary. When false negatives are more expensive, a high-recall classifier is preferred. For example, in a loan default screening task, the cost of false negatives would result in huge losses while the cost of false positives may be additional auditing. Conversely, when false positives are more expensive, a high-precision classifier is preferred. For example, suppose a classifier is trained to detect whether the news will be relevant for trading signals, a false negative is a relatively minor noise, while a false positive might mean that important information gets missed. It is also possible to combine recall and precision into a single metric, called the F_1-*score*, which is the harmonic mean of precision and recall:

$$F_1\text{-score}(y, \hat{y}, k) = \frac{2}{\text{precision}^{-1} + \text{recall}^{-1}} = 2 \frac{\text{precision} \times \text{recall}}{\text{precision} + \text{recall}}.$$

It symmetrically represents both precision and recall in one metric. The highest possible value of F_1 is 1, with perfect precision and recall, and the lowest possible value is 0 when either precision or recall is zero.

Note that until now, all recall, precision, and the F_1-score are defined with respect to a specific class c_k. When there are multiple classes $\mathcal{K}$ (more than two classes), we average the F_1 score across all classes, called *macro-F_1*:

$$\text{macro-}F_1(y,\hat{y}) = \frac{1}{|\mathcal{K}|} \sum_{k \in \mathcal{K}} F_1(y,\hat{y},k).$$

14.4.2 Evaluating language models

As seen in skip-gram word2vec and n-gram, language models are usually not the task itself, but an auxiliary task. To evaluate the performance of the task that we ultimately care about, we would ideally evaluate the word embedding extrinsically on some real task at hand. However, these tasks are typically slow to compute. In contrast, intrinsic evaluation is task-neutral and ties closely to the language modeling task itself. We may expect that better performance on intrinsic metrics may improve extrinsic metrics, but we may also need to monitor how intrinsic performance gains carry over to the real task to avoid over-optimizing the intrinsic metric.

Perplexity is a commonly used intrinsic metric for evaluating language models. Given a test set $W = w_1, w_2, \ldots, w_N$, the perplexity on the test set is the inverse probability of the data, normalized by the number of words:

$$\begin{aligned}
\text{Perplexity}(W) &= P(w_1, w_2, \ldots, w_N)^{-\frac{1}{N}} \\
&= \sqrt[N]{\frac{1}{P(w_1, w_2, \ldots, w_N)}} \\
&= \prod_{i=1}^{N} \sqrt[N]{\frac{1}{P(w_i|w_1, \ldots, w_{i-1})}}.
\end{aligned} \quad (14.23)$$

From (14.23), the perplexity depends on what n-gram we choose. Because of the inverse, the higher the conditional probability of the entire sequence, the lower the perplexity, which is equivalent to maximizing the test set probability based on the language model. Intuitively, it measures how surprised the language model is when seeing the word sequence. A lower perplexity means that a language model is less surprised and has a better idea of what words might come next in the test set.

14.4.3 Evaluating text generation

As seen in Section 14.3.4, text generation has a wide range of applications. To evaluate the quality of the generated text, we typically compare it with human-written references. The most accurate metric is therefore human evaluation. While humans provide the best evaluations, having human evaluations is often time-consuming and expensive. As a result, automatic metrics that align well with human evaluation are often used as temporary proxies. Although they are less accurate than human evaluation, they provide a cost-effective way to improve model designs.

BLEU (Bilingual Evaluation Understudy)[14] was first introduced to evaluate the quality of the machine translation system. BLEU captures the *n*-gram word precision between the output and the reference ground truth sentences. An *n*-gram word precision is the percentage of *n*-grams in the output text that are present in any of the reference texts. However, consider a reference text *"The Federal Reserve"* and a candidate output *"The The The."* The unigram precision of this candidate is 100% while it is obviously not a good output. To resolve the issue, we use a "modified" *n*-gram precision that matches *n*-grams of the candidate text $\hat{S}$ only as many times as they are present in any of the reference texts S:

[14] Papineni, Kishore et al. "Bleu: a Method for Automatic Evaluation of Machine Translation," *Annual Meeting of the Association for Computational Linguistics* (2002).

$$p_n(\hat{S}; S) = \frac{\sum_{s \in \hat{S}} \min\left(C(s, \hat{S}), C(s, S)\right)}{\sum_{s \in \hat{S}} C(s, \hat{S})}.$$

So the modified unigram precision for the above example becomes 33%. Then, we include all modified *n*-gram precision by taking their geometric mean:

$$\text{Precision} = \exp\left(\sum_{n=1}^{N} w_n \log p_n(\hat{S}; S)\right), \quad w_n = \frac{1}{n}.$$

Typically, we compute the *n*-gram precision for unigrams, bigrams, trigrams, and 4-grams. The modified *n*-gram precision unduly gives a high score for candidate output that can be very short. Given the reference *"The Federal Reserve has increased interest rates to fight rising inflation,"* the candidate output *"The Federal"* will have a unigram and bigram precision of 1 but is inarguably too short. In order to punish candidates that are too short, we consider the *brevity penalty*

$$\text{BP}(\hat{S}, S) = \begin{cases} 1 & \text{if } r \leq c, \\ \exp(1 - \frac{r}{c}) & \text{otherwise,} \end{cases}$$

where *c* and *r* are the lengths of the candidate output and reference output respectively. Finally, the BLEU score is defined as

$$\text{BLEU}_N(\hat{S}; S) = \text{BP}(\hat{S}; S) \cdot \exp\left(\sum_{n=1}^{N} w_n \log p_n(\hat{S}; S)\right), \quad w_n = \frac{1}{n}.$$

BLEU's output is always a number between 0 and 1. This value indicates how similar the candidate text is to the reference texts, with values closer to 1 representing more similar texts.

Example 14.8 (*BLEU score for machine translation*) Consider the following two translations for the text "聯準會已調升利息來對抗通膨" and the reference translation *"The Federal Reserve has increased interest rates to fight inflation"*:

1. To combat inflation, the Fed started raising its rates.
2. Fed raises interest rates further to fight inflation.

	Unigram	Bigram	Trigram	BP
Candidate 1	4/9	0/8	0/7	$\exp(1 - 10/9)$
Candidate 2	5/8	2/7	1/6	$\exp(1 - 10/8)$

$$BLEU_{\text{candidate 1}} = \exp\left(1 - \frac{10}{9}\right) \cdot \exp\left(1 \cdot \log \frac{4}{9}\right) = \exp\left(-\frac{1}{9}\right) \cdot \frac{4}{9}$$

$$BLEU_{\text{candidate 2}} = \exp\left(1 - \frac{10}{8}\right) \cdot \exp\left(1 \cdot \log \frac{5}{8} + \frac{1}{2} \cdot \log \frac{2}{7} + \frac{1}{3} \cdot \log \frac{1}{6}\right)$$

However, the BLEU score can be overly strict, since a good translation may use alternate words or paraphrases such as *combat* and *fight* or *interest rates* and *rates*. To allow synonyms to match between the reference and candidate, more recent metrics use word embeddings to achieve this goal. The *BERTscore* algorithm[15] passes the reference and the candidate texts through BERT model and uses BERT embeddings to compute precision, recall, and F_1.

[15] T. Zhang, V. Kishore, F. Wu, K. Q. Weinberger, and Y. Artzi, "BERTscore: Evaluating Text Generation With BERT," *ICLR*, 2020.

14.5 Exercises

Exercise 14.1 *Bag-of-words*
You are given a text data set containing 1,500 consumer reviews for a variety of products on Amazon.[16] We treat each review as a bag-of-words, discarding the word order but keeping the term frequency (number of occurrences of a word in a review). To be more specific, we use a matrix X to represent the review data set, with each row corresponding to a review, each column to a word, and element X_{ij} to the term frequency of the jth word in the ith review.

[16] The original data set is available at https://archive.ics.uci.edu/ml/datasets/Amazon+Commerce+reviews+set. In this exercise, you are provided with a "cleaned" version of the data set that is easier to process.

The data matrix X is a sparse matrix with 1,500 rows and 10,000 columns and is given in `Amazon_review_spmat.dat` in the online resources. Since X is a sparse matrix, we only need to store nonzero entries of X. Each line in `Amazon_review_spmat.dat` contains three numbers: row index, column index, and value of a nonzero element. The corresponding dictionary containing 10,000 words is given in `dict.txt` in the online resources.

Before analyzing the data, let us first refine the data set. There are 10,000 words in the dictionary, including bigrams, trigrams, punctuations, and so on, most of which are not needed in our analysis. Moreover, the data set is case-sensitive. For example, "book" and "Book" are treated as two different words in the dictionary and thus correspond to two columns in X, which should be merged into one column. Please write code to complete the following tasks:

1. Merge redundant words (same word in different cases) in the dictionary and their term frequencies in the data matrix. In the new data matrix, for example, the term frequency of "book" should be the sum of "book" and "Book" from the original matrix.

2. Remove words in `RemoveWordList.txt` from the dictionary and their corresponding columns from the data matrix. Construct a new dictionary for the new data matrix.

3. Finally, analyze the data to answer the following questions: (a) How many words are there in your new dictionary? (b) How many times in total does the word "book" appear in all the reviews? Sort the words by their total term frequency in descending order and plot the term frequency distribution.

Exercise 14.2 *Topic modeling via sparse PCA*
In this exercise, we are going to apply Sparse PCA to extract topics from the Amazon review data set that was prepared in Exercise 14.1. We first compute the first sparse principal component considering problem (4.14) introduced in Section 4.3.5:

$$\min_{u,v,\lambda} \quad \|X - \lambda uv^\top\|_F,$$
$$\text{s.t.:} \quad \|u\|_2 = 1, \|v\|_2 = 1, \quad (14.24)$$
$$\|u\|_0 \le k, \|v\|_0 \le h,$$

where $X \in \mathbb{R}^{n,m}$ is the refined data matrix obtained in question 2 in Exercise 14.1; n is the number of documents and m is the number of words in the dictionary; $u \in \mathbb{R}^n$ is a vector in "document space," $v \in \mathbb{R}^m$ is a vector in "word space" and $\lambda \in \mathbb{R}$ is a coefficient; and k and h are constraints on the the cardinality of u and v. In this exercise, we do not care about the sparsity of the document space vector u, so just let $k = n$. The cardinality of v should be equal to the number of keywords per topic that we want to retrieve.

To solve problem (14.24) we can use the thresholded power iteration method presented in Section 4.3.5. In this exercise, we adopt the convergence criterion

$$\max(|u^{[i+1]} - u^{[i]}|) \le 10^{-12} \text{ AND } \max(|v^{[i+1]} - v^{[i]}|) \le 10^{-12}.$$

Once we obtain u and v, we can compute $\lambda = u^\top X v$. Vector v is the desired "word space" vector whose h nonzero elements correspond to the h most "representative" keywords for this topic. To extract the next topic, we deflate the original data matrix with this "topic component" as $X - \lambda uv^\top$ and repeat the above procedure for the new data matrix.

1. Implement the thresholded power iteration algorithm as instructed above.

2. Apply the algorithm on the refined data matrix and extract 10 topics with eight keywords per topic from the data set. List your topics in the solution and briefly comment on your result.

Exercise 14.3 *Wasserstein distance between distributions*
In machine learning, the notion of Kantorovich or Wasserstein distance has been proposed to measure the distance between (discrete) distributions. The idea rests on the cost of transporting the mass from one distribution to the other distribution. More rigorously, let $n \in \mathbb{N}$, and define two discrete probability distributions $\mu = (\mu_1, \ldots, \mu_n)$ and $\nu = (\nu_1, \ldots, \nu_n)$ with $\sum_i \mu_i = \sum_i \nu_i = 1 \ldots$. We define $C \in \mathbb{R}^{n \times n}$ to be the cost matrix where $c_{ij} \ge 0$ is the cost of transporting one unit of mass from location $i \in [1, \ldots, n]$ to location $j \in [1, \ldots, n]$. We also define the flow matrix $P \in \mathbb{R}^{n \times n}$ where $p_{ij} \ge 0$ denotes the quantity of mass to be moved from location i to location j. In addition, the flow matrix P satisfies the limit conditions:

$$P\mathbf{1}_n = \mu,$$

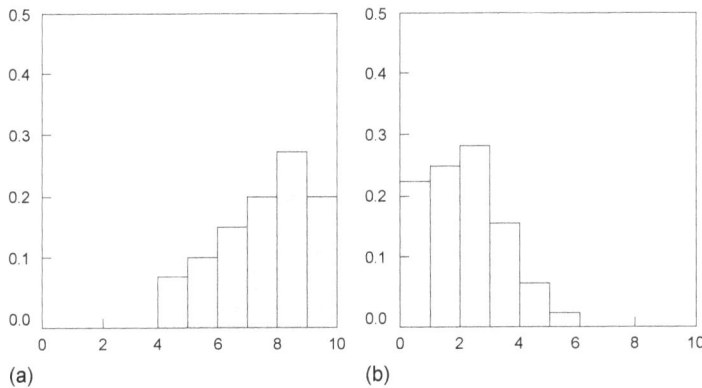

Figure 14.7 Visualization of (a) μ histogram and (b) ν histogram.

$$P^\top \mathbf{1}_n = \nu,$$

where $\mathbf{1}_n$ is a vector of 1.

1. What is the total cost of transporting the mass μ into ν by following the transportation plan P?

2. Given the cost matrix C, write the optimization problem of finding the transportation plan P with minimal total cost.

Exercise 14.4 *Document similarity*
In this exercise, we apply the idea of Wasserstein distance from Exercise 14.3 to document similarity[17] as illustrated in Figure 14.8. Assume we are provided with a *word2vec* embedding, the word travel cost c_{ij} between word i and word j is the Euclidean distance $\|x_i - x_j\|_2$ in the word embedding space. The similarity between the two documents is thus the minimum cumulative cost required to move all nonstop words from one document to the other. Implement the calculation of the Wasserstein distance and visualize the resulting flow matrix P for the provided word embedding.

[17] Kusner, Matt J. et al. "From Word Embeddings To Document Distances." *International Conference on Machine Learning* (2015).

Exercise 14.5 *Examplar selection*
Apply the optimization problem in part 3 to the tweets dataset trump-tweets.csv. We consider a matrix $X = [x^{(1)}, \ldots, x^{(m)}] \in \mathbb{R}^{n \times m}$, where $x^{(i)}$ in each column is a data point in $\mathbb{R}^n$. We would like to select a few columns, referred to as "examplars," that "best" represent the data points. In this problem, we seek to find a subset of indices $\mathcal{I} \subseteq \{1, \ldots, m\}$ such that every data point is best approximated by a linear combination of the "examplars" $x^{(i)}, i \in \mathcal{I}$:

$$\forall j \in \{1, \ldots, m\} \quad \exists (w_{ij})_{i \in \mathcal{I}}: x^{(j)} \approx \sum_{i \in \mathcal{I}} w_{ij} x^{(i)}.$$

In the above, the regression weights depend on the data point j, but not the "examplar" set $\mathcal{I}$, which is required to be the same for all the data points.

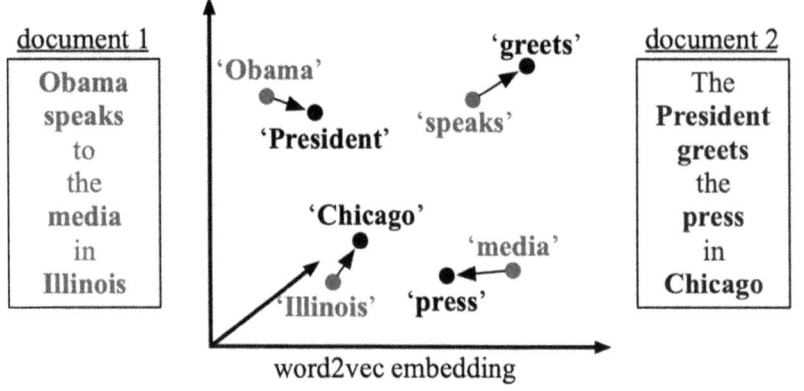

Figure 14.8 An illustration of the Wasserstein distance. All nonstop words (**bold**) of both documents are embedded into a *word embedding* space. The similarity between the two documents is the minimum cumulative distance that all words in Document 1 need to travel to exactly match Document 2.

1. Show that the condition above can be written in matrix form as $X \approx XW$, for some matrix W that you will determine. How does the number of zero rows in W relate to the cardinality of the examplar set?

2. We would like to tradeoff the size of the approximation error matrix $X - XW$, against the cardinality of the examplar set. Formulate an SOCP that accomplishes this tradeoff with a hyper-parameter λ that controls the degree of the tradeoff. *Hint:* Remember the "ℓ_1 norm trick" and apply it to some vector made up of row norms.

3. Write the problem as an unconstrained problem using the squared Frobenius norm and the ℓ_1 norm trick. Show that the solution depends only on the scalar products between data points. What are the implications of this fact, when it comes to using a different kernel for the data?

4. Apply the optimization problem in part 3 to the tweets dataset. Report your cardinality vs. approximation error tradeoff plot and cardinality/error vs. λ plot. Comment on the results, including the examplars selected for $\lambda = 1$.

Index

accuracy, 166
activation function, 219
affine recourse, 326
alternating minimization, 93
approximate risk budgeting, 313
AR, 148
ARCH, 66
ARIMA, 66
ARMA, 65
attention mechanism, 238
AUC, 175
auto-encoder, 93
autocorrelation function, 61
autoregressive process, 65, 193

backpropagation algorithm, 222
backtesting, 301
bag-of-words, 11, 374, 393
bagging, 210
balanced accuracy, 170
Bayes' rule, 33
Bayesian learning, 158
BERTscore, 392
Beta distribution, 35
bias error, 134
binary classification, 165
Binomial distribution, 35
BLEU score, 391
bootstrap, 210, 344
box plot, 25

candlesticks, 24
Cauchy–Schwartz inequality, 18
centering, 14, 108, 139
centrality measures, 353
chance constraints, 257
Chebyshev regression, 141

class imbalance, 167, 178
clearing vector, 351
clustering, 363
clusterpath, 120
community detection, 363
concave function, 248
conditional
  independence, 55
  VaR, 298
confidence ellipsoid, 58
confusion matrix, 169, 211
constraints, 248
context window, 373
convex
  function, 248
  problems, 251
  set, 248
convexity, 248
convolutional neural networks, 228
correlation networks, 342
cost function, 135
covariance approximation, 78
covariance matrix, 21
  sparse, 54, 57
cross validation, 144
cross-entropy loss, 382
curse of dimensionality, 124
custering, 111
CVX, 57
cvx, 259
cvxpy, 154, 259

data
  centering, 15
  frame, 7
  generation mechanism, 32
  imputation, 17

matrix, 9
  projection, 19
  scaling, 15
data set
  banknote authentication, 128
  car evaluation, 213
  car sales, 29
  consumer reviews, 393
  credit card, 122
  credit card fraud, 189, 212, 242
  cryptocurrencies, 14
  currency, 71
  customer churn, 212
  customer profiling, 129
  customers, 127
  customers spending, 27
  direct investments, 361
  DJI, 344
  economic growth, 150
  exchange rates, 108
  fashion, 242
  GDP, 235
  housing, 156
  loan, 8
  mtcars, 148
  NY Times headlines, 102
  NYSE, 241
  predictive maintenance, 190
  S&P 500, 28, 86, 110, 115
  S&P 500 fundamentals, 26, 122
  text classification, 188
  US gasoline, 146
  web traffic, 29
DBSCAN, 119
decision
  policy, 326
  trees, 203

deep
  learning, 227, 237
  neural networks, 384
default contagion, 349
deflation, 83
discrimination function, 166
diversification, 290
document similarity, 378, 395
dominated portfolios, 279
dyad, 13

Eckart–Young–Mirsky theorem, 50, 92
efficient
  frontier, 279
  portfolios, 279
Elastic Net regression, 140, 256
ellipsoidal uncertainty, 264
empirical cdf, 142
ensemble methods, 209
entropy, 204
estimator
  MAE, 77
  MAP, 77
  MMSE, 76
examplar, 395
expected shortfall, 298
extractive summarization, 257
extreme portfolios, 278

F_1 score, 170, 389
factor models, 48, 94, 286
false
  negative, 166
  positive, 166
features, 7
  similarity, 18
feedforward neural networks, 217
fictitious default algorithm, 351
fundamental theorem of linear algebra, 197

GARCH, 68
Gaussian
  distribution, 36
  kernel, 199, 213
Gaussian distribution, 159
generative
  learning, 240
  process, 376

gradient descent, 269
graph, 338
  edge, 338
  node, 338
  path, 341
  vertex, 338
  walk, 341
graphical lasso, 57
gross return, 274
group lasso, 257

halfspaces, 171
heatmap, 26
hierarchical clustering, 121
hinge loss, 104, 183
histogram, 25
hold-out method, 144
Huber
  loss, 136
  regression, 141
hyperplanes, 171

in-sample loss, 143
index tracking, 315
information gain, 205
initialization (of NN), 227
inner product, 11
interval uncertainty, 185, 264, 320
investment networks, 348
iterative dichotomizer, 203

k-means, 112
k-medoids, 117
kernel
  matrix, 197
  trick, 196
kernel function, 198

ℓ_1 regression, 140
Laplace prior, 162
Laplace smoothing, 377
Laplacian, 118
Laplacian noise, 162
large language models, 383
Lasso regression, 138, 162
law of cosines, 18
learning rate, 222
least squares, 107, 136, 152
  robust, 266

liability networks, 349
likelihood, 33, 159
line plot, 24
line search, 269
linear
  programs, 252
  regression, 131
  transaction costs, 286
linear matrix inequality, 258
linkage function, 121
Lipschitz continuous gradient, 269
LMI, see linear matrix inequality
log-return, 274
logistic classifier, 170
long short-term memory, 234
look-back window, 301
loss function, 33, 173, 178, 225
low-rank approximation, 48, 50, 78, 87, 91, 107, 108
LP, see linear programs
LS, see least squares

MAE loss, 136
Mahalanobis distance, 59
Mahalanobis uncertainty, 320
MAP, 159
margin of separation, 180
market impact, 289
Markowitz's model, 273
matrix
  adjacency, 338
  co-occurrence, 372
  completion, 105
  Laplacian, 338
  orthogonal, 74
  positive semidefinite, 75
  primitive, 355
  symmetric, 73
maximum margin classifier, 181
mean squared error, 160
mean/variance portfolio optimization, 255, 273
median, 143
median risk, 309
mini-batch, 222
minimum spanning tree, 343
misclassification error, 166
Moore–Penrose pseudoinverse, 160
moving average, 64
MSE loss, 136

multi-class classifier, 175, 190
multi-period
  decision problems, 323
  portfolio optimization, 327

n-gram, 379
naive Bayes, 375
NARX networks, 234
node centrality, 352, 353
nonlinear SVM, 199
nonnegative factorization, 103
norm-bounded uncertainty, 266, 267
normal equations, 138
norms, 12

one-hot vector, 372
one-vs-all classifier, 177
optimal
  set, 246
  solution, 246
  value, 246
optimization problem, 246
out-of-sample loss, 144
outliers, 57, 59, 70
ownership networks, 348

pagerank, 358
partial covariance, 259
payment networks, 348
PCA, see principal component analysis
Pearson correlation, 40, 342
penalty parameter, 145
perceptron, 179
perceptron loss, 173
performance metrics, 168
perplexity, 390
Perron–Frobenius theorem, 354
point estimates, 76
pooling, 232
portfolio, 40, 273
  diversification, 42
  two-asset, 43
portfolio optimization, 277
  in practice, 301
  robust, 318, 333
posterior, 33
power iteration algorithm, 90, 101, 359
pre-trained models, 383

precision, 168
precision matrix, 37, 56
  sparse, 56
prices, 274
principal component analysis, 82
  regularized, 98
  robust, 98
  sparse, 99
principal direction, 83
prior, 32, 160
pro-rata rule, 351
production problem, 260
projection
  networks, 348
  on a plane, 23
  onto a subspace, 22
  theorem, 20
proximal gradient algorithm, 270

QP, see quadratic programs
quadratic programs, 255
quantile regression, 142

random
  forests, 210
  walk, 62
recall, 168
rectified linear unit, 219
recurrent neural networks, 232, 384
reduced-error pruning, 209
regression model, 132
regularization, 136, 152, 184, 226
residual sum-of-squares, 160
ResNets, 237
returns, 10, 274
Ridge regression, 137, 161
risk
  aversion, 278
  budgets, 311, 337
  parity, 312, 334
risk/return tradeoff, 277
RNN, see recurrent neural networks
robust
  least squares, 266
  linear regression, 162
  LP, 262
  optimization, 260
  portfolio optimization, 304, 333
  QP, 265
  SVM, 187

ROC, 175

safe feature elimination, 100
sample
  covariance matrix, 37, 45
  quantile, 142
scaling, 139
scatterplot, 25
scenario uncertainty, 263, 265, 320
score plot, 24
SDP, see semidefinite programs
second-order cone programs, 256
semi-variance, 300
semidefinite programs, 258
sensitivity, 168
Seq2Seq models, 237
Sharpe ratio, 292
short-selling constraints, 255, 288
short-term financing, 253
shorting, 275
shrinkage estimator, 53
sigmoidal function, 134
silhouette, 115
simple return, 274
singular value decomposition, 88, 92, 373
skip-gram, 381
SOCP, see second-order cone programs
softmax, 220
sparse
  PCA, 99, 393
  precision matrix, 56, 71
  SVM, 253
  tracking, 317
specificity, 168
spectral
  clustering, 118
  theorem, 74
spectral clustering, 127
spherical uncertainty, 264, 267, 320
stochastic gradient descent, 222
subgradient, 151
sum of k-largest elements, 291
support vector machine, 179, 182
  robust, 185
support vectors, 201
SVD, see singular value decomposition
SVM, see support vector machine

text
    classification, 385
    generation, 387
text analytics, 371
text classification, 375
TF-IDF, 378
time series, 61, 70, 91
topic modeling, 393
training set, 144
transaction costs, 305
transformers, 386
true
    negative rate, 168
    positive rate, 168

type I error, 166
type II error, 166

unbiased estimator, 135
uncertainty, 260
    on covariance matrix, 321
universal approximation, 221

validation set, 144
value-at-risk, 296
VaR, *see* value-at-risk
variance
    directional, 20, 39, 82
    error, 135
    explained, 46, 85
    maximization, 83
    sample, 21
    total, 39
vectors, 9
vertex degree, 338
vocabulary, 372
volatility, 174

Wasserstein distance, 394
white noise, 62
word2vec, 381
worst-case risk, 259

For EU product safety concerns, contact us at Calle de José Abascal, 56–1°,
28003 Madrid, Spain or eugpsr@cambridge.org.

www.ingramcontent.com/pod-product-compliance
Lightning Source LLC
LaVergne TN
LVHW060054080526
838200LV00084B/202